# Electrical Machines and Transformers

# ELECTRICAL MACHINES
# and TRANSFORMERS
# Principles and Applications

**Peter F. Ryff**
*Ryerson Polytechnical Institute*

**David Platnick**

**Joseph A. Karnas**
*Ryerson Polytechnical Institute*

*PRENTICE-HALL, INC., Englewood Cliffs, N.J. 07632*

*Library of Congress Cataloging-in-Publication Data*

RYFF, PETER F.
   Electrical machines and transformers.

   Bibliography: p. 436
   Includes index.
   1. Electric machinery.   2. Electric transformers.
I. Platnick, David, d. 1982.   II. Karnas, Joseph A.,
III. Title.
TK2000.R93 1987        621.31′042        86-15094
ISBN 0-13-247222-8

Editorial/production supervision and
   interior design: *Erica Orloff*
Cover design: *20/20 Services Inc.*
Manufacturing buyer: *Carol Bystrom*

Printed in the United States of America

10  9  8  7  6  5  4  3  2  1

ISBN   0-13-247222-8   025

PRENTICE-HALL INTERNATIONAL (UK) LIMITED, *London*
PRENTICE-HALL OF AUSTRALIA PTY. LIMITED, *Sydney*
PRENTICE-HALL CANADA INC., *Toronto*
PRENTICE-HALL HISPANOAMERICANA, S.A., *Mexico*
PRENTICE-HALL OF INDIA PRIVATE LIMITED, *New Delhi*
PRENTICE-HALL OF JAPAN, INC., *Tokyo*
PRENTICE-HALL OF SOUTHEAST ASIA PTE. LTD., *Singapore*
EDITORA PRENTICE-HALL DO BRASIL, LTDA., *Rio de Janeiro*

# CONTENTS

## CHAPTER 3   DC Generators                                    56

## CHAPTER 4   DC Motors                                        86

## CHAPTER 5 Starting and Control of DC Motors 126

## CHAPTER 6 Electric Power Generation 147

# PREFACE

This textbook deals with the fundamental principles underlying the operation of electrical machines and transformers, which include relays and contactors, single-phase and three-phase transformers, and dc and ac electrical machines. Classical machines texts have paid very little attention to machines in fractional and subfractional horsepower sizes. A respectable amount of material in this book is devoted to small specialty machines. It is recognized that an understanding of these devices is necessary for nonmachine specialists who utilize motors in their design.

The authors, having taught electrical machines courses at the engineering technology level for nearly two decades, have felt an urgent need for a concise, up-to-date introductory textbook on this subject. A new textbook should be relevant in content and useful in current applications to the present generation of engineering technology graduates and engineers. Many of the detailed analysis and descriptions provided in the older books on electrical machines are unnecessary and in fact may be detrimental to developing in the uninitiated students curiosity about and interest in electrical machines and their applications.

As a consequence, there is a pressing demand for an introductory textbook which is lucid, easy to read, interesting, highly illustrative, and containing many solved example problems. In addition, a modern textbook should contain a variety of sample computer programs written in the popular BASIC language and have a good selection of questions and problems with answers for use in the classroom or tutorials.

The requirements and needs stated above are fulfilled by this book. It has been written primarily for use in a first-level machines course in the three-year associate degree program and four-year bachelor of technology program in engineering technology. The book will also serve for electrical machines courses offered to non-electrical engineering students in universities and will be a foundation for students of electrical technology who elect to pursue further studies in advanced machines and electrical power systems.

Much of the material in this book has been developed from a two-sequence semester course in electrical machines taught to all students of electrical engineering technology at Ryerson Polytechnical Institute. Obviously, more material is presented than was treated in the typical curriculum. This leaves detailed content and order of subject matter to the desires, flexibility, and enthusiasm of individual instructors, as governed by local circumstances.

There are 14 chapters in this book. Browsing through it for a couple of hours and reading the contents should enable an instructor to outline a diversity of courses with differing contents and sequences to fill particular needs. It is assumed that the student has completed or is concurrently taking an introductory electric circuits course. No mathematical sophistication beyond this level is required; The emphasis is on physical understanding as the basis for the derived equivalent machine circuit diagrams. All theory has been developed using SI units, for simplification and clarity of presentation of the derived equations. Should instructors desire to use U.S. customary units, they will find all the necessary conversion factors listed in Appendix B.

To help the reader understand the basic principles of electromagnetic devices, a reasonably comprehensive study of electromagnetics is presented in Chapter 1, which provides a suitable basis for the remainder of the textbook. Chapters 2 through 5 deal with the physical construction and operating characteristics of direct-current machines. Starting and control techniques for dc motors using solid-state devices are presented with appropriate diagrams. The remainder of the book concentrates on alternating-current machines, including transformers. After discussing electric power generation in Chapter 6, the synchronous generator is discussed in Chapter 7, followed by single-phase transformers in Chapter 8 and three-phase transformers and connections in Chapter 9. Chapter 10 deals with three-phase induction motors. Single-phase motors are discussed in Chapter 11, and Chapter 12 deals with synchronous motors. Chapter 13 gives a comprehensive overview of starting and control of ac motors, with emphasis on automatic starting techniques and solid-state speed control. Finally, Chapter 14 is devoted to special-purpose devices such as linear induction machines, special fractional-horsepower synchronous motors, servomotors, and eddy current clutches.

We strongly believe that the practical aspect of this textbook is more beneficial than pure theory for the majority of students and should be enjoyed by both students and teachers. As a result, this book is concerned primarily with principles and applications to electrical machines and transformers.

The untimely death of Dr. David Platnick in October 1982 was a great loss of a co-author and dear colleague. His sound judgment and lucid writing style throughout the first five chapters and portions of the last two chapters had been developed and partially completed by him. They have been preserved in essence and idea by the third author, who completed the first five chapters and who was instrumental in all aspects of the finished work.

*P. F. Ryff*

*J. A. Karnas*

# 1

# ELECTROMAGNETIC PRINCIPLES

## 1-1 INTRODUCTION

The dynamic age of electricity began with the work of Hans Christian Oersted (1771–1851), who demonstrated in the year 1819 that a current-carrying conductor produced a magnetic field. This was the first time that a relationship was shown to exist between electricity and magnetism. His discovery set off a chain of experiments all across Europe which culminated in the discovery by Michael Faraday (1791–1867) of his law of electromagnetic induction in 1831. Faraday showed that it was possible to produce an electric current by means of a magnetic field. This led, in a very short time, to the development of electrical generators, motors, and transformers, and opened up our modern electrical era.

All electromagnetic devices make use of magnetic fields in their operation. These magnetic fields may be produced by permanent magnets or electromagnets. Magnetic fields are created by alternating- and direct-current sources to provide the necessary medium for developing generator action and motor action. Throughout this book we will be studying the application of magnetic fields to electromechanical energy conversion processes as demonstrated in rotating electric machinery. Also, transformers provide energy transfer from one electric circuit to another via the changing magnetic field. It will become apparent that there is both transfer and storage of energy in the magnetic fields of the various electromagnetic devices. Hence all electromagnetic devices are constructed with appropriate magnetic circuits, as will be described in subsequent chapters.

The oldest magnetic instrument is a suspended permanent magnet, called a *compass*. We can define a *magnetic field* as a region in space in which a compass needle is acted upon.

In a region where there are no large magnetic objects, the compass needle points in a general north-south longitudinal direction, with the "north" pole of the compass pointing to the earth's north magnetic pole. However, we know that similar to the law of electric charges, unlike magnetic poles attract and like magnetic poles repel. In spite of the fact that the attracting poles of the compass and earth must be of opposite magnetic polarity, this north-seeking pole of the compass is defined as the *north pole*. Similarly, it would be correct to describe the other (unmarked) pole of the compass as the south-seeking pole. For brevity this pole is called the *south pole*.

It is well known that a bar of iron can be magnetized by placing it in contact with a strong magnet. By observing the direction of the compass needle at many points around the magnetized bar, a map of the magnetic field can be traced. A map of these lines can be obtained by the familiar method of sprinkling iron filings on a sheet of paper held over the magnetized bar. When this is done, the pattern of Fig. 1-1 is produced.

The map of Fig. 1-1 should not be interpreted too literally. The iron filings are just a local manifestation of the direction of the magnetic field at that point in space. Each particle of iron has in effect become a small magnet and is aligned with the magnetic field of the larger magnet (the magnetized bar). Although this map seems to show "lines of force," the lines do not actually exist in space. They can, however, be conceptualized and treated as if they had physical reality. This visualization of magnetic lines of force which was developed by Faraday will be of great value in our understanding of electromagnetic principles.

### Properties of Magnetic Lines of Force

The following properties may be ascribed to magnetic lines of force:

*Property 1.* Magnetic lines of force are directed from north to south outside a magnet. The direction is determined by the north pole of a small magnet held in the field.

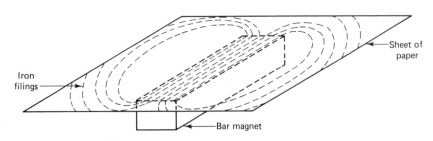

Iron filings

Sheet of paper

Bar magnet

**Figure 1-1** Magnetic field pattern near a magnet.

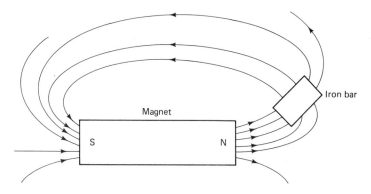

**Figure 1-2** Magnetic field distortion.

*Property 2.* Magnetic lines of force are continuous.

*Property 3.* Magnetic lines of force enter or leave a magnetic surface at right angles.

*Property 4.* Magnetic lines of force cannot cross each other.

*Property 5.* Magnetic lines of force in the same direction tend to repel each other.

*Property 6.* Magnetic lines of force tend to be as short as possible.

*Property 7.* Magnetic lines of force occupy three-dimensional space extending (theoretically) to infinity.

These properties can be seen in the field map of Fig. 1-2. Because of properties 1 and 2, we must assume that there is a magnetic field within the bar; the direction of the field is south to north. The lines are perpendicular to the magnetic surface because of property 3. The lines spread out because of properties 4 and 5, and they assume their shape because of the interaction of properties 4, 5, 6, and 7. Because of property 6, if the lines "find" an easy magnetic path (e.g., through iron), they will prefer this to a more difficult path—through air, as seen in Fig. 1-2, which is the field map around a magnet when a piece of iron is brought near it. The iron bar is an "easier" path than the air, hence the lines tend to concentrate around this part of the circuit. We say that the reluctance of the iron is less than the reluctance of air, hence the iron is an easier path for the flux lines. Reluctance of a magnetic circuit may be described as magnetic resistance which tends to oppose the establishment of magnetic flux lines.

## Magnetic Field Produced by Current-Carrying Conductor

A magnetic field is always associated with a current-carrying conductor, as illustrated in Fig. 1-3. Exploring the magnetic field by means of a compass, we observe the following:

1. The magnetic field is strongest perpendicular to the current direction.

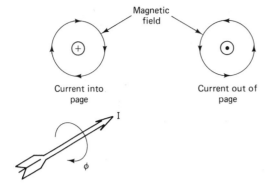

Magnetic field

Current into page

Current out of page

**Figure 1-3** Direction of magnetic field around a current-carrying conductor.

2. As we traverse a path around the conductor, we find that the magnetic field is always tangent to the direction of current flow. We can trace a path around the conductor so that continuous magnetic lines of force surround the conductor.

3. If we reverse the direction of current flow, the direction of the magnetic field also changes.

4. The field is strongest near the wire and decreases as we move farther from it. (We can obtain a measure of field strength by trying to deflect the magnet needle from the position it has assumed in the field. At a point where the field is strong, it will be more difficult to deflect it than at a point where it is weak.)

5. If we look at a single current-carrying conductor end on, and draw it as in Fig. 1-3, where the symbol $\oplus$ indicates current flowing into the page, it is easier to draw the magnetic field. If we reverse the current, we have the symbol $\odot$ for current coming out of the page, and we have the situation depicted in Fig. 1-3. The dot and cross symbols, respectively, represent the head and tail of an arrow.

6. If we grasp the conductor with our right hand, the thumb pointing in the direction of the current, our fingers will point in the same direction as the north pole of the compass. This method of determining the directions of current flow in a conductor and the surrounding lines of force is called *Ampère's right-hand rule.*

### *Practical Magnetic Circuits*

If we construct a coil of many turns, we can increase the magnetic field strength very greatly, as shown in Fig. 1-4. We can also increase the magnetic field strength by increasing the magnitude of current in the coil. This coil provided with excitation current is called a *solenoid.* Thus we see that the magnetic field strength is proportional to both the number of turns and the current.

We can determine the direction of the magnetic field in a cylindrical coil of many turns of insulated wire by using our right hand. If we grasp the coil with our right hand with the fingers pointing in the direction of the current, the thumb

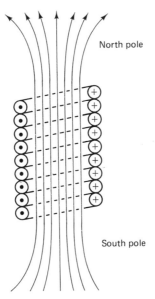

**Figure 1-4** Magnetic field inside a long solenoid.

will point in the direction of the north pole. This method of determining directions of current flow in a coil and magnetic fields of force is another form of Ampère's right-hand rule. André Marie Ampère (1775–1836), pursuant to the experimental work of Oersted, developed extensively the foundations of electromagnetic theory. Refer to Fig. 1-4. Several practical magnetic circuits are illustrated in Fig. 1-5.

## 1-3 ELECTROMAGNETIC RELATIONSHIPS

### Magnetic Lines of Force

The "quantity of magnetism" which exists in a magnetic field is the magnetic line of force, or more simply, the *magnetic flux*. In the SI system magnetic flux is measured in units called webers, abbreviated Wb, and its symbol is $\phi$ (the Greek lowercase letter phi). Those readers not familiar with the International System of Units (Système International d'Unités, abbreviated SI) should refer to Appendix B for a description of all units that will be encountered in this book. In addition, Appendix B contains the more common conversion factors for dealing with the other systems of units. The weber is defined in terms of an induced voltage, so that the definition of the unit will be postponed until we study electromagnetic induction. Although there is no actual flow of magnetic flux, we will consider flux to be analogous to current in electric circuits.

### Magnetic Flux Density

The total magnetic flux that comes out of the magnet is not uniformly distributed, as can be seen in Fig. 1-2. A more useful measure of the magnetic effect is the

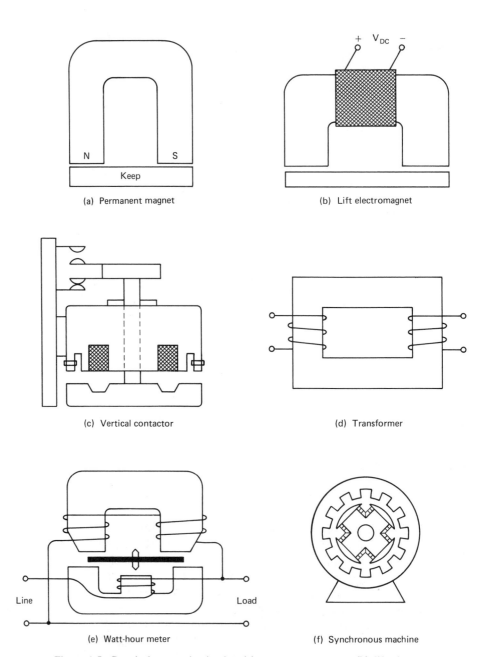

**Figure 1-5** Practical magnetic circuits: (a) permanent magnet; (b) lift electromagnet; (c) vertical contactor; (d) transformer; (e) watthour meter; (f) synchronous machine.

magnetic flux density, which is the magnetic flux per unit cross-sectional area. We will consider two equal areas through which the magnetic flux penetrates at right angles near one end of the permanent magnet along its centerline. From the illustration it becomes apparent that there is a greater amount of magnetic flux passing through an area that is nearer the magnet pole. In other words, the magnetic flux density increases as we approach closer to the end of the magnet. However, it must be noted that the magnetic flux density inside the magnet is uniformly constant. Magnetic flux density is measured in units of tesla (T) and is given the symbol $B$. One tesla is equal to 1 weber of magnetic flux per square meter of area. We can state that

$$B = \frac{\phi}{A} \tag{1-1}$$

where    $B$ = magnetic flux density, T
         $\phi$ = magnetic flux, Wb
         $A$ = area through which $\phi$ penetrates perpendicularly, m$^2$

**EXAMPLE 1-1**

The total magnetic flux out of a cylindrical permanent magnet is found to be 0.032 mWb. If the magnet has a circular cross section and a diameter of 1 cm, what is the magnetic flux density at the end of the magnet?

**SOLUTION**

The total flux = $0.032 \times 10^{-3}$ Wb, cross-sectional area of magnet:

$$A = \frac{\pi D^2}{4} = \frac{\pi (0.01)^2}{4} = 78.53 \times 10^{-6} \text{ m}^2$$

$$B = \frac{\phi}{A} = \frac{0.032 \times 10^{-3}}{78.53 \times 10^{-6}} = 0.407 \text{ T}$$

Note that this magnetic flux density exists only at the immediate end of the magnet. As we move away from the end of the magnet, the magnetic flux spreads out, and therefore the magnet flux density decreases.

**Magnetomotive Force**

We have seen that an increase in the magnitude of current in a coil or a single conductor results in an increase in the magnetic flux. If the number of turns in a coil are increased (with the current remaining constant), there is an increase in magnetic flux. Therefore, the magnetic flux is proportional to the products of amperes and turns. This ability of a coil to produce magnetic flux is called the *magnetomotive force*. Magnetomotive force is abbreviated MMF and has the units of ampere-turns (At). The magnetomotive force is given the symbol $F_m$. Strictly speaking, the units of MMF are amperes because turns are dimensionless quantities. However, from a pedagogical standpoint, we prefer and shall use throughout this book the units of ampere-turns (At) for MMF.  We may write

$$F_m = NI \tag{1-2}$$

where    $F_m$ = magnetomotive force (MMF), At
          $N$ = number of turns of coil
          $I$ = excitation current in coil, A

Magnetomotive force in the magnetic circuit is analogous to electromotive force in an electric circuit.

### EXAMPLE 1-2

The coil in Fig. 1-6 has 1000 turns wound on a cardboard toroid.  The mean (or average) diameter $D$ of the toroid is 10 cm, and the cross section is 1 cm.  The total magnetic flux in the toroid is 3 $\mu$Wb when there is an excitation current of 10 mA in the coil.

(a) What is the magnetic flux when the current is increased to 20 mA?
(b) What is the magnetic flux density within the coil when the current is 20 mA?

### SOLUTION

(a) If we double the current to 20 mA, then

$$F_m = NI = 1000 \times 20 \times 10^{-3} = 20 \text{ At}$$

and $\phi$ must double to 6 $\mu$Wb.

(b) For a toroid, the magnetic flux is assumed to be uniform across the interior cross-sectional area of the coil.  From Eq. (1-1),

$$B = \frac{\phi}{A} = \frac{6 \times 10^{-6}}{(\pi/4)\,(1 \times 10^{-2})^2} = 76 \text{ mT}$$

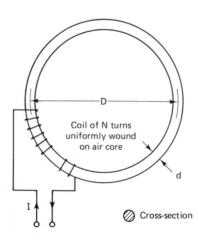

Coil of N turns
uniformly wound
on air core

I

Cross-section

**Figure 1-6**   Toroid coil.

## Magnetic Reluctance

In Example 1-2 we have seen that doubling the driving force (MMF) in the circuit results in a doubling of the output quantity (magnetic flux). We consider this ratio of MMF to magnetic flux:

$$\frac{F_m}{\phi} = R_m \tag{1-3}$$

where $\quad F_m = NI$, the MMF, At
$\quad\quad\quad \phi =$ magnetic flux, Wb
$\quad\quad\quad R_m =$ reluctance of the magnetic circuit, At/Wb

Transposing, we have

$$F_m = R_m\phi$$

which shows us that the magnetic flux is directly proportional to the magnetomotive force. This equation represents *Ohm's law of magnetic circuits*. The proportionality factor $R_m$ is called the reluctance of the magnetic circuit and is, obviously, analogous to resistance in an electric circuit. Assuming that a coil has fixed turns and a constant excitation current, the amount of magnetic flux produced will depend on the material used in the core of the coil. A much larger amount of flux can be produced in an iron-core coil than in an air-core coil. Thus we see that the reluctance of the magnetic circuit depends on the material properties of the magnetic circuit. For our purposes, the materials are classified as either magnetic or nonmagnetic. Only the ferrous (irons and steels) group of metals, including cobalt and nickel, are magnetic materials. All other materials, such as air, insulators, wood, paper, plastic, brass, and bronze, including vacuum, are nonmagnetic materials. The strength and pattern of the magnetic field in nonmagnetic materials would be identical to that of air or vacuum (free space). In our discussions we will assume that the magnetic properties of air and vacuum are the same. We consider some of the peculiar characteristics of magnetic materials in subsequent sections.

The reluctance of a homogeneous magnetic circuit may be expressed in terms of its physical dimensions and magnetic property as follows:

$$R_m = \frac{l}{\mu A} \tag{1-4}$$

where $\quad R_m =$ reluctance of the magnetic circuit, At/Wb
$\quad\quad\quad l =$ average or mean length of the magnetic path, m
$\quad\quad\quad A =$ cross-sectional area of the magnetic path, m$^2$
$\quad\quad\quad \mu =$ absolute (or total) permeability of the magnetic path, H/m

Reluctance is in essence magnetic resistance, that is, the property of a magnetic circuit which is reluctant or unwilling to set up magnetic flux.

### Permeability

Permeability is the magnetic property that determines the characteristics of magnetic materials and nonmagnetic materials. The permeability of free space and nonmagnetic materials has the following symbol and constant value in SI units:

$$\mu_0 = 4\pi \times 10^{-7} \text{ H/m}$$

As we can see, the reluctance of magnetic materials is much lower than that of air or nonmagnetic materials. From the inverse relationship of reluctance and permeability, we determine that the total permeability of magnetic materials is much greater than that of air. However, the value of permeability varies with the degree of magnetization of the magnetic material and, of course, the type of material. Since the permeability of magnetic materials is variable, we must employ magnetic saturation ($B$–$H$) curves to perform magnetic circuit calculations. Permeability in magnetic circuits is somewhat analogous to conductivity in electric circuits.

### EXAMPLE 1-3

In Fig. 1-6 we assume that the magnetic flux is practically uniform in the cross-sectional area of the toroid. The mean path length is 0.314 m and the cross-sectional area through which the flux exists is $78.5 \times 10^{-6}$ m². Calculate the number of ampere-turns required to set up magnetic flux of 1 Wb.

### SOLUTION

The reluctance of the homogeneous magnetic circuit is

$$R_m = \frac{l}{\mu_0 A} = \frac{0.314}{4\pi \times 10^{-7} \times 78.5 \times 10^{-6}} = 3.18 \times 10^9 \text{ At/Wb}$$

$$F_m = R_m \phi = 3.18 \times 10^9 \times 1.0 = 3.18 \times 10^9 \text{ At}$$

This is obviously a very large number and we may conclude that the path reluctance is very high. This means that it is comparatively difficult to establish a large magnetic flux in air. For this reason, when we need high flux densities, it becomes necessary to use materials having high values of permeability (such as iron or steel) for large portions of the magnetic paths.

### Magnetic Field Intensity

One other important magnetic quantity is the magnetomotive force gradient per unit length of magnetic circuit, or more commonly, the magnetic field intensity. Its symbol is $H$ and from the definition,

$$H = \frac{F_m}{l} \tag{1-5}$$

the unit is ampere-turns per meter (At/m). The former name for magnetic field intensity was magnetizing force. We have seen that more ampere-turns (MMF) are required to set up the same magnetic flux in magnetic circuits of air than in iron of similar configuration. Hence the magnetic field intensity for the air path is much larger than for the iron path. In the toroid of Fig. 1-6, a magnetomotive force of 10 At acts along the mean path of 0.314 m. The magnetic field intensity is

$$H = \frac{NI}{l} = \frac{10}{0.314} = 31.8 \text{ At/m}$$

Equation (1-5) transposed, $H = NI$ is one form of Ampère's circuital law applied to a simple magnetic circuit. Magnetic field intensity in magnetic circuits is analogous to potential or voltage gradient in electric circuits.

We can derive a useful relationship for magnetic circuits by summarizing the equations developed so far.

$$B = \frac{\phi}{A}$$

$$F_m = NI$$

$$\phi = \frac{F_m}{R_m}$$

$$R_m = \frac{l}{\mu A}$$

$$H = \frac{F_m}{l}$$

Thus

$$B = \frac{\phi}{A} = \frac{F_m}{R_m A} = \frac{Hl\mu A}{lA} = \mu H \quad \text{or} \quad B = \mu H \quad (1\text{-}6)$$

Equation (1-6) shows that the magnetic flux density is directly dependent on both permeability and magnetic field intensity. Only in air or free space is the permeability ($\mu_0$) constant, and thus a linear relationship between $B$ and $H$ exists. In the next section we consider ferromagnetic materials in which the absolute permeability is not a constant but depends on the degree of magnetization.

## 1-4 MAGNET CIRCUITS

A toroid of homogeneous magnetic material, such as iron or steel, is wound with a fixed number of turns of insulated wire as shown in Fig. 1-6. The magnetic flux ($\phi$) and the excitation current ($I$) are related by Eq. (1-6):

$$B = \mu H = \frac{\phi}{A}$$

Thus

$$\frac{\phi}{A} = \mu\,\frac{NI}{l} \qquad \text{that is,} \qquad \phi = \text{(constant)} \times \mu I$$

where the constant is $NA/l$. At the outset, the sample of ferromagnetic material in the toroid was totally demagnetized. In experimental measurements, the excitation current is varied and the corresponding values of magnetic flux recorded. Then the calculated values of $B$ and $H$ are plotted on linear scales as illustrated in Fig. 1-7.

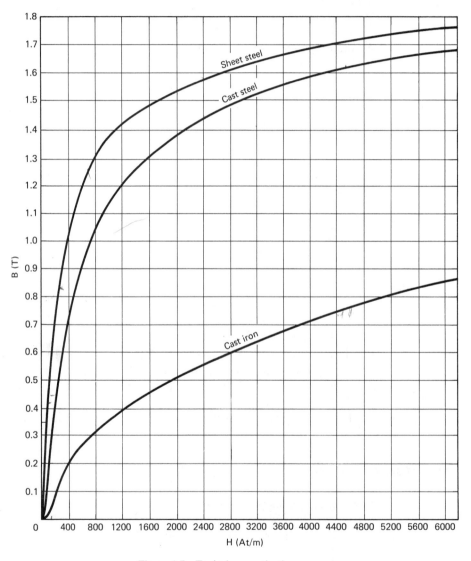

**Figure 1-7** Typical magnetization curves.

## Magnetization (B−H) Curve

Typical magnetization or $B-H$ curves for sheet steel, cast steel, and cast iron are plotted in Fig. 1-7. The nonlinear relationship between magnetic flux density $B$ (teslas) and magnetic field intensity $H$ (ampere-turns per meter) is illustrated. It is observed that the magnetic flux density increases almost linearly with an increase in the magnetic field intensity up to the knee of the magnetization curve. Beyond the knee, a continued increase in the magnetic field intensity results in a relatively small increase in the magnetic flux density. When ferromagnetic materials experience only a slight increase in magnetic flux density for a relatively large increase in magnetic field intensity, the materials are said to be *saturated*. Magnetic saturation occurs beyond the knee of the magnetization curve.

The characteristic of saturation is present only in ferromagnetic materials. An explanation of magnetic saturation is based on the theory that magnetic materials are composed of very many tiny magnets (magnetic domains) that are randomly positioned when the material is totally demagnetized. Upon application of a magnetizing force ($H$), the tiny magnets will tend to align themselves in the direction of this force. In the lower part of the magnetizing curve, the alignment of the randomly positioned tiny magnets increases proportionately to the magnetic field intensity until the knee of the curve is reached. Beyond the knee of the curve, fewer tiny magnets remain to be aligned, and therefore large increases in the magnetic field intensity result in only small increases in magnetic flux density. When there are no more tiny magnets to be aligned, the ferromagnetic material is completely saturated. In the saturation region of the curve, the magnetic flux density increases linearly with magnetic field intensity, just as it does for free space or nonmagnetic materials. From the origin of the $B-H$ curve there is a slight concave curvature beyond which is the essentially linear region. We shall see that the nonlinear characteristics of the magnetization curve have practical implications in the operation of electrical machines.

## Relative Permeability

Transposition of Eq. (1-6) gives the absolute permeability as the ratio of the magnetic flux density to the corresponding magnetic field intensity:

$$\mu = \frac{B}{H}$$

Thus we can obtain the values of absolute permeability of ferromagnetic materials from the magnetization ($B-H$) curves. Another method of obtaining the absolute permeability would be to take the slope (differential) of the curve at various points. Although the differential method may be more realistic, for our purposes in this book, the simpler method of ratios to obtain the absolute permeability will be acceptable.

If we wish to compare the permeability of magnetic materials with that of air, we may use the relative permeability $\mu_r$, which is defined by the equation

$$\mu_r = \frac{\mu}{\mu_0} \qquad (1\text{-}7)$$

where   $\mu$ = absolute permeability of the material, H/m
$\mu_0 = 4\pi \times 10^{-7}$ H/m = permeability of free space
$\mu_r$ = relative permeability

From the typical magnetization curves of Fig. 1-7, we can calculate the value of absolute and relative permeabilities for any magnetic operating condition. When we do this we observe that the value of relative permeability is not a constant but obtains a maximum value at about the knee of the $B$–$H$ curve.

### EXAMPLE 1-4

Calculate the absolute and relative permeabilities of cast steel operating at magnetic flux densities of 0.7 T and 1.0 T.

### SOLUTION

From the saturation curve for cast steel, the values of $H$ are 400 At/m and 800 At/m, respectively. The absolute permeabilities are:
For 0.7 T:

$$\mu = \frac{B}{H} = \frac{0.7}{400} = 1.75 \times 10^{-3} \text{ H/m} \quad \text{or} \quad \text{T/At/m}$$

For 1.0 T:

$$\mu = \frac{1.0}{800} = 1.25 \times 10^{-3} \text{ H/m}$$

The relative permeabilities are:
For 0.7 T:

$$\mu_r = \frac{\mu}{\mu_0} = \frac{1.75 \times 10^{-3}}{4\pi \times 10^{-7}} = 1392.61$$

For 1.0 T:

$$\mu_r = \frac{1.25 \times 10^{-3}}{4\pi \times 10^{-7}} = 994.72$$

Thus we see that cast steel has at least 1000 times more ability to set up magnetic flux lines than do nonmagnetic materials.

### Series and Parallel Magnet Circuits

By definition, a series magnetic circuit contains magnetic flux, which is common throughout the series magnetic elements. These series magnetic elements may consist of composite sectors of ferromagnetic materials of different lengths and cross-sectional areas, and of air gaps. The simplest series magnetic circuit would be of a toroid of homogeneous material and the steel core of a transformer. More complex series circuits which contain air gaps are illustrated in Fig. 1-5.

Parallel magnetic circuits are defined by the number of paths that the magnetic flux may follow. Any of these paths or branches may consist of composite sectors of magnetic materials, including air gaps. A detailed calculation for a typical parallel magnet circuit is demonstrated in Section 1-5.

### Electric Circuit Analogs

In our discussion so far, we note the following analogous relationships between magnetic quantities and electric quantities:

| Electric circuit | Magnetic circuit |
|---|---|
| $E$ (volts) | $F_m$ ($NI$ ampere-turns) |
| $I$ (amperes) | $\phi$ (webers) |
| $R$ (ohms) | $R_m$ (ampere-turns/weber) |
| $\rho = \dfrac{1}{\sigma}$ (conductivity) | $\mu$ (henries/meter) |

We can draw useful electrical analogs for the solution of magnetic circuit problems. In an electrical circuit the driving force is the voltage, the output is the current, and the opposition to establishing current is the resistance. In the same way, the driving force in the magnetic circuit is the magnetomotive force, the output is the magnetic flux, and opposition to establishing the flux is the reluctance.

Thus we have for the magnetic circuit of Fig. 1-8a the analogous electric circuit and the analogous magnetic circuit in Fig. 1-8b and c, respectively. The iron and air portions of the magnetic circuit are analogous to the two series resistors of the electric circuit. Analogous to the electric circuit, the magnetomotive force must overcome the magnetic potential drops of the two series reluctances in accordance with Kirchhoff's voltage law applied to magnetic circuits. Therefore,

$$F_m = R_{m_{\text{iron}}} \phi + R_{m_{\text{ag}}} \phi \qquad (1\text{-}8)$$

is the equivalent magnetic-potential-drop equation. Since the permeability of ferromagnetic materials (iron) is a variable depending on the state of magnetization, we must use the $B-H$ curves to obtain the magnetic field intensity if the magnetic flux density is available. Hence we can calculate the MMF drop for the iron from Eq. (1-5) as follows:

$$F_{m_{\text{iron}}} = H_{\text{iron}} l_{\text{iron}} \qquad \text{At} \qquad (1\text{-}9)$$

Finally, the general MMF-drop equation for series magnetic circuits is modified for calculation purposes to the following form:

$$F_m = H_{\text{iron}} l_{\text{iron}} + \frac{l_{\text{ag}} \phi}{\mu_0 A_{\text{ag}}} \qquad (1\text{-}10)$$

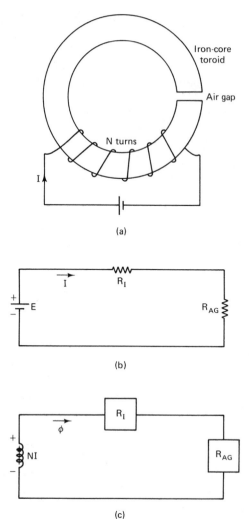

**Figure 1-8** Iron-core toroid with air gap: (a) Magnetic circuit; (b) analogous electric circuit; (c) analogous magnetic circuit.

Given the physical parameters of the series magnetic circuit and the value of magnetic flux or magnetic flux density, the required magnetomotive force can be calculated in a straightforward manner using Eq. (1-10).

The general principles of electric circuits embodied in Ohm's and Kirchhoff's laws are applied as analogous equivalents to parallel magnetic circuits. With the presence of air gaps, most complex magnetic circuits are solved using the series–parallel equivalent analogs. In analogous equivalents, Kirchhoff's current law for magnetic circuits states that the sum of magnetic fluxes entering a junction or node is equal to the sum of magnetic fluxes leaving the junction or node. Needless to say, magnetic flux must not be perceived as flowing.

### Fringing and Leakage Flux

In a series magnetic circuit containing an air gap, there is a tendency for the air-gap flux to spread out (i.e., to create a bulge). This spreading effect, termed *fringing*, reduces the net flux density in the air gap.

   *Leakage flux* is that flux in a magnetic circuit which is not useful or effective. Since a large amount of leakage flux requires a greater magnetomotive force, the designer of electromagnetic devices must minimize this ineffective flux.

### Magnetic Core (Iron) Losses

It will be shown later that the magnetic flux within the armature of dc machines changes direction as rotation occurs past the magnetic field poles. This change in direction of the armature magnetic flux is effectively an alternating flux. This results in *core losses*, which are treated in more detail in Chapter 8. Magnetic core losses consist of hysteresis losses and eddy-current losses.


## 1-5 MAGNETIC CIRCUIT CALCULATIONS

We have seen that magnetic circuits may be represented by electric circuit analogs. Thus the methods of solution for series and parallel electric circuits may be applied to magnetic circuit problems. Typically, we will be required to calculate the magnetomotive force, flux, or permeability for some given conditions. The major difference between the two types of circuits is the nonlinear characteristics of ferrous magnetic materials. Thus it is necessary to make use of $B-H$ curves and graphical methods.

### Series Magnetic Circuits

Example 1-5 illustrates the method of solution for a simple one-material series circuit.

### EXAMPLE 1-5

The circuit of Fig. 1-9 is a magnetic core made of cast steel. A coil of $N$ turns is wound on it. For a flux of 560 $\mu$Wb, calculate the necessary current, neglecting any fringing effects. The cross-sectional area $A$ is constant.

### SOLUTION

$$N = 550 \text{ turns}$$

$$l_1 = 20 \text{ cm} = 20 \times 10^{-2} \text{ m}$$

$$l_2 = 12 \text{ cm} = 12 \times 10^{-2} \text{ m}$$

$$A = 4 \text{ cm}^2 = 4 \times 10^{-4} \text{ m}^2$$

$$\phi = 560 \times 10^{-6} \text{ Wb}$$

$$B = \frac{560 \times 10^{-6}}{4 \times 10^{-4}} = 140 \times 10^{-2} = 1.4 \text{ T}$$

For $B = 1.4$ T, $H = 2200$ At/m (from the $B-H$ curve of Fig. 1-7). The average or mean length of the magnetic path is $20 + 12 + 20 + 12 = 64$ cm $= 0.64$ m. Therefore,

$$Hl = NI = 2200 \times 0.64 \text{ At}$$

$$I = \frac{2200 \times 0.64}{550} = 2.56 \text{ A}$$

### Parallel Magnetic Circuits

Figure 1-10a shows a parallel magnetic circuit. There are $NI$ ampere-turns on the center leg. The flux that is produced by the MMF in the center leg exists in the center leg and then divides into two parts, one going in the path *afe* and the other in the path *bcd*. If we assume for simplicity that *afe* = *bcd*, the flux is distributed evenly between the two paths. Now

$$\phi_g = \phi_{afe} + \phi_{bcd} \tag{1-11}$$

where
$\phi_g$ = flux in portion $g$
$\phi_{afe}$ = flux in portion *afe*
$\phi_{bcd}$ = flux in portion *bcd*

Equation (1-11) is actually the analog of Kirchhoff's current law, but now we can say that the amount of flux entering a junction is equal to the amount of flux leaving the junction.

Another observation that we may make on this circuit is that the MMF drops around a circuit are the same no matter what path we take. Thus the MMF drop

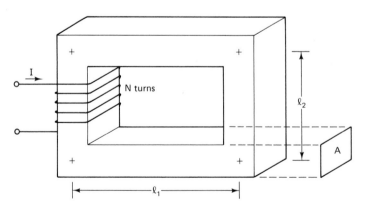

**Figure 1-9**  Magnetic circuit for Example 1-5.

Electromagnetic Principles   Chap. 1

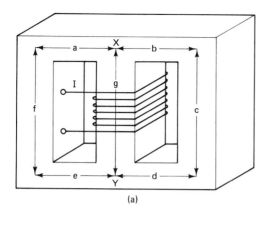

(a)

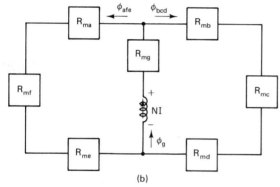

(b)

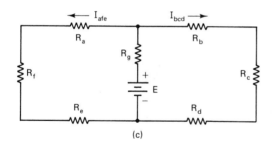

(c)

**Figure 1-10** Magnetic core with center leg: (a) Magnetic circuit; (b) equivalent magnetic circuit; (c) analogous electric circuit.

around *afe* must be equal to the MMF drop around *bcd*. This can be stated more precisely as

$$H_a l_a + H_f l_f + H_e l_e = H_b l_b + H_c l_c + H_d l_d \tag{1-12}$$

The drop in MMF around either path *afe* or *bcd* must also be equal to the MMF drop along path *g*. But *g* also has an "active source," the *NI* ampere-turns of the coil. The actual MMF existing between *X* and *Y* is the driving force *NI* minus the drop $H_g l_g$ in path *g*. Then we can write

$$\begin{aligned}(NI - H_g l_g) &= H_a l_a + H_f l_f + H_e l_e \\ &= H_b l_b + H_c l_c + H_d l_d\end{aligned} \tag{1-13}$$

Sec. 1-5    Magnetic Circuit Calculations                                    **19**

Again we can draw analogous magnetic and electrical circuits as in Fig. 1-10b and c. For Fig. 1-10b we may write

$$NI - R_{mg}\phi_g = \phi_{bcd}(R_{mb} + R_{mc} + R_{md}) \tag{1-14}$$
$$= \phi_{afe}(R_{ma} + R_{mf} + R_{me})$$

and in Fig. 1-10c we may write

$$E - R_g I_g = I_{bcd}(R_b + R_c + R_d) \tag{1-15}$$
$$= I_{afe}(R_a + R_f + R_e)$$

In the analogous magnetic circuit, note that $NI$ is drawn in series with $R_{mg}$, although physically the coil surrounds the central magnetic path.

**EXAMPLE 1-6**

In Fig. 1-10a, the following dimensions apply:

$$l_g = l_f = l_c = 12 \text{ cm}$$

$$l_a = l_b = l_e = l_d = 14 \text{ cm}$$

$$A_a = A_b = A_c = A_d = A_e = A = 1 \text{ cm}^2$$

$$A_g = 3 \text{ cm}^2$$

The material is sheet steel. The flux density in the center leg is 0.9 T. Calculate the MMF required to produce this flux density.

**SOLUTION**

The total flux in the center leg is $0.9 \times 3 \times 10^{-4} = 2.7 \times 10^{-4}$ Wb. The flux divides into two parts, the left-hand path through *afe* and the right-hand path through *bcd*. The flux density in path $g$ is $B_g = 0.9$ T and therefore $H_g = 320$ At/m. The flux density in section $a$ is

$$B_a = \frac{2.7 \times 10^{-4}}{2 \times 1 \times 10^{-4}} = 1.35 \text{ T}$$

and therefore

$$H_a = 950 \text{ At/m}$$

$$H_a = H_b = H_c = H_d = H_e = H_f$$

Therefore,

$$NI = H_g l_g + H_a(l_a + l_f + l_e)$$
$$= 320 \times 12 \times 10^{-2} + 950(14 + 12 + 14) \times 10^{-2}$$
$$= 38.4 + 380 = 418.4 \text{ At}$$

## EXAMPLE 1-7

We can add one more degree of complexity to the circuit of Fig. 1-10a. In Fig. 1-11 we cut an air gap in the center leg, and the air gap is 1.5 mm wide. All other dimensions remain unchanged and the flux density in the center leg is still 0.9 T. Find the number of ampere-turns on the center leg required to produce this flux density.

### SOLUTION

We can still use the equivalent-circuit concept as shown in Fig. 1-10b, the only difference being that $NI$ is now in series with two reluctances in the center path, the air gap and the steel in leg $g$.

$$(NI) - \text{(MMF drop in air gap)} - \text{(MMF drop in section } g)$$

$$= \text{MMF drop in section } b + c + d$$

$$= \text{MMF drop in section } a + f + e$$

In the center leg, the flux density is still fixed at 0.9 T. Therefore,

$$B_g = 0.9 \text{ T}$$

The MMF drop per unit length in the center steel section is still $H_g = 320$ At/m, as before. Therefore,

$$\text{MMF drop in leg } g = 320(12 - 0.15) \times 10^{-2} = 37.92 \text{ At}$$

The MMF drop across the air gap is found from

$$F_{m\text{gap}} = H_{\text{gap}} l_{\text{gap}}$$

For air

$$\mu = \mu_0 = 4\pi \times 10^{-7} \text{ Wb/(At/m) or H/m}$$

Therefore,

$$H_{\text{gap}} = \frac{0.9}{4\pi \times 10^{-7}} = 7.16 \times 10^5 \text{ At/m}$$

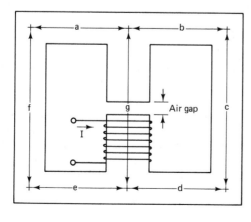

**Figure 1-11** Parallel magnetic circuit with air gap.

The MMF drop across the gap

$$NI_{gap} = 7.16 \times 10^5 \times 1.5 \times 10^{-3} = 1.074 \times 10^3 \text{ At}$$

Noting that the MMF drop across the path *afe* is still 380 At, as before,

$$NI - (37.92 + 1074) = 380 \text{ At}$$

$$NI = 1491.91 \text{ At}$$

By adding a very small air gap, the MMF required has increased by a factor of 3.57. This is because the reluctance of the air is so high and the reluctance of unsaturated steel is very low. This, in turn, is the reason why the largest part of a magnetic circuit is usually in iron and only a small portion is in air.

There are many combinations of circuits and coils that can be solved by this method, and it is suggested that the reader do several of the problems at the end of this chapter. The inverse problem—given the MMF $F_m$, find the resulting flux—is considerably more complicated. The complication arises from the fact that the relationship between $B$ and $H$, the permeability, is not a constant. This leads to a trial-and-error method that will not be dealt with here.

## 1-6 PERMANENT MAGNETS

Permanent magnets are commonly used as compasses and magnetic lifts. Today, there is a substantial increase in the application of permanent magnets for electromagnetic devices such as instruments, magnetic clutches and brakes, loudspeakers, and relays, as well as small generators and motors.

Modern permanent-magnet materials are alloys composed of nickel, aluminum, and iron, described by the trade name Alnico. Current research has developed rare-earth materials for permanent magnets having extremely high values of residual flux density. A wide variety of powdered-composition permanent magnets called *ferrites* are useful for relatively high-frequency applications. The composition materials of ferrites are usually barium and ceramic.

## 1-7 ELECTROMAGNETIC FORCES

By the interaction of magnetic fields produced in electromagnetic devices, mechanical forces are developed which may do useful work. Electromagnetic forces fall into two general classifications: (1) the magnetic tractive force, and (2) the force on a conductor. There are many examples of forces acting in electromagnetic fields. An electromagnet used to separate ferrous from nonferrous material is one, the deflection of an electron beam in a cathode-ray tube is another, and the action of an electric motor is a third. A fourth example is the attraction of an armature to an electromagnet, such as in relays, contactors, and lift magnets.

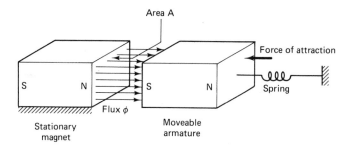

Area A

Force of attraction

S N S N

Spring

Flux $\phi$

Stationary
magnet

Moveable
armature

**Figure 1-12**   Force of attraction in magnetic circuits.

### Magnetic Tractive Force

We will consider the forces of attraction acting in an air gap between parallel surfaces.   In Fig. 1-12 the magnetic tractive force $F_{MT}$ between the two pieces of magnetic material of cross-sectional area $A$ having a flux density of $B$ tesla is expressed by

$$F_{MT} = \frac{B^2 A}{2\mu_0} \qquad \text{newtons (N)} \qquad (1\text{-}16)$$

This is the force that acts in the case of an electromagnet used for lifting large sheets of steel or in a magnetic ore separation.   The work done or energy expended in moving an armature a distance $l$ is given by

$$W_{MT} = F_{MT}l = \frac{B^2 A l}{2\mu_0} \qquad \text{newton-meters (N · m) or joules (J)} \qquad (1\text{-}17)$$

### EXAMPLE 1-8

Let us calculate the current required to lift a large cast-iron plate using the electromagnet of Fig. 1-13.   We will assume that since the magnet and the plate are both rough surfaces, we have an equivalent air gap of about 1.5 mm.   The plate has a mass of 400 kg.

### SOLUTION

The force required to lift the magnet is

$$\text{Total force} = 2 \times \text{force per pole}$$

$$= 2 \times \frac{B^2 A}{2\mu_0} = \frac{B^2 A}{\mu_0}$$

The force weight is

$$F = ma$$

$$= 400 \times 9.80 = 3920 \text{ N}$$

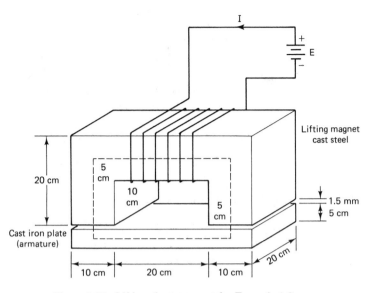

**Figure 1-13** Lifting electromagnet for Example 1-8.

Therefore,

$$3920 = \frac{B^2(0.10 \times 0.20)}{4\pi \times 10^{-7}}$$

and

$$B^2 = \frac{4\pi \times 10^{-7} \times 3920}{0.02} = 0.25$$

$$B = 0.5 \text{ T}$$

$$NI = Hl_{cs} + Hl_{CI} + \frac{B(2l_{ag})}{\mu_0}$$

$$= 250 \times 0.6 + 1950 \times 0.35 + \frac{0.5 \times 3.0 \times 10^{-3}}{4\pi \times 10^{-7}}$$

$$= 150 + 682.5 + 119.4$$

$$= 952 \text{ At}$$

Thus if $N = 1000$ turns, $I = 0.95$ A.

In general we observe that almost all of the ampere-turns are usually consumed by the relatively small air gap. In a practical case, since leakage flux and fringing have been neglected, increasing the value of the current by about 20% would probably yield a satisfactory solution.

## Force on a Conductor

Ampère demonstrated in 1820 that there is a magnetic field associated with a conductor carrying current. When placed in a transverse magnetic field, this conductor experiences a force that is proportional to (1) the strength of the magnetic field, (2) the magnitude of current in the conductor, and (3) the length of the conductor in, and perpendicular to, the magnetic field. In SI units, the electromagnetic force developed on the conductor carrying current in a magnetic field $B$ is given by

$$F = BlI \qquad \text{newtons} \tag{1-18}$$

Much use will be made of this important equation in subsequent chapters. Also, the direction of the force developed is discussed in Chapter 2.


## REVIEW QUESTIONS

**1-1.** What is a magnetic field?

**1-2.** List the important properties of magnetic lines of force.

**1-3.** List the important characteristics of a magnetic field produced by a current-carrying conductor.

**1-4.** State Ampère's right-hand rule.

**1-5.** Ferrous metals are magnetic; copper, aluminum, brass, and so on, are not magnetic. Explain why this is so.

**1-6.** State several advantages of electromagnets and solenoids.

**1-7.** Name the important magnetic quantities with their symbols that have the following units: webers, teslas, ampere-turns, ampere-turns per weber, henries per meter, or webers per ampere-meter.

**1-8.** State Ohm's law of magnetic circuits.

**1-9.** Define magnetic reluctance and state the factors involved in it.

**1-10.** Explain how a magnetization or $B-H$ curve is produced.

**1-11.** Why are there linear and nonlinear portions to this curve?

**1-12.** What is the significance of magnetic saturation? What disadvantages does it possess?

**1-13.** Why does relative permeability of ferromagnetic materials vary with the degree of magnetization?

**1-14.** Define series and parallel magnetic circuits. State the analogous electric circuit laws that would apply to the calculations of both types of magnetic circuits.

**1-15.** Explain the difference between fringing flux and leakage flux.

**1-16.** Name the two components of magnetic core (iron) losses.

**1-17.** Explain how electromagnetic forces are developed.

**1-18.** State the factors involved in the magnetic tractive force and in the force development on a current-carrying conductor.

## PROBLEMS

**1-1.** A coil of 1000 turns is wound on an air-core toroid as shown in Fig. 1-14, where $D_i$ = 4 cm and $D_s$ = 0.5 cm. If the flux in the cross section is 0.8 $\mu$Wb, calculate $I$. Assume that the flux is confined to the inside of the coil and is uniformly distributed across the cross section.

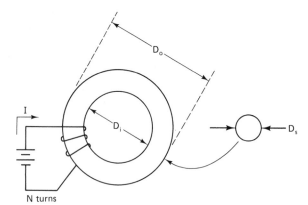

**Figure 1-14** Magnetic circuit for Problem 1-1.

**1-2.** A coil of 1000 turns is wound on an air-core toroid as shown in Fig. 1-14. The current in the coil is 5 A. $D_o$ = 7 cm, and $D_i$ = 5 cm. Calculate the flux density inside the coil, assuming that it is uniformly distributed over the coil cross section.

**1-3.** The toroid shown in Fig. 1-14 has a circular cross section and a cast iron core. The inner diameter is 4 cm, the outer diameter is 6 cm. If the flux in the core is 0.059 mWb, find the current in the coil.   500 TURNS

**1-4.** Consider the toroid shown in Fig. 1-14, given that $D_i$ = 13 cm and $D_s$ = 1.5 cm. The cast steel material has the magnetic characteristics of the $B$–$H$ curve shown in Fig. 1-7. If the coil has 150 turns, calculate $I$ so that the magnetic flux density is 1.5 T.

**1-5.** An air gap 1 mm wide is cut in the toroid of Problem 1-4. What is the current now required to produce flux density of 1.5 T?

**1-6.** A coil having 200 turns is wound on the toroid of Fig. 1-14. A 1.5-mm air gap is cut in the cast steel and a current of 2 A is passed through the coil. $D_i$ = 13 cm and $D_s$ = 1.5 cm. Assuming no leakage flux, calculate the flux density in the air gap.

**1-7.** Calculate the MMF needed to produce a flux of 1.3 × 10$^{-3}$ Wb in the air gap of Fig. 1-15.

**1-8.** A magnetic circuit consists of a cast steel yoke which has a cross-sectional area of 2 cm$^2$ and a mean length of 12 cm. There are two air gaps, each 0.2 mm long. Calculate the ampere-turns required to produce a magnetic flux of 50 $\mu$Wb in the air gaps.

**1-9.** A magnetic circuit has the dimensions shown in Fig. 1-16 (all dimensions in centimeters except as indicated). Find $I$ if the air-gap flux is 645 $\mu$Wb.

**1-10.** A magnetic core is made up of two parts, a sheet steel portion plus a cast steel portion (Fig. 1-17). The effective length of the cast steel is 8 cm, and of the sheet steel, 20 cm. The core has a uniform cross section of 15 cm$^2$. Find the number of ampere-turns of the coil required to produce a flux of 2.1 mWb in the core.

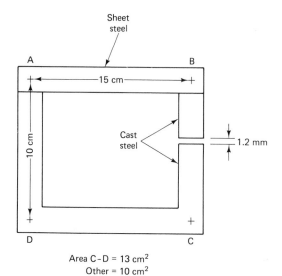

Area C-D = 13 cm²
Other = 10 cm²

**Figure 1-15** Magnetic circuit for Problem 1-7.

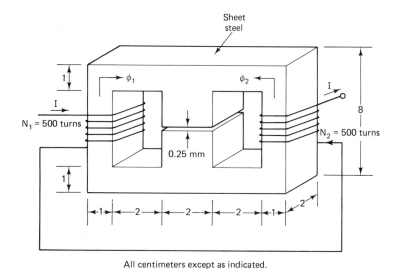

All centimeters except as indicated.

**Figure 1-16** Magnetic circuit for Problem 1-9.

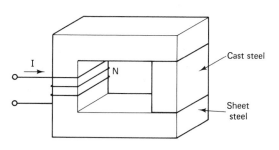

**Figure 1-17** Magnetic circuit for Problem 1-10.

**1-11.** A magnetic circuit made of cast steel has the dimensions shown in Fig. 1-18. The cross-sectional area is 4 cm² for all parts in the magnetic path. The air gap is 1.5 mm wide. Find the number of ampere-turns required on the center leg to produce a flux of 0.52 mWb in the center leg.

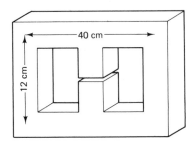

**Figure 1-18** Magnetic circuit for Problem 1-11.

**1-12.** The shunt-field winding of a dc two-pole machine has 1200 turns shown in Fig. 1-19. The magnetic flux path has a net cross-sectional area of 200 cm². The iron portion has a mean length of 50 cm, and there are two air gaps, each 0.1 cm in length. The magnetization curve for the iron in the circuit is:

| B (Wb/m²) | H (At/m) |
|-----------|----------|
| 1.0 | 350 |
| 1.2 | 650 |
| 1.4 | 1250 |

Draw the magnetization curve.
For the two-pole machine shown in Fig. 1-19, find the shunt-field current required to set up a flux of 0.02 Wb in each air group. Neglect all leakage and fringing effects.

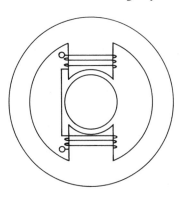

**Figure 1-19** Magnetic circuit for Problem 1-12.

**1-13.** A mild steel bar 0.4 m long and $4 \times 10^{-4}$ m² in cross section is bent to form a circle but leaving a uniform air gap. A 60-Ω 10,000-turn coil wound on the bar is excited by a 12-V battery. Find the length of the air gap required to give a flux of $4 \times 10^{-4}$ Wb in the bar. Neglect leakage and fringing fluxes. Assume a relative permeability of 1000 for mild steel at this flux density.

**1-14.** The electromagnet has a core and armature made of sheet steel (Fig. 1-20). A flux density of 1.42 T produces a force that closes the gap. Assume that the armature moves parallel to the core.
 (a) What current is required to close the gap?
 (b) What force is exerted to close the gap (i.e., what force is exerted to overcome the spring)?
 (c) After closure, to what value may the current be reduced and still maintain closure? Assume no air-gap reluctance after closure.

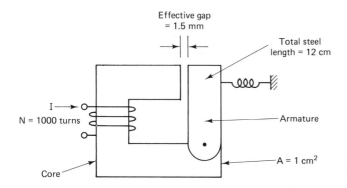

**Figure 1-20**  Magnetic circuit for Problem 1-14.

**1-15.** The shunt-field winding of a dc two-pole machine has 1000 turns (500 on each pole) and a total resistance of 250 $\Omega$. The equivalent total series reluctance of the iron part of the magnetic circuit (pole core, yoke, and armature core) is $2 \times 10^4$ At/Wb. There are two air gaps, each of length 1.5 mm, and a cross-sectional area of 200 cm² Neglecting leakage and fringing effects, find the magnetic flux in the air gap when the field winding is connected to a 125-V supply.

**1-16.** The shunt-field winding of a dc four-pole machine has 1000 turns (250 turns on each pole) and a total resistance of 100 $\Omega$. The equivalent total series reluctance of all the iron parts of the magnetic circuit (pole cores, yoke, and armature core) is 30,000 At/Wb. There are four air gaps, each of length 2.0 mm, and a cross-sectional area of 300 cm². Neglecting leakage and fringing effects, find the magnetic flux density in the air gap when the field winding is connected to a 125-V supply.

**1-17.** A circular ring of cast steel has an internal diameter of 3.0 in. The cross section of the ring has a diameter of 0.5 in. It has 500 turns of insulated wire wrapped around it. Determine the current required for a magnetic flux of 0.125 mWb:
 (a) If the ring is continuous.
 (b) If there is an air gap of 0.04 in.

# 2

# PRINCIPLES of DC MACHINES

## 2-1 INTRODUCTION

As noted at the beginning of Chapter 1, in 1819, H. C. Oersted demonstrated that a current-carrying coil produced a magnetic field. In 1821, Michael Faraday demonstrated electromagnetic rotation of a conductor and a magnet. In effect, Faraday was the first scientist to develop the basis of motor action and hence the first electric motor. In 1831, Faraday made his second great discovery—that it was possible to produce an electric current from a magnetic field. On the basis of his experiments, it was feasible to design both electric generators and electric motors, and it is fair to say that the electrical age really began with the work of Faraday.

## 2-2 GENERATED VOLTAGE EQUATION

We can demonstrate the results of Faraday's discoveries in a very simple, straightforward experiment using the layout suggested in Fig. 2-1. A conductor of length $L$ is moving along a set of parallel rails at a constant velocity of $v$ meters per second perpendicular to a magnetic field of flux density $B$ (tesla) which points into the paper as indicated by the plus sign. It is observed that the meter indicates a generated voltage

$$E_{XY} = BLv \qquad \text{volts} \qquad (2\text{-}1)$$

having the polarity shown. This equation is valid for a system in which $B$, $L$, and $v$ are all mutually perpendicular and is called the *generator equation* or the *voltage–speed equation*.

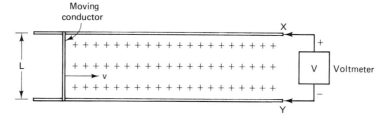

**Figure 2-1** Voltage generated by a moving conductor. +, Magnetic field into page; flux density $B$.

### EXAMPLE 2-1

If a conductor 0.5 m in length moves at a velocity of 10 m/s perpendicular to a magnetic field of flux density 0.1 T as in Fig. 2-1, find the generated voltage.

**SOLUTION**

From Eq. (2-1),

$$E_{XY} = 0.1 \times 0.5 \times 10 = 0.5 \text{ V}$$

## 2-3 DIRECTION OF GENERATED VOLTAGE

If the magnetic field direction is reversed (i.e., is out of the page), the voltage polarity is reversed and $E_{xy}$ is $-0.5$ V. If we keep the original field direction and move the conductor from right to left, we will find that $E_{XY}$ is $-0.5$ V.

Let us return to the original conditions as shown in Fig. 2-1 and connect a 2-$\Omega$ resistor in series with an ammeter as shown in Fig. 2-2. As before, the generated voltage is 0.5 V and the current read by the ammeter is

$$I = \frac{0.5}{2} = 0.25 \text{ A}$$

in the direction shown. We can determine the polarity of generated voltage and the current by the use of the right-hand rule. The rule can be described as follows. Imagine that the flux lines are elastic bands and that the moving conductor bends

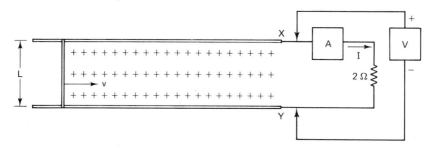

**Figure 2-2** Moving conductor generator connected to a resistor load.

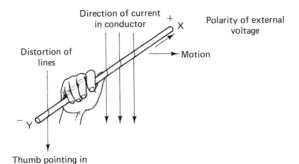

Direction of current
in conductor
Polarity of external
voltage

Distortion of
lines

Motion

Thumb pointing in
direction of current

**Figure 2-3** Illustrating Ampère's right-hand rule.

them as it moves forward. Then grasp the conductor with the right hand with the fingers following the curvature of the magnetic field and pointing in the direction of the field. The thumb will point in the direction of conventional current flow within the conductor.

In Fig. 2-2 or 2-3 the current is attempting to flow from $Y$ to $X$ *within* the conductor, and this produces a voltage of polarity $E_{XY}$ at the voltmeter. The voltmeter "sees" the externally generated voltage of 0.5 V with point $X$ positive with respect to point $Y$.

## 2-4 GENERATOR PRINCIPLES

Equation (2-1) describes a system in which $L$, $B$, and $v$ are always at right angles. If we modify the apparatus of Fig. 2-1 by tilting the track, we have the system shown in side view in Fig. 2-4b.

We now find that the generated voltage is less than that obtained previously. When we examine the motion, we see that the velocity perpendicular to the flux has decreased. In fact, if $\theta$ were 90°, we would have zero generated volts, and if $\theta$ is 45°, we have $0.5 \times 0.707 = 0.3535$ V. We can see that the velocity perpendicular to the magnetic field is given by

$$v_p = v \cos \theta \qquad (2-2)$$

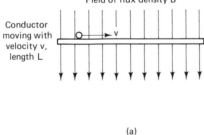

Field of flux density B

Conductor
moving with
velocity v,
length L

(a)

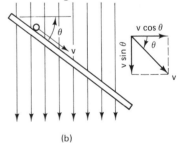

(b)

**Figure 2-4** Conductor moving at an angle to a magnetic field: (a) side view of Fig. 2-1; (b) side view with track tilted.

where  $v_p$ = velocity perpendicular to the magnetic field, m/s

 $v$ = velocity of conductor, m/s

 $\theta$ = angle between $v$ and $v_p$

This equation is useful if the angle between $v$ and $B$ changes, and this of course is what happens if we move a conductor in a circle in a uniform magnetic field as shown in Fig. 2-5. Figure 2-5a shows a one-turn coil being rotated in a magnetic field, and this represents the configuration of an actual machine. For the moment, we are interested in the mechanism of voltage generation in a single

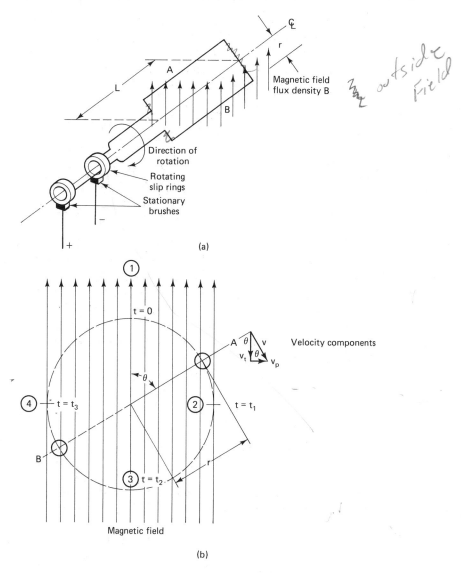

**Figure 2-5** Voltage generated in a coil: (a) coil rotating in a field; (b) sectional view of conductors $A$ and $B$ of coil ($A$, $B$, conductors of length $L$).

Sec. 2-4  Generator Principles

conductor so that we may consider the situation depicted in Fig. 2-5b. Here we have a conductor of length $L$ moving at a radius $r$ meters in a magnetic field of flux density $B$ teslas. The tangential velocity of the conductor is $v$ meters per second. When $\theta = 0$, the velocity perpendicular to the field is $v \cos \theta = v \cos 0 = v$ meters per second, so that at this instant of time, the generated voltage is $e_G = BLv$ volts. At time $t_1$, the conductor reaches the horizontal position shown, and $\theta$ is 90° or $\pi/2$ radians. At this point, from Eq. (2-2),

$$e_G = BLv \cos 90° = 0$$

We may note also that at $t_1$, the conductor is moving parallel to the magnetic field; hence it is not cutting any flux lines and thus the generated voltage must be zero. We can extend this concept further by noting that at any time, the velocity $v$ is always tangent to a circle of radius $r$, and that the velocity $v$ can be considered to be made up of two components,

$$v_p = v \cos \theta \quad \text{and} \quad v_l = v \sin \theta \tag{2-3}$$

which are perpendicular to and parallel to the magnetic field, respectively. Since voltage is generated only by the velocity perpendicular to the field, the voltage generated is

$$e_G = BLv_p = BLv \cos \theta \tag{2-4}$$

as we would expect from Eq. (2-2), and the velocity parallel to the field generates no voltage.

Note that the generated voltage in this case is represented by a lowercase $e$. In this instance $e$ is a constantly changing quantity, since $\theta$ is always changing. We must be careful to distinguish between voltages that do not change with time, such as the voltage of a battery or the constant-velocity voltage represented by Eq. (2-1), and the varying-velocity voltage represented by Eq. (2-4).

If the conductor has a tangential velocity of $v$ meters per second, and it is at a radius of $r$ meters, it travels a distance of $2\pi r$ meters in one revolution and the time per revolution is

$$T = \frac{2\pi r}{v} \quad \text{s} \tag{2-5}$$

and

$$v = \frac{2\pi r}{T} \quad \text{m/s} \tag{2-6}$$

where $T$ is the time required for one revolution and is called the *period*, in seconds. The frequency of revolution in hertz (cycles per second) is

$$f = \frac{1}{T} \quad \text{Hz} \tag{2-7}$$

and the angular velocity is

$$\omega = 2\pi f \quad \text{rad/s} \tag{2-8}$$

Substitution in Eq. (2-4) yields

$$e_G = BL2\pi rf \cos \theta \qquad (2-9)$$

where $e_g$ is the voltage at any instant of time.

It is often convenient to express an angle in radians rather than in degrees. Equation (2-9) is equally valid for $\theta$ in radians or degrees, but we must be careful to express it in the correct units. The angular velocity $\omega$ radians per second is related to the angle $\theta$ by

$$\theta = \omega t \qquad (2-10)$$

and we can then substitute Eq. (2-10) into Eq. (2-9) and obtain

$$e_G = BL2\pi rf \cos \omega t \qquad (2-11)$$

At the end of one revolution

$$\theta = 2\pi \qquad \text{and} \qquad t = T = \frac{1}{f}$$

Substitution in Eq. (2-10) gives

$$2\pi = \omega T$$

$$\omega = \frac{2\pi}{T} = 2\pi f$$

and Eq. (2-11) becomes

$$e_G = BLr\omega \cos \omega t \qquad \text{V} \qquad (2-12)$$

where $\omega$ is the angular velocity in radians per second.

This is the voltage induced in one conductor as it spins through a complete revolution starting from a vertical position. As noted, in a practical generator conductors always come in pairs to form complete turns, and several turns are usually joined in series to form complete coils. For a single-turn coil, Eq. (2-12) becomes

$$e_G = 2BLr\omega \cos t \, \omega t \qquad \text{V}$$

and for a coil of $N$ turns,

$$e_G = 2BLr\omega N \cos \omega t \qquad \text{V} \qquad (2-13)$$

In a practical machine, conductors are arranged in pairs or groups of pairs.

The voltage generated by a moving conductor is collected by a pair of slip rings that touch a set of stationary brushes. The voltage generated between the slip rings of Fig. 2-5a as a function of time is shown in Fig. 2-6. The equation of the voltage wave is

$$e_G = E_{\max} \cos \omega t$$

$$= E_{\max} \cos 2\pi ft$$

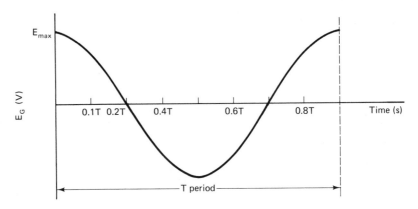

**Figure 2-6**   Generated voltage as a function of time.

where $\quad e_G$ = generated voltage, V
$E_{max}$ = $2BLr\omega N$ = maximum generated voltage, V
$\omega$ = angular velocity, rad/s
$t$ = time, s
$f$ = angular frequency, Hz

Note that this is the voltage of two coil sides in series and that it is twice the voltage of one conductor acting alone.   The quantity $T = 1/f$ is the period of the wave and is the time required for one revolution of the coil.

### EXAMPLE 2-2

A one-turn coil is arranged to spin in a magnetic field at a rotational frequency of 3600 r/min.   It is 0.25 m long and has a radius of 0.3 m.   The magnetic field is uniform and has a flux density of 1.1 T.   Tabulate the generated voltage versus time.

### SOLUTION

The generated voltage at any instant of time is given by Eq. (2-13).   We have

$$B = 1.1 \text{ T}$$

$$L = 0.25 \text{ m}$$

$$r = 0.3 \text{ m}$$

The rotational frequency is 3600 r/min.

$$f = \frac{3600}{60} = 60 \text{ r/s}$$

$$\omega = 2\pi f = 2\pi \times 60 = 377 \text{ rad/s}$$

$$e_G = 2 \times 1.1 \times 0.25 \times 0.3 \times 377 \times 1 \times \cos 377t$$

$$= 62.2 \cos 377t = 62.2 \cos \theta$$

$$\theta = 377t$$

| $t$ (ms) | $\theta$ (rad) | $\cos \theta$ | $e_g$ (V) |
|---|---|---|---|
| 0 | 0 | 1.000 | 62 |
| 2 | 0.759 | 0.730 | 45 |
| 4 | 1.508 | 0.069 | 43 |
| 6 | 2.25 | $-0.630$ | $-39$ |
| 8 | 3.01 | $-0.990$ | $-61$ |
| 10 | 3.76 | $-0.814$ | $-50$ |
| 12 | 4.51 | $-0.201$ | $-12$ |
| 14 | 5.27 | 0.530 | 33 |
| 16 | 6.02 | 0.966 | 6 |
| 18 | 6.77 | 0.875 | 54 |
| 20 | 4.52 | 0.326 | 2 |

## 2-5 GENERATION OF UNIDIRECTIONAL VOLTAGE

From the results of Example 2-2 we can see that it is comparatively simple to generate an alternating voltage. However, the generation of a unidirectional voltage, or one of constant polarity and magnitude, is a bit more complex. The method of obtaining a unidirectional voltage—one that does not change polarity—requires the use of a switch or commutator. In its simplest form the commutator consists of two half-cylinders of copper separated by a nonconducting material such as mica. Each segment of copper is connected to a coil side as shown in Fig. 2-7, and the whole commutator rotates with the coil. Two stationary carbon brushes make contact with the copper, so that at any given time one commutator segment is in contact with one brush, except for a short period of time when the insulation is at the brush centers and the brushes each touch both commutator sections. Figure 2-8 is a cutaway view of the coil viewed from the commutator end as the coil rotates.

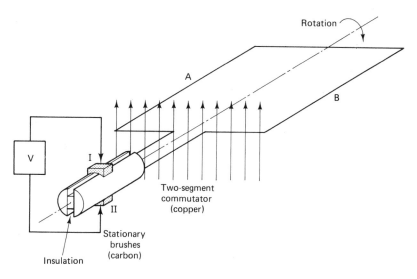

**Figure 2-7** Commutator connection.

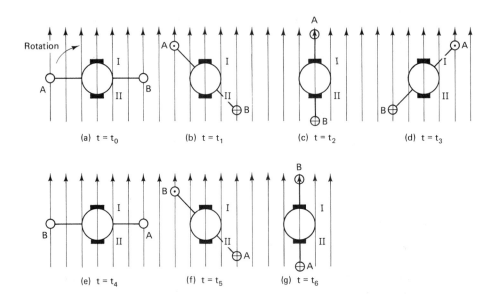

**Figure 2-8** Time sequence of voltage generation: (a) $t = 0$; (b) $t = t_1$; (c) $t = t_2$; (d) $t = t_3$; (e) $t = t_4$; (f) $t = t_5$; (g) $t = t_6$.

At $t = 0$ the angle $\theta$ is zero, and the voltage produced in the coil is zero. As the coil rotates, the voltage changes sinusoidally until the coil is vertical, as shown at time $t_2$ when $\theta$ is 90°, and the voltage is *maximum*. The voltmeter in Fig. 2-7 indicates $E_{max}$ volts and this is the voltage between brushes I and II, or $e_{I\text{-}II}$. At time $t_4$ the coil sides $A$ and $B$ are both moving parallel to the magnetic field; hence no voltage is induced in either one and the brushes can short-circuit the commutator segments without causing any harm. As the coil continues to rotate, coil side $B$ comes into contact with commutator segment I, and coil side $A$ comes in contact with commutator segment II. The voltage generated in $B$ is now beginning to go positive, and by time $t_6$, $B$ reaches the vertical position the generated voltage is a maximum. Coil side $B$ is in contact with brush I, so that voltage $e_{I\text{-}II}$ at this instant of time is exactly the same as it was at time $t_2$.

The commutator and brush arrangement allows the machine always to have the upper brush in contact with a coil side which is producing a voltage that is positive with respect to the bottom brush. The conductor that is in contact with brush I is always positive with respect to the conductor in contact with brush II, so that the voltage of brush I is always positive to brush II.

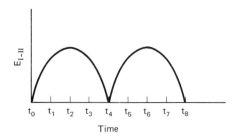

**Figure 2-9** Output voltage of commutated coil.

Principles of DC Machines    Chap. 2

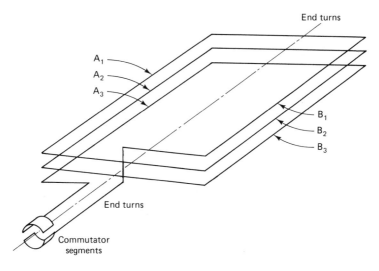

Figure 2-10  Three-turn coil.

## 2-6  MULTITURN COIL

The resulting voltage shown in Fig. 2-9 is obviously unidirectional, but it can hardly be called pure dc. It has a very large ripple component, and as we have seen from Example 2-1, a single turn does not produce a very large voltage. To increase the generated voltage and decrease the ripple, we can add more turns to the coil and more coils spaced around the periphery of the machine rotor.

Figure 2-10 shows a three-turn coil which would produce three times the voltage of a single-turn coil. In an actual machine the coil sides $A_1$, $A_2$, and $A_3$ are located in a slot in an iron core as shown in Fig. 2-11, and coil sides $B_1$, $B_2$, and $B_3$ are located in another slot on the opposite side of the core. If coil sides $A_n$ are located under a north pole, coil sides $B_n$ will be located under a south pole. This arrangement shows how it is possible to increase generated voltage, but it does not do anything to reduce the ripple.

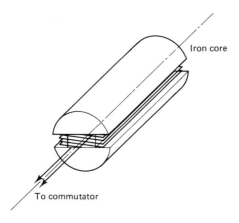

Figure 2-11  Coil sides in iron core.

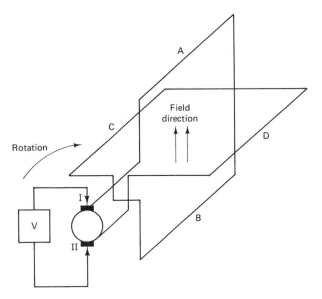

**Figure 2-12** Rotor position at $t = 0$.

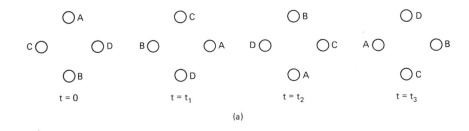

(a)

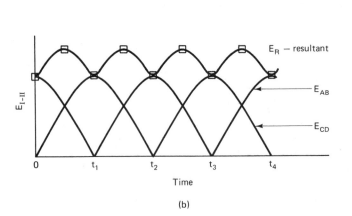

(b)

| Time | A | B | C | D |
|------|------|------|------|------|
| 0 | 0 | 180° | 270° | 90° |
| $t_1$ | 90° | 270° | 0 | 180° |
| $t_2$ | 180° | 0 | 90° | 270° |
| $t_3$ | 270° | 90° | 180° | 0 |
| $t_4$ | 0 | 180° | 270° | 90° |

**Figure 2-13** (a) Position of coil sides during one revolution; (b) voltage generated by out-of-phase coils.

Principles of DC Machines Chap. 2

## 2-7 MULTICOIL WINDING

The method used to reduce the ripple may be illustrated in an elementary form by considering the arrangement of Fig. 2-12. Here there are two coils, $A-B$ and $C-D$, which are arranged on the rotor so that they are always at right angles to each other. Figure 2-13a shows successive positions of the coil ends in space. The voltage generated by the individual coils is shown in Fig. 2-13b. The voltage in coil $A-B$ starts off at a maximum at $t = 0$ and decreases to zero at $t = t_1$, and the shape is that of a cosine. The voltage in coil $C-D$ starts at zero and increases to a maximum at $t = t_1$ and then decreases to zero at $t_2$. The waveshapes are of course identical, but there is a phase displacement between them. Every event that occurs in coil $A-B$ happens $t_2$ seconds later in coil $C-D$, and $t_2$ seconds corresponds to a phase of 90° on a sine wave, so that $E_{C-D}$ is 90° out of phase with $E_{A-B}$.

If we add $E_{A-B}$ and $E_{C-D}$ point by point by electrically connecting the two voltages in series, we obtain a winding and the resultant voltage is $E_R$, shown in Fig. 2-13b.

## 2-8 WINDING DIAGRAM

We can also understand the commutation mechanism by considering the layout of Fig. 2-14. This is an elementary form of winding diagram. It is formed by cutting the machine in half through the interpolar space and spreading it out. Then we can clearly see the distribution of coil sides in space. Winding diagrams are drawn for one specific time, and in the case of Fig. 2-14a, we assume that this is at the time when coil $A-B$ is at the center of a pole face. In the diagram, coil side $A$ is under the north pole, and is connected to commutator segment $X$ which at $t = 0$ is in contact with brush I. Coil side $B$ is under a south pole and the end is connected to coil side $C$, in the space between poles. Coil side $D$ is in the opposite interpolar space and it is connected to commutator segment $Y$, which at $t = 0$ is in contact with brush II.

At $t = 0$, the voltage generated in the individual coil sides can be determined by the right-hand rule. The internal current direction is as indicated by the arrows, and the voltage is positive from brush I to brush II. As a coil moves, the polarity change can be seen clearly by following the sequence illustrated in Fig. 2-14. Again, we can see that the polarity of $E_{I-II}$ is always the same, and a voltage is produced which now has a similar ripple. By adding many coils in series spaced around the periphery it is possible to generate several hundred volts with a very small ripple, as shown in Fig. 2-15.

The winding shown in Fig. 2-12 is, of course, a very elementary one, as is the diagram of Fig. 2-14a. In actual machines, all the windings are made up of prefabricated coils of many turns, and all the coils are connected in series at the commutator bars.

Figure 2-16 shows a typical individual coil made up of many turns. Only the beginning and end are brought out from the insulating cover, and these ends are

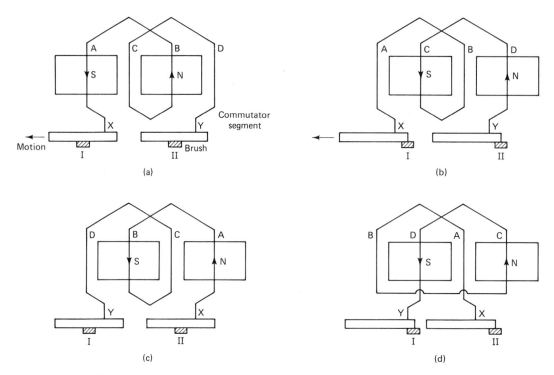

**Figure 2-14** Elementary winding diagram for Fig. 2-12 (a wave winding): (a) $t = 0$; (b) $t = t_1$; (c) $t = t_2$; (d) $t = t_3$.

connected to commutator segments as shown in Fig. 2-18. Coils are normally arranged so that the opposite edges are approximately 180° apart, so that if one side is under a north pole, the opposite side is under a south pole.

As noted in Fig. 2-11, coils sit in slots in the rotor iron. Coils are arranged in pairs for electrical and mechanical balance, and a coil is mounted so that each side occupies the top of one slot and the bottom of the opposite slot. Then a second coil is assembled in the same slot to occupy the top and the bottom of the

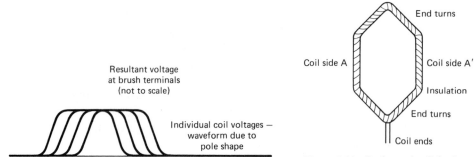

**Figure 2-15** Resultant voltage for many coils in series.

**Figure 2-16** Preformed coil for lap winding.

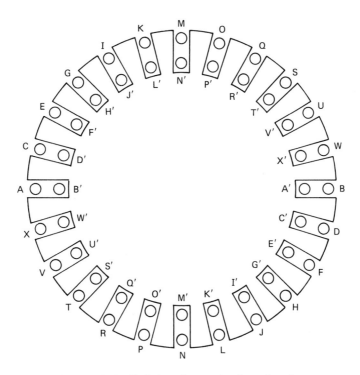

**Figure 2-17** End view of rotor showing coil ends.

next opposite slot. Figure 2-17 shows the coil arrangement. Note that in Fig. 2-16 the parts of the coil marked "end turns" do not play any significant part in the voltage-generation mechanism since they are not under the poles. They serve merely to connect the sides of the coil and in an electrical-generating sense are really passive components.

To complete the picture of winding diagrams, Fig. 2-17 shows the arrangement of pairs of coils in slots. For clarity, the connections to the commutator segment are not shown, but the connections are shown in Fig. 2-18. A solid line indicates the coil side in the top of a slot, while a dashed line indicates the side located in the bottom of a slot. By tracing the coil connections between brushes, it may be seen that there are really two paths between the brushes.

Figure 2-19 shows how the complete commutator works. The coils form one continuous set of small generators connected in series. If we imagine the rotor turning counterclockwise as suggested in Fig. 2-19, the coils on the left side of the brush axis have their trailing edge positive with respect to their leading edge, while the coils on the right of the brush axis have their leading edge positive with respect to the trailing edge. In effect, we have two parallel sets of small series-connected generators, both of which produce a resultant voltage, making brush 1 positive with respect to brush 2. As the machine rotates an individual coil passes the brush axis. When it passes the axis, its polarity is reversed so that all individual generators are always correctly connected.

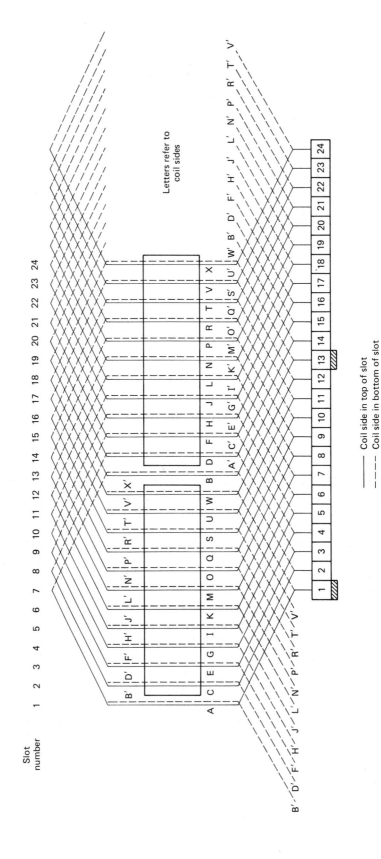

**Figure 2-18** Complete winding diagram (lap winding). Solid line, coil side in top of slot; dashed line, coil side in bottom of slot.

44

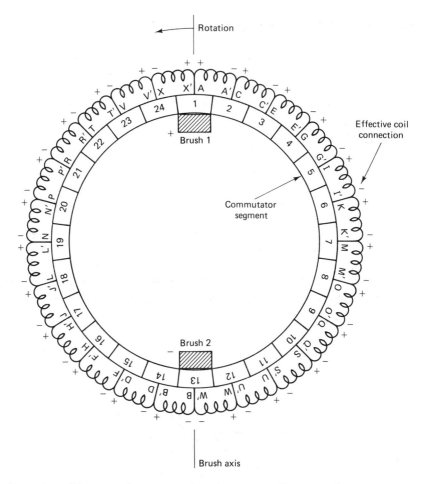

**Figure 2-19** Diagrammatic representation of armature coils connected to commutator segments (wave winding).

## 2-9 MOTOR ACTION

Motor action is the development of force by a current-carrying conductor within a magnetic field in which all or a portion of the magnetic lines of force must be perpendicular to the conductor. From experiments, the magnitude of the mechanical force developed is proportional to the current in the conductor within the magnetic field, and the strength (magnetic flux density) of the field [see Eq. (1-18)]. In order to develop continuous torque and hence rotation as required for motor operation, the current-carrying conductors when moving from a north field pole to a south field pole, and vice versa, must reverse the direction of current. This change of direction of current within the conductors is accomplished by the commutator and brushes. Note that the brushes supply the rotor winding (armature circuit) with unidirectional current from an external dc source.

It is important to be able to determine the direction of the developed force as illustrated in Fig. 2-20a. If the current flowing in the conductor is as shown here, the mechanical force developed will be to the left. However, if either the polarity of the magnetic fields or the direction of current flow were reversed, the force developed would be to the right.

The direction of the developed force in motor action can be determined as follows. Associated with the current in the conductor is a field that interacts with the main field flux of the magnetic poles (Fig. 2-20). The direction of the associated magnetic lines in the conductor are determined by Ampère's right-hand rule. This interaction results in a strengthening of the magnetic flux lines on the right (i.e., both sets of lines of force are aiding or strengthening each other). However, on the left side, the two sets of magnetic lines of force oppose each other, resulting in a net weakened field. As a consequence, the current-carrying conductor is being forced from the strengthened net field to the weakened net field (to the left) by this interaction of the two magnetic fields. This may be considered analogous to the stretching and tensioning of elastic bands when hurling an object.

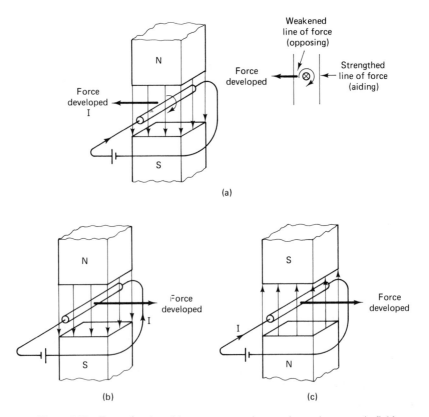

**Figure 2-20** Force developed by current-carrying conductor in magnetic field.

A simple method of determining the direction of the force acting on the conductor is the use of Fleming's left-hand rule for a motor. With this rule, the thumb, index finger, and the first (middle) finger of the left hand must all be positioned at right angles to each other. With the index finger pointing in the direction of the magnetic field and the first finger pointing in the direction of current flow, the thumb will indicate the direction of the force developed. The direction of the maximum developed force is always perpendicular to both the direction of the magnetic field and the axis of the conductor.

It must be noted that current-carrying conductors in a generator are rotated within a magnetic field, resulting in a force and torque that the prime mover must overcome. This opposing torque in a generator is called the *counter* or *retarding torque*. The torque developed by current-carrying conductors (coils) in a magnetic field (permanent magnets or electromagnets) has useful application also in measuring instruments.

We can develop the "motor equation" from an experiment using the configuration shown in Fig. 2-21, which is very similar to that of Fig. 2-1 except that now a voltage is applied to the device rather than obtaining a voltage from it. A spring balance is used to measure the force on a conductor of length $L$ carrying a current $I$ at right angles to a magnetic field having a flux density $B$. Then the force tending to move the conductor is found to be

$$F_M = BLI \qquad (1-18)$$

where
$$F_M = \text{mechanical force, N}$$
$$B = \text{flux density, T}$$
$$L = \text{length of conductor, m}$$
$$I = \text{current, A}$$

From the right-hand rule (Section 1-2) we know the direction of the magnetic field created by the current around the conductor, as indicated in Fig. 2-21. We may consider the flux lines as aiding each other on the side of the conductor toward the spring scale, and as opposing each other on the other side. The force acts in

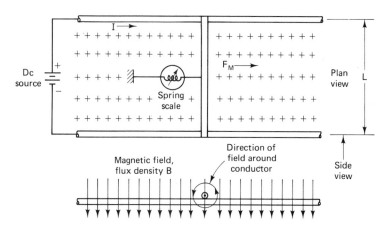

**Figure 2-21** Force acting on a conductor.

the direction from the stronger field concentration toward the weaker, as shown by the arrow labeled $F_M$.

## 2-11 FORCE, POWER, AND TORQUE

The work done when a force $F_M$ is applied to move a body a distance $d$, as shown in Fig. 2-22, is

$$W = F_M \times d \tag{2-14}$$

where    $W$ = work done, J
$F_M$ = force applied, N
$d$ = distance, m

Power is the rate at which work is done, so that

$$P = \frac{W}{t} = \frac{F_M \times d}{t} = F_M \times v \tag{2-15}$$

where    $P$ = power expended in doing work, W
$t$ = time, s
$v$ = velocity of the body, m/s

In the study of motors, we are interested primarily in circular motion. In Fig. 2-23 a body is constrained to move in a circle of radius $r$ while a force $F_M$ is applied that is tangent to the circle. The tangential velocity of the body is $v$. The power required to move the body is then $F_M \times v$, as in Eq. (2-15).

Now the time required for one revolution is the period $T_p$ and the number of revolutions per second or frequency $f$ is then $1/T_p$. In one revolution the body will travel $2\pi r$ meters, so that the work done in one revolution

$$W = F_M \times d = F_M \times 2\pi r$$

The time taken is $T_p$ seconds, so that the power required to move the body is

$$P = \frac{W}{T_p} = F_M \times 2 \times \pi \times \frac{r}{T_p} = F_M \times 2\pi r \times f$$

If the frequency is expressed in radians per second,

$$\omega = 2\pi f \tag{2-16}$$
$$P = F_M \times r \times \omega$$

where $\omega$ is the angular velocity in radians per second. The quantity $F_M \times r$ is called the torque $T$, so that we can write

$$P = T \times \omega \tag{2-17}$$

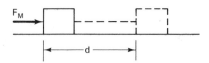

**Figure 2-22** Force applied to a body.

Principles of DC Machines    Chap. 2

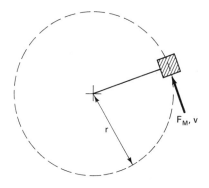

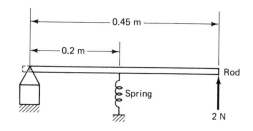

**Figure 2-23** Body constrained to move in a circle.

**Figure 2-24** Forces acting on a spring–rod system.

where $T$ is the torque in newton-meters. Torque is the turning moment and is the product of force and distance, but it is independent of velocity. A force can exert a torque even when standing still.

### EXAMPLE 2-3

A force of 2 N is applied to the end of a thin rod that is pivoted at one end and constrained by a spring, as in Fig. 2-24. The rod is 0.45 m long and the spring is located 0.2 m from the pivot.

(a) What is the torque exerted by the force applied?
(b) What is the force exerted by the spring?

### SOLUTION

(a) $T = F_M \times d$
$\quad = 2 \times 0.45 = 0.9 \text{ N} \cdot \text{m}$
(b) The applied force produces a counterclockwise torque at the end of the rod which is resisted by the spring. The spring itself produces a force in a clockwise direction which balances the applied force. Both torques must be equal, because the rod does not move; then

$$0.2 \times F_S = 0.45 \times F_m = 0.45 \times 2$$

$$F_S = \frac{0.45 \times 2}{0.2} = 4.5 \text{ N}$$

where $F_S$ is the spring force.

The relationships that have been developed thus far in this section are all mechanical ones. Equation (1-18) relates electrical and mechanical quantities, so that it is possible to calculate such things as mechanical power output, knowing electrical power input.

Sec. 2-11    Force, Power, and Torque

**49**

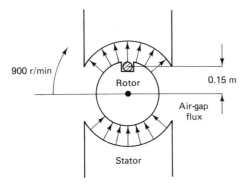

900 r/min

Rotor

0.15 m

Air-gap
flux

Stator

**Figure 2-25**  Conductor in a dc rotor.

**EXAMPLE 2-4**

Figure 2-25 is a cross section of a dc motor showing one conductor in a slot. The motor is turning at 900 r/min and the armature current is 4 A. The conductor is located 0.15 m from the center of the rotor and the average flux density in the air gap is 0.8 T. The effective length of the conductor into the page is 0.35 m. Calculate the torque exerted by this conductor.

**SOLUTION**

The force exerted by the conductor is

$$F_M = BLI = 0.8 \times 0.35 \times 4 = 1.12 \text{ N}$$

$$T = 1.12 \times 0.15 = 0.168 \text{ N} \cdot \text{m}$$

## 2-12 *CONSTRUCTION OF DC MACHINES*

Up to this point we have discussed the principles of dc voltage generation and motor action from a theoretical point of view. To correlate these ideas with physical reality, we now examine the physical construction of dc machines. Any modern dc machine can be run as a motor or a generator and the physical structure is identical in both cases. The dc machines can be manufactured in sizes from the milliwatt range found in hand calculators to the kilowatt or even megawatt range found in industrial drives, but again the same basic electromechanical components will be found in all ranges.

The two major parts of any machine are the stationary component, the stator, and the rotating component, the rotor. In the case of most dc machines the stationary part supports the fixed magnetic field windings. The rotating part either receives or delivers electrical power, depending on whether it is a motor or a generator, respectively. The armature is the part of the machine that delivers or receives power, so that in the case of a dc machine the rotor is the armature.

Figure 2-26 shows the physical arrangement of the dc machine components. The rotor consists of a steel core containing many layers of thin laminations which are punched with slots on the edge. The laminations are insulated from each other and assembled onto the rotor shaft. The armature conductors fit into the armature

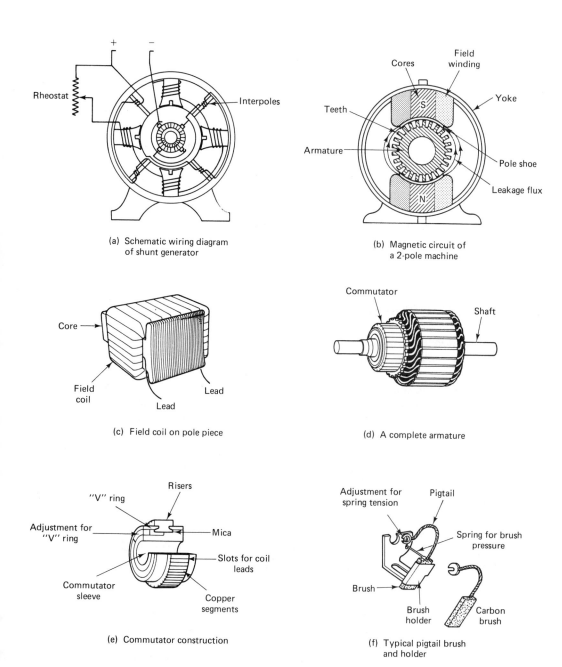

(a) Schematic wiring diagram of shunt generator

(b) Magnetic circuit of a 2-pole machine

(c) Field coil on pole piece

(d) A complete armature

(e) Commutator construction

(f) Typical pigtail brush and holder

**Figure 2-26**  Dc machine components: (a) schematic wiring diagram of shunt generator; (b) magnetic circuit of a two-pole machine; (c) field coil on pole piece; (d) complete armature; (e) commutator construction; (f) typical pigtail brush and holder.

slots and are insulated from the core. The armature conductors are arranged in coils, and the coil ends are brought out to commutator segments, as explained previously. The commutator is also fitted onto the rotor shaft, and the whole assembly is supported by a set of bearings.

The bearings support the rotor in the center of the stator, and the rotor and stator are separated by a small air gap. The magnetic field is created by a set of windings on the stator acting on a path which includes both steel and the air gap. The field windings are arranged electrically in series or parallel with the armature windings. Depending on the type of winding, it may consist of a few turns of heavy wire or many turns of fine wire. In the first case the field winding will be in *series* with the armature and in the second case in *parallel* or *shunt*. If both series and shunt windings are on the machine, it may be connected as a compound motor or generator. The properties of the various configurations are examined in subsequent chapters.

## 2-13 ARMATURE REACTION

In addition to series- and shunt-field windings, a dc machine may also have interpole and commutating windings. These windings help to prevent magnetic field distortion and aid commutation. Armature reaction is the distortion of the main magnetic field due to the current in the armature. Figure 2-27a shows a cutaway section of the armature with brushes in their normal position. Note that the conductors immediately under the brushes are moving parallel to the magnetic field and have no voltage induced in them. This is the ideal point to switch the current in a conductor since there will be no sparking in that particular conductor. However, recall that the rotor currents themselves form a coil, which in turn produces a magnetic field in the direction shown. This field combines with the main magnetic field shown in Fig. 2-27b. The result is the distorted magnetic field of Fig. 2-27c. The conductors undergoing commutation are no longer moving parallel to the magnetic field and hence have a voltage induced in them. The result is that switching occurs in the conductor under the brush while it has a voltage induced in it and sparking occurs. In dc motors, the effects of armature reaction cause the magnetic neutral to advance in the direction of rotation under load conditions.

The problems associated with armature reaction can be overcome by several means. One method is to shift the brush axis until the brushes are on the magnetic neutral axis. However, since the amount of armature reaction depends on the armature current, this would require constantly shifting the brushes. In addition, reversing the machine direction would also require brush shifting. To some extent the problem can be alleviated by properly shaping the pole pieces. A better solution, found on large and medium-size machines and in the better grade of small machines, is to use commutating or interpole windings. These are located between the main poles, as shown in Fig. 2-26, and are connected in series with the armature. They carry armature current; hence the field they produce is directly proportional to load and nearly exactly balances the effect of armature reaction.

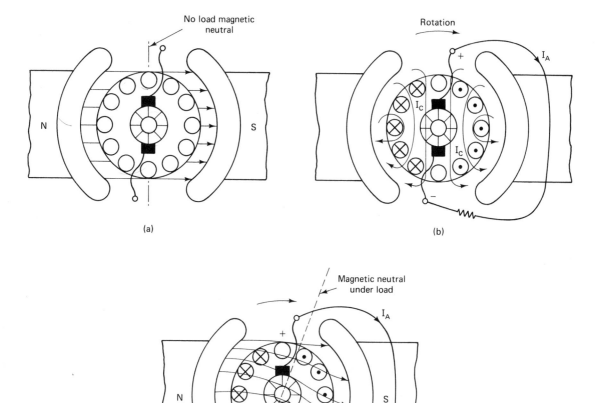

**Figure 2-27** Magnetic flux dc generator: (a) magnetic flux of main field poles only; (b) magnetic flux of armature rotor only; (c) actual magnetic flux in generator under load.

Another winding that may be found on very large machines is called a *pole face* or *compensating winding*. As the name suggests, this winding consists of conductors physically located in the pole face. It carries armature current in the direction opposite to the immediately adjacent armature conductors. These windings also help to counteract the effect of armature reaction. This is an expensive type of winding configuration and is generally found only in large machines.

It should be understood that from an external user's point of view, the commutating winding and the pole face winding only appear as added armature resistance. They produce a better machine with better voltage and commutating characteristics, but the windings do not change the machine's operating principles.

A fan is normally an integral part of a dc rotor. Heating invariably occurs in any machine due to $I^2R$, magnetic, and mechanical losses in the machine. To keep the machine cool and prevent deterioration of the components due to overheating, a fan is normally built on the shaft and this forces air over the rotor and stator, helping to keep them cool. On some machines it is necessary because of commutator sparking to use a completely enclosed structure and no air can circulate in the generator. In this case the generator may be oversized and the extra surface area will aid in cooling, or some type of special cooler may be added as an auxiliary.

## REVIEW QUESTIONS

**2-1.** State the two basic principles of electrical machines developed by Michael Faraday.

**2-2.** In your own words, explain how voltage is induced in a moving conductor in a magnetic field. State the factors that would determine the magnitude and polarity of this generated voltage.

**2-3.** Write out the generated-voltage equation and clearly identify each of the factors.

**2-4.** Determine the instantaneous voltage equation for a rotating coil of $N$ turns. Why is the waveshape of this voltage sinusoidal?

**2-5.** What determines the frequency of the generated voltage?

**2-6.** Explain clearly how a pair of brushes and a two-segment commutator connected to a rotating coil in a magnetic field produce a unidirectional generated voltage.

**2-7.** A practical dc machine consists of multiturn coils, forming a multicoil winding assembled in the rotor. Explain the advantages of such a winding.

**2-8.** What is a winding diagram? List the useful information provided by such a diagram.

**2-9.** Explain the construction of a wave winding. What are its advantages?

**2-10.** Explain the construction of a lap winding. What are its advantages?

**2-11.** Explain how motor action arises. How is the direction of the developed force determined?

**2-12.** The physical construction of dc machines consists of a stator and a rotor. Describe the functions of each and the various elements that are part of the assembly of each.

**2-13.** What is armature reaction, and what are its detrimental effects?

**2-14.** What are the purpose and location of the following?
  (a) An interpole or commutating winding.
  (b) A compensating winding.
  (c) A series field.

## PROBLEMS

**2-1.** A metal bar 1.24 m long has a resistance of 3 $\Omega$. It is placed on a set of parallel conducting rails which are in a magnetic field of flux density 0.9 T as shown in Fig. 2-28. The rails may be assumed to be ideal conductors and they are connected to an 8-V power supply. Determine the magnitude and direction of the force acting on the conductor.

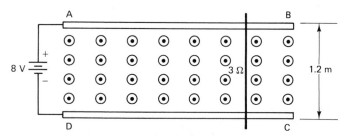

**Figure 2-28** Moving metal bar on conducting rails in magnetic field. Magnetic field perpendicular to page.

**2-2.** In Fig. 2-28, the length *AB* is 2.2 m. The total flux enclosed by the area *ABCD* is 0.02 Wb. Find the force acting on the conductor and the direction in which it acts.

**2-3.** The average flux density surrounding a conductor 0.8 m long carrying a current of 25 A is 0.6 T. The conductor is in the edge of a drum of radius 0.2 m.
  **(a)** What is the torque acting on the conductor?
  **(b)** If the conductor is part of a motor rotating at 400 r/min producing 10 hp, how many effective conductors are in the motor?

# 3

# DC GENERATORS

## 3-1 INTRODUCTION

Except for special applications, the dc generator has become nearly obsolete for large-scale power generation. There are still some important applications, but ac power generation is virtually universal and the use of solid-state devices makes the conversion to reliable dc power quite economical. Historically, the dc generator is the oldest commercial electrical machine and the operating principles are still important. In particular, the idea of generated electromotive force (EMF), which appears in motors as counter EMF, is of fundamental importance.

## 3-2 GENERATOR EQUIVALENT CIRCUIT

It was pointed out in Chapter 2 that a conductor rotating in a magnetic field produces a generated voltage. By careful design we can produce a practically pure dc voltage. This voltage is proportional to the flux in the field, the dimensions of the machine, and the rotational speed. This relationship can be written

$$E_G = K\Phi\omega \qquad \text{volts} \tag{3-1}$$

where     $E_G$ = generated voltage, V
           $\Phi$ = total air-gap flux, Wb
           $\omega$ = rotational speed, rad/s
           $K$ = constant, depending on the machine dimensions

The generated voltage may be expressed in terms of the total air-gap flux, the speed of rotation, and the number of effective armature conductors between the brush terminals, that is,

$$E_G = (\phi P) \left(\frac{\text{r/min}}{60}\right) \frac{Z}{a} = (\phi P)(\text{r/s}) \frac{Z}{a}$$

$$= (\phi P)(2\pi \text{ rad/s}) \frac{Z}{a} = (\phi P)(2\pi\omega) \frac{Z}{a} \quad \text{V}$$

where
$\phi$ = air-gap flux per field pole, Wb
$P$ = number of main field poles
r/min = speed of rotation
$Z$ = number of effective armature conductors between brush terminals
$a$ = number of parallel paths in armature winding between brush terminals, wave winding: $a = 2$, lap winding: $a = P$ (see Chapter 2)

Therefore, the generator constant, $K$ is equal to $P2\pi Z/a$.

This equation can be written in other forms, which are perhaps more useful. If we are dealing with a machine that has a magnetic field which does not vary, such as a field winding tied directly across the line or a permanent-magnet machine, voltage is directly proportional to speed and we can write

$$E_G = K\Phi\omega = K_G\omega \quad \text{V} \tag{3-1}$$

In some cases, the field current $I_F$ is variable and since the flux is proportional to field current below saturation,

$$E_G = K\Phi\omega = KK_F I_F\omega = K'_G I_F\omega \quad \text{V} \tag{3-2}$$

$K_G$ and $K'_G$ are generator constants that depend on the size of the machine. They can be readily determined experimentally, as all that is required is a measurement of generated voltage, field current, and speed. Figure 3-1 is the equivalent-circuit representation of an ideal generator. $I_F$ is the field current that produces the flux $\Phi$. The prime mover is an external mechanical power source such as a steam

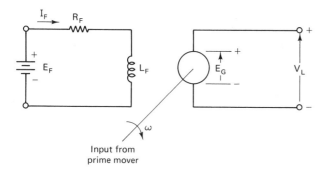

Input from prime mover

**Figure 3-1** Ideal dc generator.

turbine or a waterwheel turbine and it causes the shaft to turn at ω radians per second.  Now

$$\Phi = K_F I_F \tag{3-3}$$

where     $\Phi$ = air-gap flux, Wb
           $I_F$ = field current, A
           $K_F$ = constant

Note that $K_F$ is constant only if there is no magnetic saturation present.  If there is saturation, $\Phi$ is no longer proportional to $I_F$.  Let us postpone for the moment the problem of saturation since we consider it in detail later in the chapter.

 As users of electrical machines, we are not interested in the internal characteristics of a generator, but we are interested in how it behaves externally. Equation (3-1) gives us the relationship between generated voltage and speed.  If the field current is held constant, $\Phi$ will be constant and $E_G$ will vary with ω.

 As an example, consider the following problem.


## EXAMPLE 3-1

 A dc generator connected as in Fig. 3-1 is driven by a steam turbine and has an open-circuit output voltage of 125 V at 1800 r/min.

 (a) What is the value of $K_G$?
 (b) If $I_F$ is 10 A, what is the value of $K_G'$?
 (c) If the field current is held constant, what is the generated voltage at 2600 r/min?
 (d) If the field current is held constant, at what speed must the generator be driven to produce 100 V?
 (e) What would be the output voltage if the field current were increased to 15 A at 1800 r/min, assuming no saturation?

## SOLUTION

 (a) $E_G$ = 125 V and ω = 2π × 1800/60 = 188.5 rad/s.  Therefore,

$$K_G = \frac{E_G}{\omega} = \frac{125}{188.5} = 0.663 \text{ V/rad/s}$$

 (b) $E_G$ = 125, ω = 188.5 rad/s, $K_G' I_F$ = 125/188.5 = 0.66.  Therefore,

$$K_G' = \frac{0.66}{10} = 0.066 \text{ V/rad/s/A}$$

 (c) $E_G = K_G \omega$ = 0.66 × 2π × 1600/60 = 111 V.
 (d) 100 = 0.66ω; ω = 150.8 rad/s = 1440 r/min.
 (e) If $I_F$ = 15 A, ω = 188.5 and $E_G = K_G' \omega I_F$ = 0.0663 × 188.5 × 15 = 187.5 V.

 The relationship between the voltage generated on open circuit and speed, for various fixed values of field current, is plotted in Fig. 3-2a.

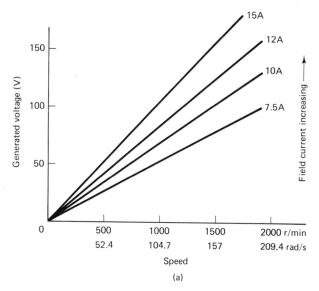

(a)

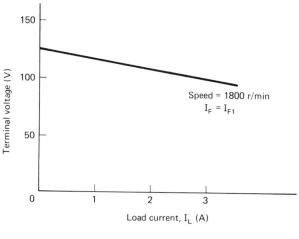

(b)

**Figure 3-2** (a) Open-circuit voltage characteristic; (b) voltage variation with load of a dc generator.

## 3-3 INTERNAL RESISTANCE

Figure 3-2a shows how the generated voltage varies with speed. If the speed is constant and the generator drives a load, we might expect that the output voltage would remain constant. However, it is easily verified that the output voltage drops as the output current changes. $V_L$ is the voltage across the load, and we can see from Fig. 3-2b that as $I_L$ increases, $V_L$ drops. We can account for this drop by making a more complete equivalent-circuit model for the generator of Fig. 3-1.

Figure 3-3 shows a circuit model for a generator that includes an internal resistance $R_A$ in the armature circuit. This armature resistance accounts for the internal voltage drop due to the current flowing in the armature. In addition to

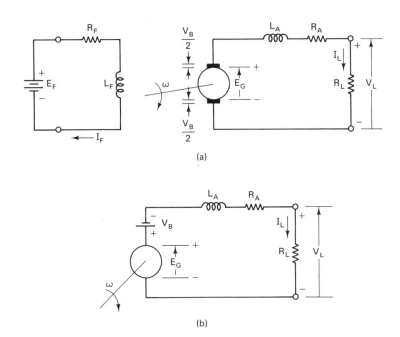

(a)

(b)

**Figure 3-3**  (a) Equivalent circuit of generator (separately excited); (b) alternative representation of rotor circuit.

the armature resistance drop, there is a voltage drop across the brushes. This drop is usually of the order of 1.5 to 4 V and is approximately constant, not changing with current. We can now write an equation for the circuit shown in Fig. 3-3a, in which $E_G$ is the generated voltage,

$$E_G = V_B + I_L R_A + I_L R_L \qquad (3\text{-}4a)$$
$$= V_B + I_L R_A + V_L$$

where    $E_G$ = generated voltage, V
           $V_B$ = brush voltage drop, V
           $I_L$ = load current, A
           $R_A$ = armature resistance, $\Omega$
           $R_L$ = load resistance, $\Omega$
           $V_L$ = load voltage, V

*Note:* For completeness, the equivalent circuit includes the armature inductance $L_A$. Since we are dealing with dc conditions, there is no voltage drop across $L_A$ and it can be neglected in our calculations.

The voltage drop across the brushes is of the order of 1.5 to 4 V. It is approximately constant with load, so that it can be modeled using a constant-voltage device, namely a battery $V_B$, as in Fig. 3-3b. In some cases the voltage drop is represented by an increase in the resistance of $R_A$, and this is most convenient when the armature current is fixed. In any event, this voltage drop is usually quite small and is very often neglected.

Hence we can then rewrite Eq. (3-4a)

$$E_G = I_L R_A + I_L R_L \tag{3-4b}$$
$$= I_L R_A + V_L$$

From Eq. (3-4a) it is possible to calculate the internal resistance of a generator.

## EXAMPLE 3-2

A 100-kW generator supplies a 50-kW load at 125 V. If the generator speed is held constant and the load is removed, the generator voltage rises to 137 V. Determine the effective armature resistance:

(a) Assuming no brush voltage drop.
(b) Assuming a total brush voltage drop of 1.8 V.

## SOLUTION

(a) For no voltage drop,

$$\text{load power} = 50,000 \text{ W}$$

$$\text{load voltage} = 125 \text{ V}$$

$$\text{load current} = I_L = \frac{50,000}{125} = 400 \text{ A}$$

$$V_L = 125 \text{ V} \qquad E_G = 137 \text{ V}$$

From Eq. (3-4b),

$$I_L R_A = E_G - V_L$$

$$R_A = \frac{E_G - V_L}{I_L} = \frac{137 - 125}{400} = 0.03 \ \Omega$$

(b) For a voltage drop of 1.8 V, from Eq. (3-4a),

$$I_L R_A = E_G - V_B V_L$$

$$R_A = E_G - V_B - V_L$$

$$= \frac{137 - 1.8 - 125}{400} = 0.0255 \ \Omega$$

## 3-4 SATURATION EFFECTS

Equation (3-3) states that the air-gap flux is proportional to field current. However, if the generator speed is held constant while the field current is varied, the curve of Fig. 3-4 is obtained. This curve is called the *no-load saturation curve* and it may conveniently be divided into three parts: the linear, transition, and saturation regions. In the linear region, the output voltage is directly proportional to field

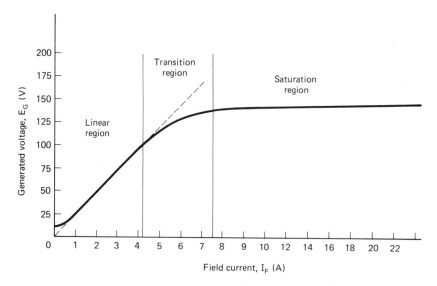

**Figure 3-4** No-load saturation curve.

current, so that 2 A of field current produces a generated voltage of 50 V, while 3 A produces 75 V, and 4 A produces 100 V. A change of 1 A produces a voltage change of 25 V in all cases. In the transition region, there is no longer a simple relationship between current change and voltage change. Going from 4 A to 5 A, the output voltage changes by only 20 V and between 5 A and 6 A by about 12 V. The straight-line relationship between current and voltage changes no longer exists; 1 A of field current no longer causes a change of 25 V.

Above about 7.5 A, the output voltage change with increasing field current is negligible. It requires a change of nearly 10 A to vary the output voltage about 6 V. In this region, the iron of the magnet circuit is in saturation and it is no longer possible to increase flux significantly by increasing field current.

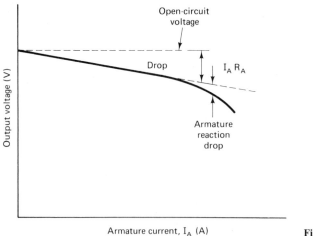

**Figure 3-5** Effect of armature reaction.

At the low end of the linear region there is actually a small nonlinear portion. If the field current is reduced to zero, there is still a small generated voltage. This is due to the residual magnetism in the iron, which is not totally demagnetized when the current is zero. Figure 3-5 shows the effects of internal resistance and armature reaction on the generated voltage with increasing armature current.

Generators are usually operated in the region where the magnetic field is somewhat saturated. The reason for doing this is so that the output voltage will not be too severely affected by small changes in field current. If we operate in either the transition region or the linear region, a small change in field current may produce a large change in generated voltage, and this is usually not desirable. To compensate for the effects of armature reaction, compensating windings are used as described in Chapter 2.

## 3-5 VOLTAGE REGULATION

The change in output voltage of a generator from no load to full load, divided by the full-load voltage, is called the *voltage regulation*.

$$\text{REG} = \frac{V_{NL} - V_{FL}}{V_{FL}} \tag{3-5a}$$

and

$$\%\text{REG} = \frac{V_{NL} - V_{FL}}{V_{FL}} \times 100\% \tag{3-5b}$$

where    $V_{NL}$ = open-circuit or no-load generated voltage, V
         $V_{FL}$ = full-load terminal voltage, V

Regulation is an important parameter in the performance of generators, as it provides us with a measure of how constant the output voltage is with load. If the regulation is too high, devices that are attached to the generator may be subjected to widely varying terminal voltages and may not operate satisfactorily.

### EXAMPLE 3-3

A 5-kW 125-V dc generator has an internal armature resistance of 0.2 Ω.

(a) What is the output voltage of the generator on open circuit?
(b) What is the output voltage if the machine is delivering 2.5 kW?

Assume that the machine is connected as in Fig. 3-3a and that the field current and speed are held constant. Armature reaction effects and brush voltage drops may be neglected.

### SOLUTION

In Fig. 3-3a the output voltage $V_L$ changes with load but $E_G$ remains fixed.

(a) $V_L$ is given as 125 V under full load (rated value) while delivering 5 kW. Therefore,

$$5000 = V_L \times I_L = 125 \times I_L$$

$$I_L = \frac{5000}{125} = 40 \text{ A}$$

$$E_G = 0.2 \times 40 + 125 = 133 \text{ V}$$

(b) In this case the output power is 2.5 kW. The output voltage is unknown as well as the output current. But we do know that

$$V_L \times I_L = 2500 \text{ W} \qquad (3\text{-}6)$$

and

$$E_G = I_L R_A + V_L \qquad V \qquad (3\text{-}7)$$

$$133 = I_L \times 0.2 + V_L$$

Therefore,

$$I_L = \frac{2500}{V_L} \qquad A$$

and substitution in Eq. (3-7) yields

$$133 = \frac{2500}{V_L} \times 0.2 + V_L$$

$$= \frac{500}{V_L} + V_L$$

Therefore,

$$133 \, V_L = 500 + V_L^2$$

$$V_L^2 - 133 V_L + 500 = 0$$

$$V_L = \frac{133 \pm \sqrt{133^2 - 4 \times 500}}{2}$$

$$= 129 \quad \text{or} \quad 4 \text{ V}$$

Our obvious choice is $V_L$ of 129 V across the load. To confirm that this is true, let us substitute in Eq. (3-6) for $I_L$. If $V_L = 129$ V,

$$I_L = \frac{2500}{129} = 19.4 \text{ A}$$

and if $V_L = 4$ V,

$$I_L = \frac{2500}{4} = 625 \text{ A}$$

Since the machine is rated at 40 A, at half-power output the load current must obviously be 19.4 A.

## 3-6 OUTPUT-VOLTAGE VARIATION WITH LOAD

As we have seen, the output voltage of a separately excited generator can be controlled by varying a rheostat in the field winding. A curve of output voltage as a function of load current is very often of interest, as this tells us how the voltage changes and we can then determine if this voltage variation is tolerable. In Example 3-3 we have solved for one point on the load characteristic curve. It is convenient to write a computer program that will give us this information for many points.

## 3-7 VOLTAGE BUILDUP IN A SHUNT GENERATOR

Figure 3-3a shows a generator connected with the field supplied from a source $E_F$ that is not connected to the armature. This is called a *separately-excited generator*, for obvious reasons. It is quite practical to connect the field directly across the armature as in Fig. 3-6, thus eliminating the extra power supply. Note that the armature must now supply both the load and the field winding, so that

$$I_A = I_F + I_L \tag{3-8}$$

where  $I_A$ = armature current, A
$I_F$ = field current, A
$I_L$ = load current, A

Again we may write

$$E_G = V_B + I_A R_A + I_L R_L$$
$$= V_B + I_A R_A + V_L \tag{3-9}$$

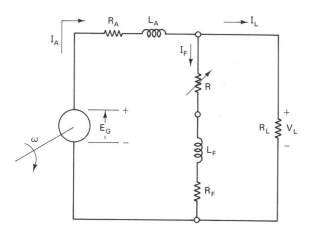

**Figure 3-6** Shunt generator with field control rheostat.

which is similar to Eq. (3-4) except that $I_L$ has been replaced by $I_A$ in the second term on the right. The connection shown in Fig. 3-6 is a self-excited or shunt-excited or simply a shunt generator.

In the shunt generator, the output voltage is dependent on the field current, and the field current is in turn dependent on the generated voltage. If we neglect the brush voltage drop and assume for the moment that the resistance of the rheostat is zero, we may write

$$E_G = I_A R_A + I_F R_F \tag{3-10}$$

and

$$E_G = K\Phi\omega = KK_G I_F \omega = K_G' I_F \omega \tag{3-11}$$

Equation (3-10) is merely Kirchhoff's voltage law around the field winding loop. Equation (3-11) is the generated-voltage law which is directly related to Faraday's law. Equation (3-11), on the other hand, assumes the existence of a current which produces a flux that is large enough to produce $E_G$ when $\omega$ is large enough.

If the resistance $R$ of Fig. 3-6 is too high, $I_F$ will never be large enough to produce a magnetic flux which, in turn, is large enough to produce $E_G$. It is necessary to determine a maximum value of the field circuit resistance that will allow the generator voltage to "build up."

Figure 3-6 shows a shunt-connected generator with a rheostat connected in series with the shunt field. The rheostat is adjusted to produce 120 V on open circuit, at which point the field current $I_F$ is 1 A. The armature resistance is 0.2 $\Omega$ and the shunt-field resistance is 40 $\Omega$, but the resistance of the rheostat is unknown. Figure 3-7 is a plot of generated voltage for the generator versus field current.

Now, since $V_L$ is 120 V and $I_F$ is 1 A, the total field circuit resistance $R + R_F$ is given by

$$R + R_F = \frac{120}{1} = 120\ \Omega$$

Since

$$R_F = 40\ \Omega$$

$$R = 80\ \Omega$$

then

$$V_L = I_F \times (R + R_F)$$

or

$$\frac{V_L}{I_F} = R + R_F \tag{3-12}$$

But Eq. (3-12) represents a straight line through the origin having a slope of $R + R_F$. This is shown in Fig. 3-7. Now if $R$ is decreased so that the output voltage

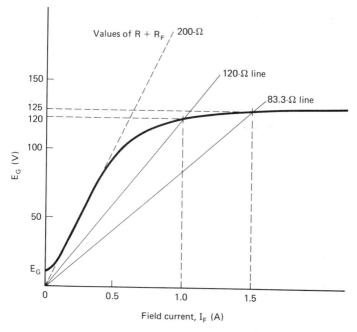

**Figure 3-7** Effect of changing total field circuit resistance.

is 125 V, the field current is 1.5 A, as can be seen in Fig. 3-7. Then

$$R + R_F = \frac{125}{1.5} = 83.3 \ \Omega$$

and

$$R = 83.3 - 40 = 43.3 \ \Omega$$

Decreasing $R$ has shifted the point of intersection to the right. If we increase $R$, the point shifts to the left. However, we can shift so far to the left that no intersection can occur, as shown by the dashed line labeled 200 Ω. The maximum value that we can use is called the *critical resistance* $R_c$ and occurs when the resistance line is just tangent to the curve. The slope of the line can be determined from Fig. 3-7 and is

$$R_c + R_F = \frac{120}{0.6} = 200 \ \Omega$$

$$R_c = 200 - 40 = 160 \ \Omega$$

Obviously, in practice a smaller resistance is desirable, and it should be chosen to given an operating point on the curve within the saturated region. The machine should not necessarily be operated too deeply in the saturation region, as this wastes power in the field resistance.

The output voltage variation with load of a shunt generator is somewhat more difficult to calculate than that of a separately excited generator. In the separately excited generator, the field current $I_F$ is constant; hence for a fixed speed the generated EMF is constant. However, for a shunt generator, as the load current increases the armature voltage drop $I_A R_A$ increases and the voltage across the field winding decreases, decreasing the field flux. The decrease in field or air-gap flux will cause a further decrease in $E_G$ and in $V_F$. However, the amount of change of flux with change in field voltage and current cannot be readily calculated without detailed knowledge of the magnetic properties of the generator. This information is not usually available from the manufacturer; hence these calculations are not easily made. In practice, a set of actual performance curves can be obtained experimentally, and these will provide a complete picture of the machine performance over its entire range.

The point $E_G'$ on the no-load saturation curve of Fig. 3-7 is extremely important for the buildup of voltage in a shunt generator. It represents the output voltage when no field current is present. In Fig. 3-7, if $R$ is set at, say, 90 Ω and the machine is rotating at rated speed, a small voltage is initially produced due to the rotor conductors cutting the residual magnetic field. This voltage will produce a small current in the field winding, which in turn produces a small flux. If this flux is in the correct direction, it will aid the residual field and produce a larger field current. This field current will again produce flux to aid the existing flux and a large voltage is rapidly built up. The process is limited, however, by saturation in the magnetic field, since after a while additional current will not produce significant additional flux. The operating point is the intersection of the no-load saturation curve of the generator and the straight line representing the total resistance in the field circuit.

From the preceding argument, it should be obvious that it is possible to connect a dc generator so that it will not build up a voltage. If the machine has been used and the field is connected, there will probably not be any problem. However, if the field leads have been disconnected from the armature, the machine may be reconnected with the wrong polarity. The voltage that is being built up may produce a flux in opposition to the residual flux. The two fluxes will tend to cancel each other and thus the generator will not build up its voltage. Reversing the armature connections may cure the problem but if it does not, it becomes necessary to "flash" the field. This procedure is to connect an external dc voltage source to the field winding in order to produce a slight residual magnetic field. The generator is then connected in the normal manner and it will build up properly.

Throughout this discussion it is assumed that the shunt generator is rotating in the proper direction and turning at rated speed. Changing the direction or speed may prevent voltage buildup in the generator. In fact, for a fixed value of field circuit resistance there is a corresponding critical speed below which voltage buildup cannot occur.

## 3-8 COMPOUND GENERATORS

In a shunt generator the load voltage decreases as load current increases. When a series-field winding is appropriately connected to the armature winding or in series with the load, the series-field flux can be the same magnetic polarity as the main field flux (i.e., currents in both windings produce magnetic fluxes that aid each other). This series-field flux is proportional (without saturation) to the armature or load current. By choosing the appropriate relative size of shunt- and series-field windings, it is possible to design a compound generator in which the output voltage is relatively independent of load conditions; that is, the load current may change over a wide range (say, 0 to 100%) while the load voltage will change over a comparatively small range (say, 5%). Figure 3-8a is called a long-shunt connection, and Fig. 3-8b is a short-shunt connection. The shunt location with respect to the armature determines whether the connection is "short" or "long." There is very little difference in performance between the short- and long-shunt connection. The difference is due to the fact that there is a small voltage drop across the series winding, which means that the voltage across the shunt field differs in both slightly for the same terminal voltage, $V_L$.

If the series and shunt fields are connected so that the fluxes aid each other, the compound generator is said to be *cumulatively compounded*; if the fluxes oppose each other, the connection is called *differentially compounded*. The optimum degree of compounding is called *flat*. A flat compound generator has the same output voltage at full load as at no load. An undercompounded generator has a higher voltage at no load than at full load, and an overcompounded generator has a higher voltage at full load than at no load.

All these characteristics are shown in Fig. 3-9, where it is assumed that all the generators have the same output voltage at 100% full load. It is obvious why the differentially compound generator is not used, since the voltage regulation is far too great. An ordinary shunt-connected characteristic is shown for comparison,

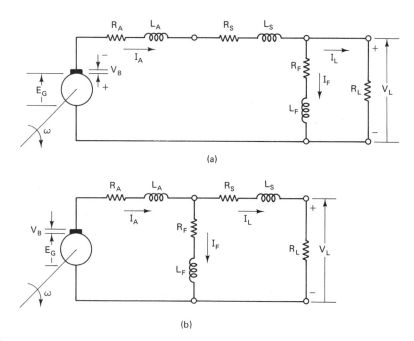

(a)

(b)

**Figure 3-8** (a) Long-shunt generator; (b) short-shunt generator.

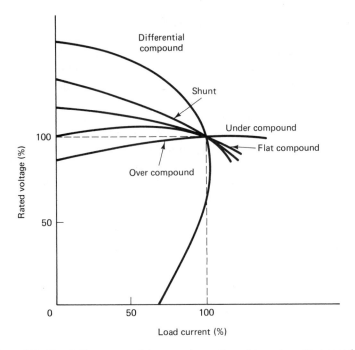

**Figure 3-9** Regulation curves of dc generators, all machines at constant speed.

and it is assumed that there would be negligible difference if the individual machines were either short- or long-shunt connected.

From Fig. 3-8 it can be seen that the armature must produce all of the current required for the load and field so that

$$I_A = I_F + I_L \tag{3-13}$$

For a long-shunt generator,

$$E_G = V_B + I_A R_A + I_A R_S + I_L R_L \tag{3-14a}$$

For a short-shunt generator,

$$E_G = V_B + I_A R_A + I_L R_S + I_L R_L \tag{3-14b}$$

**EXAMPLE 3-4**

A 50-kW 250-V generator has a series field resistance of 0.01 $\Omega$, an armature resistance of 0.012 $\Omega$, and a shunt-field resistance of 120 $\Omega$. Find the generated voltage if the machine is delivering rated output power and is connected:

   (a) Long shunt.
   (b) Short shunt.
Neglect brush voltage drop.

**SOLUTION**

(a) The long-shunt connection is shown in Fig 3-8a.

$$I_A = I_F + I_L$$

$$I_L = 50 \times \frac{10^3}{250} = 200 \text{ A}$$

$$I_F = \frac{250}{120} = 2.08 \text{ A}$$

Therefore,

$$I_A = 200 + 2.08 = 202.08 \text{ A}$$

$$E_G = 250 + 0.01 \times 202.08 + 0.012 \times 202.08 = 254.44 \text{ V}$$

(b) The short-shunt connection is shown in Fig 3-8b.

$$I_L = 50 \times \frac{10^3}{250} = 200 \text{ A}$$

$$V_F = 250 + 200 \times 0.01 = 252 \text{ V}$$

$$I_F = \frac{252}{120} = 2.08 \text{ A}$$

$$I_A = 200 + 2.08 = 202.08 \text{ A}$$

$$E_G = 252 + 202.08 \times 0.012 = 254.42 \text{ V}$$

## 3-9 VOLTAGE CONTROL

Figure 3-9 shows the output voltage versus load current characteristic of the various types of dc generators for various fixed configurations of these machines. It may be that a particular generator has a suitable voltage at one point on its curve but is not suitable at another, and it may be necessary to modify its behavior. For a compound generator, the various methods of control possible are shown in Fig. 3-10. There are three rheostats shown which can be used to change the output voltage—$R'_A$, $R'_S$, and $R'_F$. All three rheostats would normally not be used; the diagram is being used to illustrate the possible control modes.

$R'_A$ is a resistor in series with the armature. The armature current flowing in $R'_A$ has a voltage drop of $I_A R'_A$, so that the voltage across the load is $E_G - I_A(R'_A + R_A)$, neglecting any brush drop. In order to set $V_L$ at a particular value, $E_G$ is therefore set higher by the required amount. This particular method of control, however, has two disadvantages: (1) it requires a rheostat having a large power capacity, and (2) the losses in the rheostat are large, therefore decreasing the generator efficiency. The problem arises because the total armature current

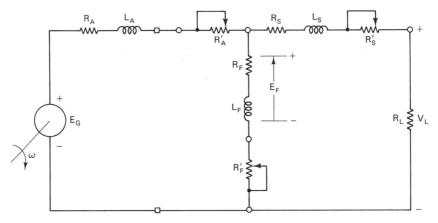

**Figure 3-10**   Possible generator control modes.

flows in $R'_A$, and if this current is high, the $I^2_A R'_A$ loss can be very high.   Note that $I_A R'_A$ voltage drop causes the voltage $E_F$ to change and hence the current $I_F$, as well as the flux $\Phi_F$ in the shunt field, also changes.

Control using $R'_S$ suffers from the same disadvantages as the previous configuration, since there will again be a large power loss due to $I^2_L R'_S$.   Neither of these methods is really very useful, for the reasons stated.

A practical and common method of voltage control is to use the rheostat $R'_F$ in series with the shunt field $R_F$.   The resistor normally has a high resistance and is required to carry a small current so that it can have a comparatively low dissipation.   It does not affect the overall machine efficiency as drastically as the previous arrangement.

In all cases, the voltage is being controlled by varying the field current, which in turn varies the field flux.   We can modify Eqs. (3-1) and (3-3) slightly and write

$$E = \Phi K \omega \tag{3-1}$$

$$\Phi = K_S I_S + K_F I_F \tag{3-15}$$

where
$\begin{aligned}
E_G &= \text{generated voltage, V} \\
\Phi_S &= \text{series field flux, Wb} \\
\Phi_F &= \text{shunt field flux, Wb} \\
I_S &= \text{series field current, A} \\
I_F &= \text{shunt field current, A} \\
K_S, K_F &= \text{constants relating current and flux (assuming no saturation),} \\
&\quad \text{Wb/A}
\end{aligned}$

**EXAMPLE 3-5**

Figure 3-8b shows a cumulatively compound generator with a rheostat connected in series with the shunt field.   The generator is rated at 10 kW, 250 V at 1200 r/min, $R_A = 0.35\ \Omega$, $R_S = 0.5\ \Omega$, $R_F = 40\ \Omega$, and $I_F = 5$ A at rated voltage.   Find the value of $R'_F$ and $E_G$.

**SOLUTION**

The rated load voltage is

$$V_L = 250 \text{ V}$$

The rated load current is

$$I_L = \frac{10{,}000}{250} = 40 \text{ A}$$

Further,

$$V_F' = 250 + 40 \times 0.5 = 270 \text{ V}$$

$$I_F = 5 \text{ A}$$

$$R_F + R_F' = \frac{270}{5} = 54 \ \Omega$$

Then

$$R_F' = 54 - 40 = 14 \ \Omega$$

$$I_A = 40 + 5 = 45 \text{ A}$$

Therefore,

$$E_G = 45 \times 0.35 + 270 = 285.75 \text{ V}$$

The power dissipation of $R_F$ is

$$P = 5^2 \times 14 = 350 \text{ W}$$

All the methods of voltage control outlined above use a rheostat in series with one of the generator components. A second alternative that is fairly common can be used to control the series-field current. In Fig 3-11, $R_D$ is a rheostat used

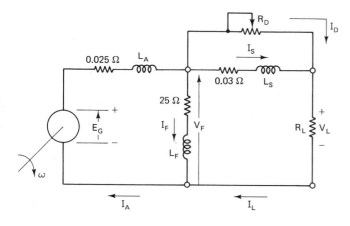

**Figure 3-11** Circuit for Example 3-6.

to divert some of the load current from flowing through $R_S$, and hence is called a *diverter*.

**EXAMPLE 3-6**

Figure 3-11 shows a generator delivering 100 A at 250 V to a load. If $R_A$ = 0.025 Ω, $R_F$ = 25 Ω, $R_S$ = 0.03 Ω, and $R_D$ = 0.09 Ω, find the generated voltage $E_G$.

**SOLUTION**

$$I_D + I_S = I_L = 100 \text{ A}$$

$$I_D R_D = I_S R_S$$

Therefore,

$$I_D = \frac{I_S R_S}{R_D} = I_S \times \frac{0.03}{0.09} = \frac{I_S}{3}$$

$$\frac{I_S}{3} + I_S = 100$$

Therefore,

$$I_S = 75 \text{ A}$$

Further,

$$I_D = 100 - 75 = 25 \text{ A}$$

$$V_F = I_D R_D + V_L = 25 \times 0.09 + 250 = 252.25 \text{ V}$$

$$I_F = \frac{252.25}{25} = 10.09 \text{ A}$$

$$I_A = 10.09 + 100 = 110.09 \text{ A}$$

$$E_G = 252.25 + 110.09 \times 0.025 = 255 \text{ V}$$

### Automatic Voltage Control

In many applications of generators, it is desirable to keep the output voltage constant over a wide range of output current. The methods we have discussed so far allow us to do this, but they require constant adjustment of one of the rheostats, either $R_F'$ or $R_D$ of Fig. 3-10 or 3-11. A more useful arrangement, which is now fairly common, is to have an automatic voltage regulator to keep the output voltage constant. In essence, a voltage regulator is a device that senses the output voltage, compares it to a fixed reference, and attempts to keep it equal to the reference. It does this by varying the field current to produce the right amount of flux to keep $V_L$ at the desired value.

## 3-10 EFFICIENCY

The primary purpose of a generator is to convert mechanical energy into electrical energy. Normally, we are interested in the efficiency of the conversion process, but we are more interested in power conversion than energy conversion. For our purpose we define *efficiency* as the ratio of output power to input power. The symbol for efficiency is the Greek lowercase letter eta $\eta$, so that

$$\eta = \frac{P_{out}}{P_{in}} \qquad (3\text{-}16)$$

The difference between output power and input power is the total power loss $P_L$, so that we may write

$$\eta = \frac{P_{out}}{P_{in}} = \frac{P_{in} - P_L}{P_{in}} = 1 - \frac{P_L}{P_{in}}$$

$$= \frac{P_{out}}{P_{out} + P_L} \qquad (3\text{-}17)$$

where
$$\eta = \text{efficiency, per unit}$$
$$P_{out} = \text{output power, W}$$
$$P_{in} = \text{input power, W}$$
$$P_L = \text{total power loss, W}$$

The input power is the mechanical power delivered to the generator shaft from a prime mover, such as a waterwheel, steam turbine, or diesel engine. Output power is the electrical power delivered to the load at the output terminals. Losses in a generator may be conveniently divided into two classes, mechanical or rotational losses and winding resistance losses.

### Rotational Losses

These losses depend on the rotational speed and may themselves be divided into two categories: (1) purely mechanical losses, and (2) electrical losses due to mechanical motion. Purely mechanical losses are those that do not depend on any electrical phenomenon, and in fact exist when the machine is electrically "dead" but is turning. These losses are bearing friction, brush mechanical losses, cooling-fan rotational losses, and windage loss. Rotational losses due to electrical effects are hysteresis and eddy current losses. *Hysteresis losses* occur in any magnetic material in which the magnetic field continually reverses direction. In the case of the dc generators, as we have seen previously, an observer on the rotor would continually see alternating north and south magnetic poles as the rotor moved. This causes the internal magnetic domains within the rotor iron to be continually reoriented. It requires a definite amount of energy to rotate these domains and this energy must be supplied from the prime mover.

Any conductor that moves in a magnetic field has a voltage induced in it, and if there is a continuous path and the voltage is of the correct polarity, a current will be induced in that conductor. The iron laminations of the rotor are themselves conductors and there are currents induced which circulate in the laminations. These losses are called *eddy-current losses*. Theoretically, hysteresis, eddy current, and purely mechanical losses can be measured individually, but usually they are measured together and considered as stray losses.

### Electrical Losses

Electrical losses in a generator occur in the brushes, the field windings, and the armature winding. The brush loss is given by

$$P_B = I_A V_B \tag{3-18}$$

where    $P_B$ = brush power loss, W
        $I_A$ = armature current, A
        $V_B$ = brush voltage drop, V

The brush loss is a constant times the current, and the constant is in the range 1.5 to 4 V. The actual voltage-drop mechanism is quite a complex physical process related to a gas discharge in a plasma. However, the actual voltage drop is usually fairly constant, and the amount of power loss is not very great, in any event.

The winding resistance losses are the $I^2R$ losses and these occur in the armature, series-field, commutating-field, and shunt-field windings.

### EXAMPLE 3-7

*Calculation of Efficiency*
The following data was measured for the circuit of Fig. 3-12:

$$R_A = 0.27 \ \Omega \qquad R_F = 53 \ \Omega$$

At 1200 r/min with switch $S$ open it requires 600 W from the prime mover

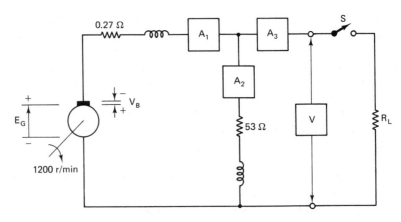

**Figure 3-12** Circuit for Example 3-7.

to turn the machine. When the machine is loaded, S is closed, the output voltage is 200 V, ammeter $A_3$ reads 75 A, and the total brush voltage drop is 1.8 V. Determine:

(a) The ammeter $A_1$ and $A_2$ readings.
(b) The generated voltage.
(c) The input power.
(d) The efficiency.

**SOLUTION**

(a) $I_L = 75$ A
$\quad V_L = 200$ V
$\quad I_F = \dfrac{200}{53} = 3.77$ A = reading of $A_2$
$\quad I_A = I_L + I_F = 75 + 3.77 = 78.77$ A = reading of $A_1$

(b) $E_G = V_B + I_A R_A + V_L$
$\qquad = 1.8 + 78.77 \times 0.27 + 200$
$\qquad = 223$ V

(c) $P_{in} = P_{out} + P_{losses}$

Losses:

Friction and windage, hysteresis, eddy current = 600 W

Brush: $I_A \times V_B\ =\ 78.77 \times 1.8$ $\qquad\qquad$ = 141.8 W

Armature: $I_A^2 R_A = 78.77^2 \times 0.27$ $\qquad$ = 1675.3 W

Field: $I_F^2 R_F\quad = 3.77^2 \times 53$ $\qquad\qquad$ = 753.3 W

$\qquad\qquad$ Total $\qquad\qquad\qquad\qquad\qquad$ = 3170.4 W

$P_{out} = 75 \times 200$ $\qquad\qquad\qquad\qquad$ = 15,000 W

$P_{in}$ $\qquad\qquad\qquad\qquad\qquad\qquad\qquad$ = 18,170.4 W

(d) Efficiency $= \dfrac{P_{out}}{P_{in}} = \dfrac{15,000}{18,170.4} = 0.8255$ or 82.55%

The power relationships with a dc generator may be clearly seen in Fig. 3-13. The actual useful output $P_{out}$ is shown at the right-hand side, and it is the power actually delivered to the load.

One can draw a complete schematic diagram of a compound generator indicating where all the losses occur. No control rheostats are usually shown, but if we are concerned with overall system efficiency, the losses due to the rheostats must be included. If we are discussing only generator efficiency, these losses are omitted.

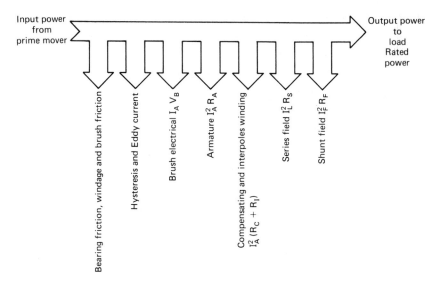

**Figure 3-13** Generator power losses.

## 3-11 TESTING DC MACHINES

The efficiency of a dc generator can be found in a straightforward manner by determining the various losses.

### Rotational Losses

These losses can be determined by driving the generator from an external motor. The motor is initially decoupled from the generator and a curve of speed versus power input obtained for the driving motor. The generator is then connected to the motor with full field excitation on the generator. A curve of speed versus power input for the combination is now run. The difference in power between the two curves, at any speed, represents the rotational losses. (There is a slight error due to increased $I^2R$ losses in the driving motor, but this is usually insignificant.)

### Field Losses

The resistance of the shunt-field, armature, series-field, interpoles, and pole-face windings can be measured using a Wheatstone bridge. The losses are then calculated as $I^2R$ losses. These measurements are made when the machine has been run long enough for it to have reached a stable operating temperature.

### Brush Losses

Brush losses are calculated on the assumption of a constant voltage drop across the brushes and a convenient value to use is 2 V. As noted previously, the error in neglecting brush drop is not normally very serious.

## EXAMPLE 3-8

Figure 3-14 is a curve of rotational speed versus load losses for a 25-kW 240-V, 1200-r/min dc compound generator.  Given that

$$R_A = 0.129 \ \Omega \qquad R_I = 0.018 \ \Omega \qquad R_C = 0.005 \ \Omega \qquad R_S = 0.02 \ \Omega$$

$$R_F = 58 \ \Omega \qquad I_A = 104 \ A \qquad I_L = 99.8 \ A$$

$$I_F = 4.2 \ A \qquad V_L = 240 \ V$$

calculate the generator efficiency

## SOLUTION

$$P_{out} = 240 \times 99.8 = 23,950 \ W = 23.95 \ kW$$

Losses:

| | |
|---|---|
| Armature: $104^2 \times 0.129$ | = 1395.3 W |
| Compensating field: $99.8^2 \times 0.005 =$ | 49.8 W |
| Interpole: $99.8^2 \times 0.018$ | = 179.3 W |
| Shunt field: $4.1^2 \times 58$ | = 974.9 W |
| Series field: $99.8^2 \times 0.02$ | = 199   W |
| Rotational losses: (from graph) | = 1000   W |
| Brush loss $= 104 \times 2$ | = $\underline{\phantom{00}208\phantom{0}}$ W |
| Total | = 4006.3 W |

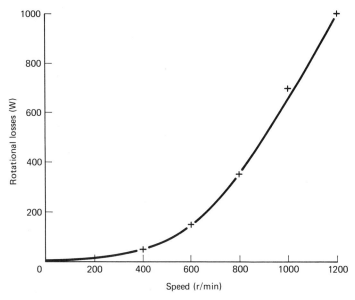

**Figure 3-14**  Rotational losses as a function of speed.

$$\text{Efficiency} = \frac{23{,}950}{23{,}950 + 4006} \times 100\% = 85.7\%$$

## 3-12 MEASURING MACHINE PERFORMANCE

Figure 3-15 shows a dc generator instrumented to make performance measurements on it. The instrumentation enables us to perform electrical measurements of the various currents in it. To make efficiency and loss measurements, we have to know the various resistance values accurately, and this requires the use of an accurate ohmmeter or bridge.

The voltage drop across the brushes is usually estimated and depends on the current, number of brushes, and to some extent the condition of the commutator. Rotational losses are accurately measured using a calibrated dynamometer. Failing this, the total rotational losses can be measured with a fair degree of accuracy by running the generator as a motor with no load and measuring the input voltage, currents, and power at several points over the operating speed range. The various $I^2R$ losses at each point are then calculated and subtracted, and the results plotted as a loss versus speed curve similar to Fig. 3-14. This method does not separate friction, windage, hysteresis, and eddy current losses. These can be separated using more sophisticated measuring techniques, but this is rarely necessary.

## REVIEW QUESTIONS

**3-1.** State the generated-voltage equation and describe each of its factors.

**3-2.** State the voltage equations associated with each of the generator constants, $K$, $K_G$, and $K'_G$. What are the fundamental assumptions in each equation?

**3-3.** Draw and label the practical equivalent circuit of a separately excited dc generator under load.

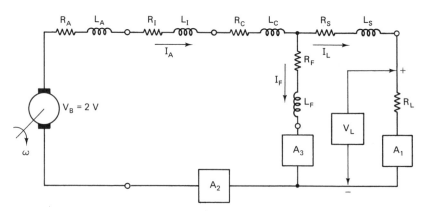

**Figure 3-15** Instrumentation for measuring dc generator performance.

**3-4.** What are the effects of magnetic saturation on a dc generator?

**3-5.** Explain the detrimental effects of armature reaction on the load characteristics of a dc generator.

**3-6.** Compensating windings are used in very large dc machines to counteract the effects of armature reaction. Explain how this is achieved. (See also Chapter 2.)

**3-7.** Interpoles or commutating windings are used in all dc machines. What is their purpose, and how are they connected? (See also Chapter 2.)

**3-8.** List the factors involved in voltage buildup of a generator.

**3-9.** Explain how the voltage builds up to a steady value in a shunt generator.

**3-10.** Define and explain critical field circuit resistance and critical speed.

**3-11.** What constitutes the internal resistance of the armature circuit? Is there justification in having it represented by a fixed resistance?

**3-12.** Explain voltage brush drop. Why is it not represented as a fixed resistance?

**3-13.** Explain why a separately excited dc generator has better voltage regulation than a self-excited dc generator.

**3-14.** Draw the connection diagrams of long-shunt and short-shunt compound generators. Label the current flowing in the series winding for each connection. How may the degree of compounding be controlled or adjusted?

**3-15.** Draw and label the regulation curves for the five types of generators, assuming that all are operating at rated terminal voltage and rated load current.

**3-16.** The total power loss consists of rotational and electrical losses. List and describe all the losses associated with the two components.

**3-17.** Draw a typical efficiency curve (efficiency versus power output) of a dc generator.

## PROBLEMS

**3-1.** The open-circuit voltage of a separately excited dc generator is 350 V when it is turning at 1800 r/min. If the excitation is held constant, what is the output voltage at 1200 r/min? At what speed would the generator run to produce 300 V?

**3-2.** A separately excited generator has the no-load saturation curve of Fig. 3-16.
(a) What is the output voltage if $I_F$ is 2, 4, and 5 A?
(b) What field current is required to produce an output voltage of 60, 120, and 160 V at 1800 r/min?

**3-3.** For the generator of Problem 3-2, draw no-load saturation curves at 1200 and 800 r/min. Determine the generated voltage at the new speeds for currents of 2.4 and 5 A.

**3-4.** The generator of Problem 3-2 has an armature resistance of 0.2 Ω and is connected to a resistive load $R_L$ as shown in Fig. 3-16. If $R_L$ is varied so that $I_L$ changes from 0 to 20 A, plot a curve of $V_L$ versus $I_L$ in steps of 5 A at 1200 r/min. No data are provided for armature reaction and brush drop effects, so that they may be neglected.

**3-5.** Problem 3-4 can readily be programmed in BASIC to yield a family of curves to give both no-load and load conditions. Assuming a brush voltage drop of 2 V, write a program to yield a regulation curve ($V_L$ versus $I_L$) in steps of 2 A at speeds of:
(a) 1200 r/min.
(b) 1000 r/min.
(c) 800 r/min.

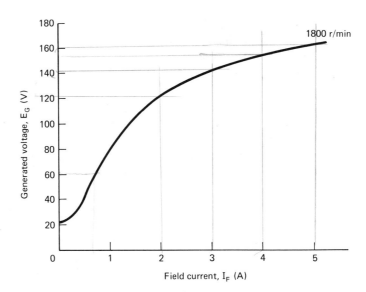

**Figure 3-16** No-load saturation curve for Problem 3-2.

**3-6.** A shunt generator has the no-load saturation curve shown in Fig. 3-17. Determine:
   **(a)** The value of the field circuit resistance if the generated voltage is 350 V.
   **(b)** The output voltage if the field circuit resistance is 60 Ω.
   **(c)** The value of the critical field resistance for a speed of 2000 r/min.

**3-7.** A generator having the no-load saturation curve of Fig. 3-17 is connected as a separately excited generator with 5 A in the field circuit. If the armature resistance is 0.16 Ω and total brush drop is 1.5 V, plot output voltage versus output current as current varies from 0 to 100 A. There are no armature reaction effects to be considered.

**3-8.** Plot the results of Problem 3-7 for the same machine at 1400 r/min.

**3-9.** A 10-kW 250-V generator has a shunt-field resistance of 125 Ω, $R_A$ of 0.4 Ω, a stray-load loss of 540 W, and a 2-V brush voltage drop. If it is running at its rated output, calculate:
   **(a)** The generated voltage.
   **(b)** The efficiency.

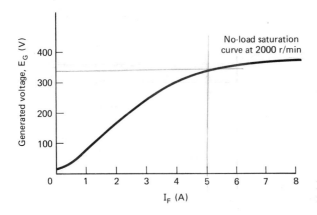

**Figure 3-17** No-load saturation curve for Problem 3-6.

DC Generators   Chap. 3

**3-10.** A shunt generator with a field resistance of 60 $\Omega$ has the no-load saturation curve of Fig. 3-17 at 2000 r/min. If $R_A$ is 0.16 $\Omega$ and the brush drop is 2 V, plot a graph of output voltage versus load current as it varies from 0 to 40 A in steps of 5 A. Assume that the field flux remains essentially constant.

**3-11.** Repeat the calculations and graph plot of Problem 3-10 for a new generator speed of 2400 r/min.

**3-12.** A separately excited dc generator is driven at 1200 r/min and the following data were recorded:

| Generated voltage, $E_G$ (V) | 5 | 26 | 50 | 76 | 98 | 131 | 153 | 170 |
|---|---|---|---|---|---|---|---|---|
| Field current, $I_F$ (A) | 0 | 0.1 | 0.2 | 0.3 | 0.4 | 0.6 | 0.8 | 1.0 |

(a) Draw the no-load curves at 1200 and 1500 r/min.
(b) The field circuit resistance is 200 $\Omega$. What is the open-circuit voltage of the machine if it is connected as a shunt generator running at 1500 r/min?
(c) What is the critical field resistance at 1000 r/min?

**3-13.** A long-shunt compound dc generator with armature, series-field, and shunt-field resistances of 0.5, 0.4, and 250 $\Omega$, respectively, gave the following readings when run at constant speed:

| Load current (A) | 0 | 10 | 20 | 30 | 40 |
|---|---|---|---|---|---|
| Terminal potential (V) | 480 | 478 | 475 | 471 | 467 |

Plot the curve of internally generated EMF against load current. Explain fully the steps by which this curve is obtained and tabulate the values from which it is plotted. Is the generator differentially or cumulatively compounded?

**3-14.** The data of Fig. 3-18 were obtained for a generator rated 6.25 kW and having a field resistance of 100 $\Omega$.
(a) What is the rated current and terminal voltage?
(b) What is the generator efficiency?

(*Hint:* On Fig. 3-18 plot a curve for a constant power output of 6.25 kW. The intersection of this curve and Fig. 3-18 will give the operating point.)

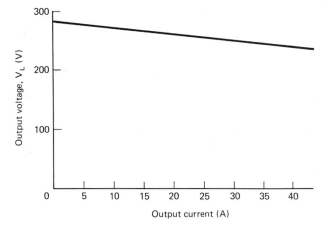

**Figure 3-18** Generator load characteristic for Problem 3-14.

**3-15.** In a 150-kW 600-V short-shunt compound generator, 645.6 V is induced in the armature when the generator delivers rated load at 600 V. The shunt-field current is 6 A. $R_{series} = 0.08\,\Omega$.

**(a)** Determine:

    **(i)** The armature resistance and shunt-field resistance. (Neglect brush voltage drop.)

    **(ii)** The voltage regulation if the EMF induced in the armature on no load is 660 V.

**(b)** Why is the EMF induced at rated load not equal to that on no load?

**3-16.** A separated excited generator has a no-load voltage of 125 V at a field current of 2.1 A when driven at a speed of 1600 r/min. Assuming that it is operating on the straight-line portion of its saturation curve, calculate:

**(a)** The generated voltage when the field current is increased to 2.6 A.

**(b)** The generated voltage when the speed is reduced to 1450 r/min and the field current is increased to 2.8 A.

**3-17.** **(a)** A shunt generator is operating at no load. A decrease of 20% in the speed results in a decrease of 40% in the voltage. Determine the percent change in the flux.

**(b)** A long-shunt compound dc generator with armature, series-field, and shunt-field resistances of 0.5, 0.4, and 250 Ω, respectively, gave the following readings when run at constant speed.

| Load current (A) | 0 | 10 | 20 | 30 | 40 |
|---|---|---|---|---|---|
| Terminal voltage, $V_T$ (V) | 480 | 478 | 475 | 471 | 467 |

    **(i)** Plot the curve of internally generated EMF against load current. Show the detailed calculations for the values obtained.

    **(ii)** If the number of turns on the shunt field and the series field are 900 and 20, respectively, calculate the total MMF produced at a load current of 25 A.

**3-18.** A 10-kW 220-V compound generator (long-shunt connected) is operated at no load at the proper armature voltage and speed, from which the stray-power loss is determined to be 705 W. The shunt-field resistance = 110 Ω, the armature resistance = 0.265 Ω, and the series-field resistance = 0.035 Ω. Assume a 2-V brush drop and calculate the full-load efficiency.

**3-19.** A 250-kW 240-V 1200-r/min short-shunt compound generator has a shunt-field resistance of 24 Ω, an armature resistance of 0.003 Ω, a series-field resistance of 0.0013 Ω, and a commutating-field resistance of 0.004 Ω. Calculate the generated EMF at full load.

**3-20.** The terminal voltage of a 200-kW shunt generator is 600 V when it delivers rated load current. The resistance of the shunt field circuit is 250 Ω, the armature resistance is 0.32 Ω, and the brush resistance is 0.014 Ω.

**(a)** Determine the EMF induced at rated current.

**(b)** The terminal voltage is 620 V at half-rated current. Determine the EMF induced.

**3-21.** The no-load saturation curve of a dc shunt generator when running at a speed of 1000 r/min is as illustrated in Fig. 3-19.

**(a)** Determine the critical field resistance at:

    **(i)** 1000 r/min.

    **(ii)** 1500 r/min.

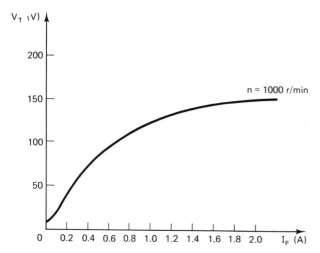

Figure 3-19   No-load saturation curve for Problem 3-21.

**(b)** If the resistance of the field coils is 100 $\Omega$, find the value of the field rheostat to set the open-circuit voltage to 125 V at a speed of 1000 r/min.

# 4
# DC MOTORS

## 4-1 INTRODUCTION

In the early years of the electrical age (about 1879), the dc motor and generator were the predominant electromechanical energy-conversion devices. The first large power station, the Pearl Street Station in New York, built in 1882, produced direct current that supplied electricity for lighting to the Wall Street area. Soon after, developments in dc motors led to their wide-scale utilization, and until the invention of the induction motor in 1887, there was no serious rival to the dc motor. However, with the development of the transformer in 1883 and the induction motor, ac power became feasible and quickly dominated the industry. Dc motors are still very much in existence and are still the primary choice in many applications involving variable-speed drives.

## 4-2 DC MOTOR PRINCIPLES

The same machine can act as either a dc generator or a dc motor. In Fig. 4-1 a dc generator is connected to a power system (in parallel with other generators) and is supplying a current $I_L$ to the system at a voltage of $V_L$. If the power system voltage is $V_L$ and the generator is producing $E_G$ volts internally, a current $I_L$ flows from the generator to the power system and

$$I_L = \frac{E_G - V_L}{R_A} \qquad (4\text{-}1)$$

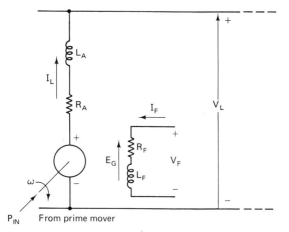

Figure 4-1 Dc generator connected to a line.

P_IN    From prime mover

where    $I_L$ = load current from generator, A
         $E_G$ = generated voltage of generator, V
         $V_L$ = power system line voltage, V
         $R_A$ = armature resistance of generator, Ω

This requires that at the given speed ω, the generator flux due to $I_F$ produces a voltage $E_G$ which is greater than $V_L$. If we decrease $I_F$ until $E_G$ is exactly equal to $V_L$, then $I_A$ becomes zero and no current flows. If we decrease $I_F$ still further, $I_A$ will become negative and current will flow into the generator. However, power flowing into a machine means that it is now acting as a motor and it will take power from the line rather than delivering power to it.


## 4-3 COUNTER EMF IN MOTORS

If a conductor moves in a magnetic field, it generates a voltage that may be calculated from Eq. (2-1) or (3-1). This voltage exists whether the conductor is in a generator or in a motor. If it is in a generator the voltage is called the generated voltage, but in a motor it is usually called the counter electromotive force (counter EMF or CEMF). Sometimes the counter EMF is described as the back EMF. Since it is a generated voltage it is proportional to the product of length, flux density, and speed. If a motor is connected to a dc line with no load on it, the counter EMF is almost exactly equal to the line voltage. Figure 4-2 shows a motor "idling" on the line; there is no load on the shaft and the only power into the motor is that required to overcome rotational losses. From Kirchhoff"'s voltage law we may write

$$V_L = I_A R_A + \frac{V_B}{2} + E_C + \frac{V_B}{2} \tag{4-2}$$

where    $V_L$ = line voltage, V
         $I_A$ = armature current, A
         $R_A$ = armature resistance, Ω

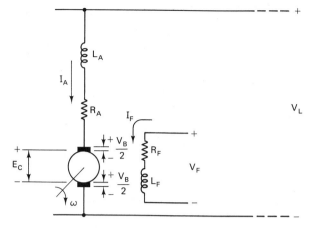

Figure 4-2 Motor idling on dc line.

$V_B$ = total brush voltage drop, V
$E_C$ = counter EMF, V

Although the generated voltage and the counter EMF are identical phenomena, to emphasize that we are dealing with motors, the symbol $E_C$ will be used instead of the symbol $E_G$.

On no load, the armature current is very small, and if we neglect the brush voltage drop,

$$V_L \approx E_C \qquad (4\text{-}3)$$

However, we do know that $E_C$ is proportional to speed and field current, so that we may write

$$E_C = K_G' I_F \omega \qquad (4\text{-}4)$$

Equations (4-3) and (4-4) can be used to calculate rotational speed. Compare Eq. (4-4) with Eq. (3-2).

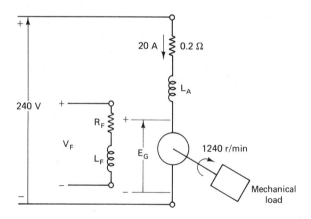

Figure 4-3 Motor of Example 4-1.

## EXAMPLE 4-1

$V_L$

Figure 4-3 shows a 240-V dc motor drawing 20 A from the line while running at 1240 r/min. If the armature resistance is 0.2 Ω, what is the no-load speed? Brush losses and stray losses may be ignored.

### SOLUTION

From Fig. 4-3 on load,

$$240 = 0.2 \times I_L + E_g$$

$$= 0.2 \times I_L + K'_G I_F \omega$$

$$= 0.2 \times 20 + K'_G I_F \times 2\pi \times \frac{1240}{60}$$

$$= 4 + 129.8 \, K'_G I_F$$

Therefore,

$$K'_G I_F = \frac{236}{129.8} = 1.818$$

At no load,                                                          *Assume*

$$I_L \cong 0$$

Therefore,

$$240 = 1.818 \times \omega$$

and

$$\omega = 132.0 \text{ rad/s} = 1260 \text{ r/min}$$

Knowing the torque and speed of a machine as discussed in Section 2-11, we can calculate the developed power, which the internal armature power converted from electrical form to mechanical form, but it is not the output power. To determine output power, it is necessary to subtract any losses that occur between the point where power is developed and the point where power is utilized (normally on the motor shaft).

## EXAMPLE 4-2

The machine described in Example 2-4 has 260 conductors in series uniformly spaced around the rotor. If 80% of the conductors are under a pole at any given time, calculate:

(a) The developed power $P_D$.
(b) The output power $P_O$ if total mechanical losses $P_M$ are 350 W.

**SOLUTION**

(a) The developed torque $T_D$ is

$$T_D = 0.168 \times 260 \times 0.8 = 34.9 \text{ N} \cdot \text{m}$$

$$P_D = 34.9 \times 900 \times 2 \times \frac{\pi}{60} = 3289 \text{ W} = 3.29 \text{ kW}$$

(b) $P_O = P_D - P_M = 3289 - 350 = 2939 \text{ W} = 2.94 \text{ kW}$

or $\qquad\qquad\qquad\qquad \dfrac{2939}{746} = 3.94 \text{ hp}$

## 4-4 MOTOR CONNECTIONS

To produce mechanical power, a motor obviously requires an armature and at least one field. The field may be connected in parallel with the armature, in series with it, or there may be two fields, one in series and one in parallel. Each of these connections will result in different motor characteristics and each will be examined in detail.

### Shunt Motors

Figure 4-4 shows the equivalent circuit for a dc shunt motor. The input voltage is $V_L$ and the input current is $I_L$. $I_L$ must supply both armature current $I_A$ and shunt field current $I_F$, and we can see from Kirchhoff's current law that

$$I_L = I_A + I_F \tag{4-5}$$

The input power is

$$P_I = V_L \times I_L \tag{4-6}$$

The field current may be calculated from

$$I_F = \frac{V_L}{R_F} \tag{4-7}$$

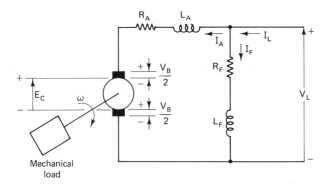

Figure 4-4  Dc shunt motor equivalent circuit.

The counter EMF may be calculated from

$$V_L = I_A R_A + V_B + E_C$$

or

$$E_C = V_L - I_A R_A - V_B \qquad (4\text{-}8)$$

For a fixed field current

$$E_C = K'_G I_F \omega \qquad (4\text{-}9)$$

where  $E_C$ = counter EMF, V
  $I_L$ = line current, A
  $I_A$ = armature current, A
  $I_F$ = shunt field current, A
  $P_I$ = input power, W
  $V_L$ = line voltage, V
  $R_A$ = armature resistance, $\Omega$
  $R_F$ = field resistance, $\Omega$
  $V_B$ = brush voltage drop, V
  $K'_G$ = armature EMF constant, V-s/A-rad
  $\omega$ = rotational speed, rad/s

In many applications using dc shunt motors, the field winding is connected across the line and $I_F$ is held constant. We may then write

$$E_C = K_G \omega \qquad (4\text{-}10)$$

Equation (4-10) applies to wound-field dc motors as well as to permanent-magnet motors. Equation (4-9) applies to wound-field machines in which the field current variation produces linear changes in the field flux.

**EXAMPLE 4-3**

In the circuit of Fig. 4-4,

$$R_A = 0.02 \ \Omega \qquad R_F = 62.5 \ \Omega \qquad V_L = 250 \ \text{V}$$

$$V_B = 2.5 \ \text{V} \qquad I_A = 200 \ \text{A} \qquad \omega = 100 \ \text{rad/s}$$

Calculate (a) $I_F$, (b) $E_C$, (c) $K'_G$, and (d) the speed when $I_A$ is 100 A.

**SOLUTION**

(a) $I_F = \dfrac{250}{62.5} = 4 \ \text{A}$

(b) $I_A = I_L - I_F = 200 - 4 = 196 \ \text{A}$

$E_C = V_L - I_A R_A - V_B$

$\qquad = 250 - 196 \times 0.02 - 2.5$

$\qquad = 243.6 \ \text{V}$

(c) $K_G' = \dfrac{E_C}{I_F \omega} = \dfrac{243.6}{4 \times 100} = 0.608$ V · s/A · rad

(d) $I_A = 100$ A

$$E_C = 250 - 100 \times 0.02 - 2.5$$

$$= 245.5 \text{ V}$$

$$\omega = \frac{E_C}{I_F K_G'} = \frac{245.5}{4 \times 0.608} = 100.9 \text{ rad/s}$$

Except for very low voltage machines, the brush voltage drop is negligible. No great loss of accuracy is incurred if it is neglected. Brush drop will be used only in problems where brush losses may be significant.

**Speed regulation.** If the shunt motor is connected as in Fig. 4-4, the field current is constant and is equal to $V_L/R_F$. As the load varies, the current $I_A$ will vary, hence $I_A R_A$ will vary and $E_C$ will vary. $E_C$ is proportional to the product of flux and speed; as flux does not vary, the speed must change with load. Specifically, from Fig. 4-4 (neglecting brush drops),

$$I_A = \frac{V_L - E_C}{R_A}$$

$$= \frac{V_L - K_G' I_F \omega}{R_A} \tag{4-11}$$

Rearranging Eq. (4-11) yields

$$K_G' I_F \omega = V_L - I_A R_A$$

$$\omega = \frac{V_L - I_A R_A}{K_G' I_F} \tag{4-12a}$$

$$= \frac{V_L - I_A R_A}{K_G} \tag{4-12b}$$

Equation (4-12) is plotted in Fig. 4-5 for a fixed value of $I_F$. It shows how speed drops off as the motor is loaded. In the case of large motors the operating speed range is only a small percentage of no-load speed, so that even at maximum load there may be only a 10 or 15% drop from the no-load value. Smaller motors, which are used primarily for position-control devices, are run near zero speed and the speed curve will drop to zero. This will be considered in detail in a later chapter.

The no-load speed $\omega_{NL}$ is given by

$$\omega_{NL} = \frac{V_L}{K_G' I_F} \tag{4-13a}$$

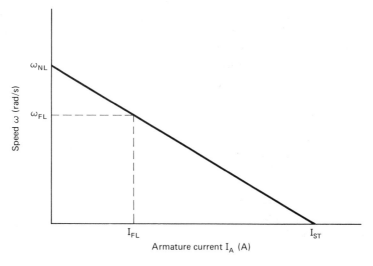

**Figure 4-5**  Speed–armature current characteristic of a dc motor.

and the full-load speed is

$$\omega_{FL} = \frac{V_L - I_{FL}R_A}{K_G'I_F}$$  (4-13b)

$I_{FL}$
*Armature at Full Load*

where

$\omega_{NL}$ = no-load speed, rad/s

$\omega_{FL}$ = rated full-load speed, rad/s

The starting current $I_{ST}$ is given by

$$I_{ST} = \frac{V_L}{R_A}$$  (4-14)

At rated load, the motor runs at rated full-load speed $\omega_{FL}$, while at no load it runs at $\omega_{NL}$. The quantity

$$REG = \frac{\omega_{NL} - \omega_{FL}}{\omega_{FL}}$$  (4-15)

is called the speed regulation and may also be written as

$$\% \ REG = \frac{\omega_{NL} - \omega_{FL}}{\omega_{FL}} \times 100\%$$  (4-16)

where     REG = speed regulation
     % REG = percent speed regulation
     $\omega_{NL}$ = no-load speed
     $\omega_{FL}$ = full-load speed

Regulation is a measure of how well a motor behaves under load. Good regulation means that the motor speed does not change very much with load, whereas a machine that has poor regulation changes speed a large amount when going from no load to full load.

## EXAMPLE 4-4

The motor of Example 4-3 has a full-load rated current of 100 A. Calculate its percent speed regulation.

### SOLUTION

No information is given for the no-load speed of the motor. However, we can make the approximation that on no load there is practically no armature current. Then

$$E_C = V_L = 250 = K'_G I_F \omega_{NL}$$

$$\omega_{NL} = \frac{E_C}{K'_G I_F} = \frac{250}{0.608 \times 4} = 102.8 \text{ rad/s}$$

$$\% \text{ REG} = \frac{102.8 - 100.9}{100.9} \times 100\% = 1.9\%$$

**Power.** The power that the motor converts from the electrical form to the mechanical form is called the *developed power* $P_D$. Developed power is the input power minus any internal $I^2R$ losses. It can be calculated directly from the equivalent circuit of Fig. 4-4 and is the power that appears across the rotor at a voltage $E_C$. It is necessary to subtract all rotational losses from the developed power to obtain the useful output mechanical power $P_O$. The expression for developed power is then

$$P_D = E_C \times I_A \tag{4-17a}$$

Also,

$$P_D = P_I - (P_F + P_A + P_B)$$

$$= V_L I_L - \left( \frac{V_L^2}{R_F} + I_A^2 R_A + V_B I_A \right) \tag{4-17b}$$

$$P_F = I_F^2 R_F = \frac{V_L^2}{R_F} \tag{4-18}$$

$$P_A = I_A^2 R_A \tag{4-19}$$

$$P_B = V_B I_A \tag{4-20}$$

$$P_O = P_D - P_R \tag{4-21}$$

where    $P_F$ = shunt field power loss, W
           $P_A$ = armature power losses, W
           $P_I$ = input power, W
           ~~$P_L$ = total losses, W~~
           $P_B$ = brush losses, W
           $P_O$ = output mechanical power load, W
           $P_D$ = (internal) developed power, W
           $P_R$ = rotational losses, W

$P_R$ includes hysteresis, eddy current, friction, and windage losses plus any other losses not directly measurable, such as tooth ripple losses. $P_R$ is usually measured experimentally, either at a fixed speed and voltage or if necessary as a set of graphs with both speed and voltage varying.

### EXAMPLE 4-5

The following measurements were made on the motor shown in Fig. 4-4.

$$R_A = 0.45 \ \Omega \quad\quad R_F = 145 \ \Omega \quad\quad V_B = 2 \ V \quad\quad P_R = 200 \ W$$

$$V_L = 125 \ V \quad\quad I_L = 22.5 \ A \quad\quad \text{speed} = 1800 \ r/min$$

Calculate:
(a) The developed power.
(b) The output power.
(c) The output torque.
(d) The generator constant $K_G$.

### SOLUTION

$$I_F = \frac{125}{145} = 0.862 \ A$$

$$I_A = 22.5 - 0.862 = 21.64 \ A$$

Input power:

$$P_I = V_L I_L = 125 \times 22.5 = 2812 \ W$$

Losses:

Shunt field: $\dfrac{125^2}{145}$        = 107.8 W

Armature: $22.5^2 \times 0.45$ = ~~227.8~~ W    *[handwritten: 21.64², 210]*

Brush: $2 \times 21.64$       = 43.28 W

Total      *[handwritten: 361.08]* = ~~378.9~~ W    *[handwritten: 361.08]*

(a) Developed power $P_D$ = 2812 − ~~378.9~~ = ~~2433~~.1 W.    *[handwritten: 2450.92]*
(b) Output power $P_O$ = ~~2433~~.1 − 200 = ~~2233~~.1 W.    *[handwritten: 2451 ... 2251]*

*[handwritten: 3.017 HP]*

(c) Speed $= 1800 \times (2\pi/60) = 188.4$ rad/s

Output torque $T_D = 2233.1/188.4 = 11.9$ N $\cdot$ m

(d) The counter EMF is

$$E_C = 125 - 21.64 \times 0.45 - 2$$
$$= 113.26 \text{ V}$$
$$E_C = K_G\omega = K_G \times 188.4$$

therefore,        $K_G = 0.601$ V/rad/s

**Torque.**    Equation (4-17) is the relationship between current, voltage, and developed power.   $E_C$ is the counter EMF produced by the rotor.   If we substitute the relationship of Eq. (4-9) into Eq. (4-17), we obtain

$$P_D = K'_G I_F \omega \times I_A = K_G \omega I_A \tag{4-22}$$

The developed torque is then

$$T_D = \frac{P_D}{\omega} = K'_G I_F I_A = K_G I_A \tag{4-23}$$

For a fixed field current, torque is again seen to be proportional to armature current. As a matter of convenience, we may define a torque constant $K_T$ for a motor in which $I_F$ is fixed such that

$$T_D = K_T I_A \tag{4-24}$$

where $K_T$ is the torque constant in N $\cdot$ m/A.

From Eqs. (4-23) and (4-24) it is obvious that $K_G$ and $K_T$ are equal.   However, this equality is true only in the SI system of units.   From Eq. (4-23), we obtain

$$T_D = K'_T I_F I_A \tag{4-25}$$

Again, it is easily shown that $K'_T$ and $K'_G$ are always equal in SI units.

The developed torque of a motor may be expressed in terms of its physical quantities as follows:

$$T_D = ZT_C = ZRF_C = ZRBlI_C = \frac{Z}{2}(2R)BlI_C = BlI_C ND \qquad \text{N} \cdot \text{m} \tag{4-26}$$

where    $B = $ magnetic flux density per pole, T

$l = $ effective length of armature conductor, m

$I_C = I_A/a = $ current in armature conductor, A

$N = Z/2 = $ number of effective one-turn coils

$D = 2R = $ diameter of armature rotor, m

### EXAMPLE 4-6

For the motor in Example 4-5, calculate the developed torque and power at 1650 r/min.

**SOLUTION**

From the solution of Example 4-5, $K_G$ was found to be 0.601 V/rad/s. At 1650 r/min,

$$E_C = 0.601 \times 1650 \times \frac{2\pi}{60} = 103.6 \text{ V}$$

$$I_A = \frac{125 - (103.6 + 2)}{0.45} = 43.1 \text{ A}$$

$$T_D = K_T I_A = 0.601 \times 43.1 = 25.8 \text{ N} \cdot \text{m}$$

$$P_D = E_C I_A = 103.6 \times 43.1 = 4.465 \text{ kW}$$

As a check, since

$$T_D = \frac{P_D}{\omega}$$

$$\omega = 1650 \times \frac{2\pi}{60} = 172.7 \text{ rad/s}$$

$$T_D = \frac{4465}{172.7} = 25.8 \text{ N} \cdot \text{m}$$

**Torque–speed relationship.** Equation (4-12) describes the relationship between armature current and speed for a dc shunt motor with fixed field current. We know that developed torque is proportional to field current times armature current from Eq. (4-25). If we substitute the relationship of Eq. (4-25) into Eq. (4-12), we obtain

$$\omega = \frac{V_L - (T_D R_A / K'_T I_F)}{K'_G I_F}$$

$$= \frac{V_L}{K'_T I_F} - \frac{T_D R_A}{(K'_T I_F)^2} \tag{4-27a}$$

since $K'_T = K'_G$. Equation (4-27a) is a straight-line relationship between $\omega$ and $T_D$ as shown in Fig. 4-6. Note that Fig. 4-6 is plotted for one value of $I_F$.

For a fixed field current,

$$\omega = \frac{V_L}{K_T} - \frac{T_D R_A}{(K_T)^2} \tag{4-27b}$$

Equations (4-27a) and (4-27b) relate speed and torque.

We are often interested in the behavior of a motor as its speed is varied. Using a digital computer, it is a comparatively straightforward task to obtain curves of current, power, and torque as functions of speed.

Normally, we would know the motor no-load speed, armature, and shunt-field resistance. If we know the no-load speed and armature current, we can calculate $K_G$ using Eq. (4-10). Knowing $K_G$, we arbitrarily select values of armature current, say in steps of 5% rated $I_L$, and proceed with our calculations.

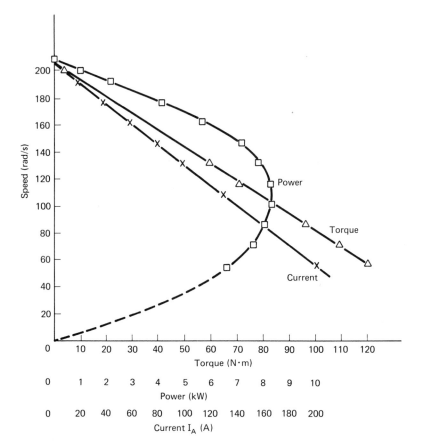

**Figure 4-6** Performance curves for motor of Program 4-1.

Knowing $I_A$, $K_G$, and $R_A$, we calculate a value of $\omega$ and $E_C$ from

$$\omega = \frac{V_L - I_A R_A}{K_G} \tag{4-12}$$

and

$$E_C = V_L - I_A R_A - V_B \tag{4-8}$$

Then using Eqs. (4-17) and (4-23), we have

$$P_D = E_C \times I_A$$

and

$$T_D = \frac{P_D}{\omega}$$

By printing out the values of $\omega$, $P_D$, and $T_D$ for each value of $I_A$, we obtain the data required. The program is shown as Program 4-1, and the results using the data of Example 4-3 are shown in Fig. 4-6.

Figure 4-7 shows a shunt motor with a rheostat $R'_F$ connected in the field circuit. By varying $R'_F$, the current $I_F$ can be changed; hence the field flux $\Phi_F$ can also be changed. If the motor is run in the linear region of its saturation curve (see Section 3-4), the flux is proportional to field current, and

$$\Phi_F = K_F I_F \tag{4-28}$$

where $\quad \Phi_F$ = field flux, Wb
$\quad\quad\quad K_F$ = field constant, WB/A
$\quad\quad\quad I_F$ = field current, A

From Fig. 4-4,

$$V_L = I_A R_A + E_C$$

$$= I_A R_A + K'_G I_F \omega \tag{4-29}$$

$$\omega = \frac{V_L - I_A R_A}{K'_G I_F}$$

## PROGRAM 4-1

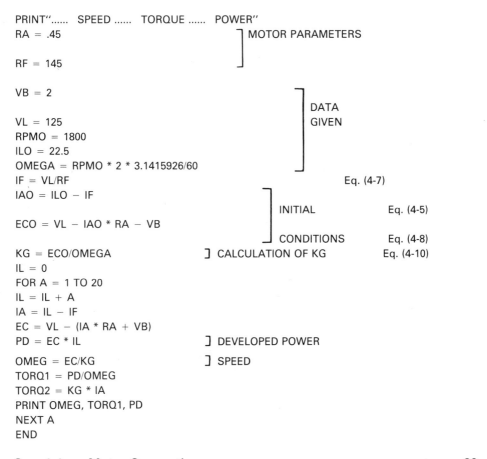

```
PRINT"...... SPEED ...... TORQUE ...... POWER"
RA = .45                          ⎤ MOTOR PARAMETERS
                                  ⎥
RF = 145                          ⎦

VB = 2                                              ⎤
                                                    ⎥ DATA
VL = 125                                            ⎥ GIVEN
RPMO = 1800                                         ⎥
ILO = 22.5                                          ⎥
OMEGA = RPMO * 2 * 3.1415926/60                     ⎦
IF = VL/RF                                    Eq. (4-7)
IAO = ILO − IF                      ⎤
                                    ⎥ INITIAL       Eq. (4-5)
ECO = VL − IAO * RA − VB            ⎥
                                    ⎦ CONDITIONS    Eq. (4-8)
KG = ECO/OMEGA              ⎤ CALCULATION OF KG     Eq. (4-10)
IL = 0
FOR A = 1 TO 20
IL = IL + A
IA = IL − IF
EC = VL − (IA * RA + VB)
PD = EC * IL                ⎤ DEVELOPED POWER

OMEG = EC/KG                ⎤ SPEED
TORQ1 = PD/OMEG
TORQ2 = KG * IA
PRINT OMEG, TORQ1, PD
NEXT A
END
```

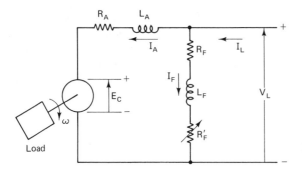

**Figure 4-7** Shunt motor with field control rheostat.

This equation states that as $I_F$ gets larger, speed decreases, and as $I_F$ gets smaller, speed increases. Furthermore, it implies that if $I_F$ goes to zero, $\omega$ will become infinitely large. A lightly loaded shunt motor in which the field current goes to zero may indeed try to increase its speed indefinitely and fly apart in the attempt. It is extremely important that the shunt-field current be maintained at all times, and in fact the shunt field is not normally fused, so as to prevent accidental loss of field in the case of the field fuse blowing. Small shunt-field machines may not run away, since the internal losses are usually high enough to keep a sufficient load on the machine, which may prevent destructive overspeeding.

**EXAMPLE 4-7**

The following data apply to the circuit of Fig. 4-13.

$$V_L = 250 \text{ V} \qquad R_F = 60 \text{ }\Omega \qquad R_A = 0.3 \text{ }\Omega \qquad R'_F = 0 \text{ to } 200 \text{ }\Omega$$

$$\omega = 120 \text{ rad/s} \quad \text{when } R'_F = 0$$

When $R'_F$ is set to zero, the line current is 50 A. If $R'_F$ is changed to 100 $\Omega$, the motor speeds up, but the load current remains the same. Calculate the new developed power, torque, and speed. Neglect brush voltage drop and assume that the motor is operating in the region where flux is directly proportional to field current.

**SOLUTION**

Initially,

$$\omega = 120 \text{ rad/s}$$

$$I_F = \frac{250}{60} = 4.17 \text{ A}$$

$$I_A = 50 - 4.17 = 45.83 \text{ A}$$

$$E_C = 250 - 45.83 \times 0.3 = 236.25 \text{ V}$$

$$E_C = K'_G I_F \omega$$

$$K'_G = \frac{236.25}{4.17 \times 120}$$

$$= 0.472 \text{ V} \cdot \text{s/A}$$

Increasing $R'_F$ will decrease $I_F$ to $I_{F_1}$:

$$I_{F_1} = \frac{250}{60 + 100} = 1.56 \text{ A}$$

The load current remains fixed, so that $E_C$ must also remain fixed but $\omega$ becomes $\omega_1$, and

$$\omega_1 = \frac{E_C}{K'_G I_{F_1}} = \frac{236.25}{0.472 \times 1.56} = 320 \text{ rad/s}$$

The output power must remain fixed since $E_C$ is fixed and $I_A$ is fixed.

$$P_D = 236.25 \times 45.83 = 10,827 \text{ W} = 10.38 \text{ kW}$$

and the new value of developed torque is

$$T_D = \frac{10.83 \times 10^3}{320} = 33.8 \text{ N} \cdot \text{m}$$

As a check on developed torque, we know that

$$T_D = K_T I_A = K_G I_A = K'_G I_{F_1} I_{A_1}$$

In Example 4-7, the load current remained constant while the speed changed. It is possible for both of these quantities to change, and the manner in which they change is dependent on the load being driven.

**EXAMPLE 4-8**

The motor of Example 4-7 is driving a load of 6000 W at 120 rad/s, with $R'_F = 0$. The shunt-field rheostat is changed and the motor speeds up to 150 rad/s and the armature current is 65 A. The rotational losses of the motor are 800 W and may be assumed constant. Calculate:

  (a) The input power at 120 rad/s.
  (b) The input power at 150 rad/s.
  (c) The rheostat resistance at 150 rad/s.

Assume that the motor is not operating in saturation.

**SOLUTION**

  (a) At 120 rad/s:

$$\text{output power} = 6000 \text{ W}$$

$$\text{rotational losses} = 800 \text{ W}$$

$$\text{developed power} = 6800 \text{ W}$$

counter EMF $= 120 \times 4.17 \times 0.472 = 236.8$ V

armature current $= \dfrac{250 - 236.8}{0.3} = 46$ A

armature power loss $= 46^2 \times 0.3 = 635$ W

field loss $= \dfrac{(250)^2}{60} = 1042$ W

input power $= 6800 + 635 + 1042 = 8477$ W

(b) At 150 rad/s:

$I_A = 65$ A

CEMF $= K'_G I_F \omega = 250 - 65 \times 0.3 = 230.5$ V

$I_F = \dfrac{230.5}{0.472 \times 150} = 3.26$ A

$P_D = 65 \times 230.5 = \cancel{24,983}$ W

*14982*

Losses:

$\qquad$ Armature: $65^2 \times 0.3 = 1268$ W

Field (including rheostat): $3.26 \times 250 = \phantom{0}815$ W

Rotational: $= \phantom{0}800$ W

Total $= \overline{2883}$ W

output power $P_O = P_D - P_R \qquad = 14{,}983 - 800$

$= 14{,}183$ W

input power $= P_O +$ losses

$= 14{,}183 + 2883 = 17{,}066$ W

$= 17.07$ kW

(c) $I_F = 3.26$ A:

$$R_F + R'_F = \frac{250}{3.26} = 76.7 \ \Omega$$

$$R'_F = 76.7 - 60 = 16.7 \ \Omega$$

**Armature control.** Figure 4-8 illustrates a second method of speed control for a shunt motor. In this method, the field circuit resistance remains fixed, but the armature current is varied using the rheostat $R'_A$. This method of control is restricted to small machines since the armature current is normally very high and

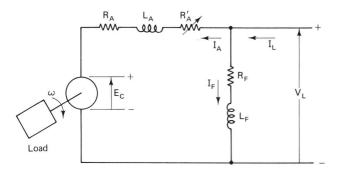

**Figure 4-8** Shunt motor with armature control rheostat.

the $I^2R$ losses in $R'_A$ would also be very high. It is particularly useful in small servomotors, where power dissipation is small and where quick time response is desirable. The armature electrical time constant $T_A$ is $L_A/R_A$, and effectively increasing $R_A$ reduces this time constant, allowing the motor to respond more quickly.

From Fig. 4-8, we have the following:

$$I_A = \frac{V_L - E_C}{R_A + R'_A} \tag{4-30}$$

$$I_F = \frac{V_L}{R_F} \tag{4-31}$$

$$I_L = I_A + I_F \tag{4-32}$$

$$E_C = K'_G I_F \omega = K_G \omega \tag{4-33}$$

In this configuration, $I_F$ is held fixed so that the second form of Eq. (4-33) will be the most useful. Many dc servomotors have permanent-magnet fields, so that this form of the equation applies to them as well.

### Series Motors

Figure 4-9 shows the connection of a series motor. In this connection, the armature and field current are identical. The series-field winding consists of a few turns of large wire having a low resistance $R_S$. For the series motor

$$I_A = I_L \tag{4-34}$$

The input power is

$$P_I = V_L I_L = V_L I_A \tag{4-35}$$

The series-field current is

$$I_S = I_A = \frac{V_L - E_C - V_B}{R_A + R_S} \tag{4-36}$$

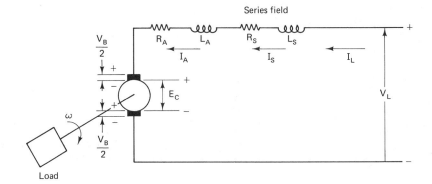

**Figure 4-9** Dc series motor equivalent circuit.

Since the brush voltage drop is small, we can ignore it and write

$$I_A = \frac{V_L - E_C}{R_A + R_S} \tag{4-37}$$

The counter EMF $E_C$ is found from

$$E_C = K'_G I_A \omega \tag{4-38}$$

But now $I_A$ and $\omega$ both vary with load. This equation assumes a linear relationship between $E_C$ and $\Phi s$, which is to say that it assumes that the iron core is not saturated.

### EXAMPLE 4-9

A series motor draws 30 A from a 250-V dc source and it is running at 90 r/min. The load changes, and the current drops to 20 A. Find the new speed. The armature resistance is 0.15 $\Omega$ and the series-field resistance is 0.12 $\Omega$.

### SOLUTION

When $I_A = 30$ A,

$$E_C = 250 - 30 \times (0.15 + 0.12) = 241.9 \text{ V}$$

$$\omega = \frac{2\pi f}{60} \qquad = \frac{2\pi \times 900}{60} = 94.2 \text{ rad/s}$$

$$K'_G = \frac{E_C}{I_A \omega} \qquad = \frac{241.9}{30 \times 94.2}$$

$$= 0.0856 \text{ V/A/rad/s}$$

$$= 0.0856 \text{ V} \cdot \text{s/A} \cdot \text{rad}$$

Now if $I_A = 20$ A,

$$E_C = 250 - 20 \times (0.15 + 0.12) = 244.6 \text{ V}$$

$$\omega = \frac{E_C}{K'_G I_A} = \frac{244.6}{(0.0856 \times 20)} = 142.8 \text{ rad/s} \quad \text{or} \quad 1364 \text{ r/min}$$

**Speed regulation.**   In Fig. 4-9, neglecting brush drop,

$$I_A = \frac{V_L - E_C}{R_A + R_S} = \frac{V_L - K_G'\omega}{R_A + R_S}$$

Then

$$I_A(R_A + R_S) = V_L - K_G'I_A\omega$$

$$K_G'I_A\omega = V_L - I_A(R_A + R_S)$$

$$\omega = \frac{V_L - I_A(R_A + R_S)}{I_A K_G'} \qquad (4\text{-}39)$$

Now $I_A(R_A + R_S)$ is the series voltage drop in the armature circuit, and is very small, so that we can approximate $\omega$ by

$$\omega = \frac{V_L}{I_A K_G'} \qquad (4\text{-}40)$$

Equation (4-39) is plotted in Fig. 4-10.   We can see from Fig. 4-10 that at very low speed, the current is very high, but that as current decreases, speed rises rapidly. In fact, the curve implies that if current approaches zero, speed will approach infinity.   This implies that there should always be a fair amount of current into a series motor, which means that it should not be run without load.   Since an unloaded series motor can run away and destroy itself, the load on a series motor should be connected directly to the shaft and should not be belt connected.

Since the speed variation with load is so great, the regulation is poor.   We cannot even determine it experimentally, as the no-load speed might be too high for safety.   The starting current, if the motor is started directly across the line, can be found from Eq. (4-39) by setting $\omega = 0$.   Then

$$I_{AS} = \frac{V_L}{R_A + R_S} \qquad (4\text{-}41)$$

where $I_{AS}$ is the starting current in amperes.

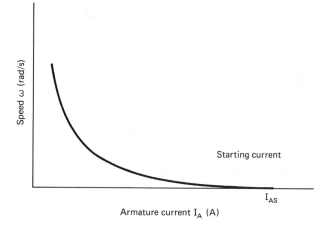

Speed $\omega$ (rad/s)

Starting current

$I_{AS}$

Armature current $I_A$ (A)

**Figure 4-10**   Series motor current–speed characteristic curve.

**Power.** The equations governing the performance of the series motor can be readily determined from Fig. 4-9. The developed power $P_D$ is given by

$$P_D = E_C I_A$$

$$E_C = V_L - I_A(R_A + R_S) - V_B$$

$$P_A = I_A^2 R_A$$

$$P_S = I_A^2 R_S$$

$$P_D = P_I - P_L = V_L I_A - I_A^2(R_A + R_S)$$

$$P_O = P_D - P_R$$

All of the definitions for the quantities above are the same as those following Eq. (4-21).

### EXAMPLE 4-10

The rotational losses for the motor of Example 4-9 are given in Fig. 4-11. Calculate the following at (a) 94 and (b) 143 rad/s:

(i) The developed power.
(ii) The output power.
(iii) The output torque.

### SOLUTION

(a) At 94 rad/s:

(i) When $I_A = 30$ A,

$$E_C = 250 - 30 \times (0.15 + 0.12) = 241.9 \text{ V}$$

$$P_D = 241.9 \times 30 = 7257 \text{ W} = 7.26 \text{ kW}$$

(ii) $P_R = 172$ W
$$P_O = 7257 - 172 = 7085 \text{ W}$$

(iii) $T_O = \dfrac{7085}{94} = 75.4$ N·m

(b) At 143 rad/s:

(i) When $I_A = 20$ A,

$$E_C = 250 - 20 \times (0.15 + 0.12) = 244.6 \text{ V}$$

$$P_D = 244.6 \times 20 = 4892 \text{ W} = 4.89 \text{ kW}$$

(ii) $P_R = 400$ W
$$P_O = 4892 - 400 = 4492 \text{ W} = 4.49 \text{ kW}$$

(iii) $T_O = \dfrac{4492}{143} = 31.4$ N·m

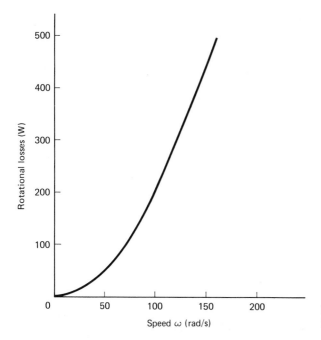

**Figure 4-11** Rotational losses as a function of speed.

**Torque.**   The developed power for a series motor is

$$P_D = E_C I_A$$

but

$$E_C = K'_G I_A \omega$$

Therefore,

$$P_D = K'_G I_A^2 \omega$$

and the developed torque

$$T_D = \frac{P_D}{\omega} = K'_G I_A^2 \qquad (4\text{-}42)$$

We know that torque is proportional to the product of current and flux, and that flux is proportional to current.   For the series motor, then, we know that

$$T_D \propto I_A \Phi \qquad T_D \propto I_A I_F \qquad T_D \propto I_A^2$$

Therefore,

$$T_D = K'_T I_A^2 \qquad (4\text{-}43)$$

Again $K'_T = K'_G$, as in the case of the shunt motor.   Then for a series motor, the developed torque is proportional to the square of the armature current.   At standstill, when the counter EMF is zero, the armature current is very high and the starting torque is extremely high.   As the motor speeds up, the counter EMF increases, the current diminishes, and the torque drops off very rapidly.

Sec. 4-4    Motor Connections                                                    **107**

**Torque–speed relationship.** From Eq. (4-39), we have

$$\omega = \frac{V_L - I_A(R_A + R_S)}{I_A K_G'}$$

Rearranging to solve for $I_A$, we obtain

$$I_A K_G' \omega = V_L - I_A(R_A + R_S)$$

or

$$I_A(K_G' \omega + R_A + R_S) = V_L$$

and

$$I_A = \frac{V_L}{K_G' \omega + R_A + R_S} \tag{4-44}$$

Substitution into Eq. (4-43) yields

$$T_D = \frac{K_T' V_L^2}{(K_T' \omega + R_A + R_S)^2} \tag{4-45}$$

Figure 4-12 shows the relationship between rotational speed and torque. Note that the series motor has a very high starting torque and low torque at high speed. As noted previously, under no-load conditions, the series motor has a tendency to run away.

Because of the high starting torque, the series motor characteristic is utilized in driving subway cars. It has been suggested that series motors would be the ideal drive for electric automobiles. However, this suggestion overlooks the poor high-speed performance of series motors. At high speeds, there would be very little torque available for acceleration and passing. This is not a serious problem on subway cars since they can be accelerated to their running speed over a fairly long period, and the problem of passing another car is nonexistent.

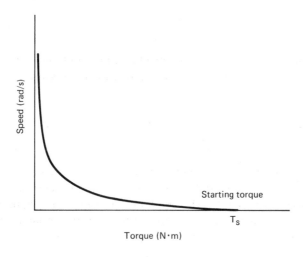

Figure 4-12 area labels: Speed (rad/s), Torque (N·m), Starting torque, $T_S$

**Figure 4-12** Torque–speed characteristic of a dc series motor.

## EXAMPLE 4-11

A series motor has a rated input power of 20 kW, 550 V, and 2000 r/min. To limit the inrush starting current to 150% of rated value, a starter resistance is needed which may be switched out of the circuit when the current is low enough. The motor armature resistance is $0.5\ \Omega$ and the series field resistance is $0.5\ \Omega$. Determine the value of the starting resistor and the starting torque. Neglect any rotational losses and brush voltage drop.

**SOLUTION**

$$\text{rated current} = \frac{20 \times 10^3}{550} = 36.36\ \text{A}$$

The maximum allowable current is

$$I_{\max} = 1.5 \times 36.36 = 54.54\ \text{A}$$

The circuit resistance on starting is

$$R_C = \frac{550}{54.54} = 10.08\ \Omega$$

The starting resistor is

$$R_{\text{ST}} = 10.08 - (0.5 + 0.5) = 9.08\ \Omega$$

The rated speed is

$$\omega = 2\pi \times \frac{2000}{60} = 209.4\ \text{rad/s}$$

$$\text{rated torque} = \frac{20 \times 10^3}{209.4} = 95.5\ \text{N·m}$$

and

$$T_D = K_T' I_A^2$$

At rated value, $T_D = 95.5$ N·m and $I_A = 36.36$ A; therefore,

$$K_T' = \frac{95.5}{36.36^2} = 7.22 \times 10^{-2}\ \text{N·m/A}^2$$

On starting, $I_A = 54.54$ A; therefore,

$$T_D = 7.22 \times 10^{-2} \times 54.54^2 = 214.7\ \text{N·m}$$

**Speed control.** To limit the starting current in a series motor, it was suggested that a series resistor might be used which could be switched out as the current decreased. This method is not suitable for continuous control since the series resistor must conduct full armature current and will have high losses. For speed control, it may be more useful to bypass some of the field current through a diverter resistor $R_D$ as shown in Fig. 4-13. As $R_D$ is decreased, $I_D$ will increase

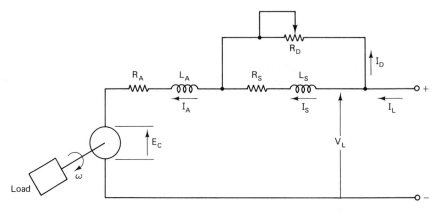

**Figure 4-13** Series motor speed control using diverter.

and $I_S$ will decrease, causing field flux to decrease and causing the motor to speed up.

From Fig. 4-13,

$$E_C = V_L - I_L \frac{R_D R_S}{R_D + R_S} - I_L R_A \qquad (4\text{-}46)$$

and

$$E_C = K'_G \omega I_S$$

Therefore,

$$K'_G \omega I_S = V_L - I_L \frac{R_D R_S}{R_D + R_S} - I_L R_A \qquad (4\text{-}47)$$

But

$$I_S = \frac{I_L R_D}{R_D + R_S} \qquad (4\text{-}48)$$

Substituting Eq. (4-48) into Eq. (4-47) gives us

$$\frac{K'_G \omega I_L R_D}{R_D + R_S} = V_L - I_L \frac{R_D R_S}{R_D + R_S} - I_L R_A$$

Solving for $\omega$, we have

$$\omega = \frac{V_L(R_D + R_S)}{K'_G I_L R_D} - \frac{R_S}{K'_G} - \frac{R_A(R_D + R_S)}{K'_G R_D} \qquad (4\text{-}49)$$

This rather formidable equation merely describes the speed versus load current relationship for a series motor with a diverter resistor in parallel with the field circuit. It will, of course, have maximum effect when the motor is operating in the linear portion of the saturation curve or near the bend. We can check the validity of Eq. (4-49) at two extremes, when $R_D = 0$ and $R_D = \infty$.

If $R_D = 0$, the first term of Eq. (4-49) becomes infinite as does the last. In fact the algebraic sum goes to infinity, which is what we would expect since the diverter takes all the line current, flux goes to zero, and speed would tend to infinity. At the other extreme, if $R_D = \infty$, we would have

$$\omega = \frac{V_L}{K'_G I_L} - \frac{R_S}{K'_G} - \frac{R_A}{K'_G}$$

$$= \frac{1}{K'_G I_L}(V_L - I_L R_S - I_L R_A)$$

which is Eq. (4-39).

## EXAMPLE 4-12

A dc series motor draws 22 A from a 240-V line while running at 840 r/min. It has an armature resistance of 0.6 $\Omega$ and a series field resistance of 0.5 $\Omega$. A diverter is to be added to the circuit so that the speed increases to 1200 r/min while the line current increases to 28 A. Find $R_D$.

### SOLUTION

Initially,

$$\omega = \frac{840 \times 2\pi}{60} = 87.96 \text{ rad/s}$$

$$I_L R_S = 22 \times 0.5 = 11 \text{ V}$$

$$I_L R_A = 22 \times 0.6 = 13.2 \text{ V}$$

$$V_L = 240 \text{ V}$$

$$87.96 = \frac{240 - 11 - 13.2}{K'_G} \times 22$$

Therefore,

$$K'_G = 0.111$$

If we call the new value of $\omega$, $\omega_N$,

$$\omega_N = 1200 \times \frac{2\pi}{60} = 125.7 \text{ rad/s}$$

Substitution into Eq. (4-49) yields

$$125.7 = \frac{240 \ (R_D + 0.5)}{0.111 \times 28 \times R_D} - \frac{0.5}{0.111} - \frac{0.6(R_D + 0.5)}{0.111 R_D}$$

$$= \frac{240}{3.1} + \frac{120}{3.1 R_D} - 4.5 - 5.4 - \frac{2.7}{R_D}$$

$$= 77.4 + \frac{38.7}{R_D} - 9.9 - \frac{2.7}{R_D}$$

$$58.2 = \frac{36}{R_D}$$

$$R_D = 1.62 \ \Omega$$

### Compound Motors

Figure 4-14 shows the possible connections of a compound motor. Figure 4-14a is a short-shunt connection, so-called because the shunt field is connected directly across the armature, while Fig. 4-14b shows a long-shunt connection, where the shunt field is a "longer" distance from the armature. Although the connections are different, the difference in performance of the two configurations is only marginal. Since the voltage drop across $R_S$ is small, the air-gap flux due to $I_F$ is practically the same in both cases.

It is possible to connect the two fields so that their fluxes aid each other, or so that they oppose each other. A compound motor in which the fields aid is said to be *cumulatively compound*, while a motor in which the fields oppose each other

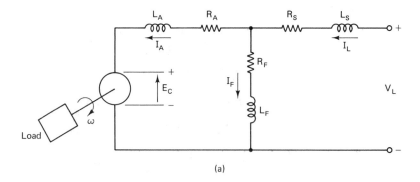

(a)

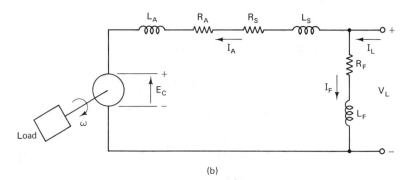

(b)

**Figure 4-14** (a) Compound motor short-shunt connection; (b) compound motor long-shunt connection.

is said to be *differentially compound*. The differentially compound motor tends to be unstable and is rarely used.

The long-shunt motor is somewhat easier to analyze than the short-shunt motor, and since there is no appreciable difference in behavior, we will confine our analysis to this configuration.

In Fig. 4-14b,

$$I_L = I_F + I_A \tag{4-50}$$

The input power is

$$P_L = V_L I_L \tag{4-51}$$

The field current is

$$I_F = \frac{V_L}{R_L} \tag{4-52}$$

Since the series-field current is the same as the armature current in the long-shunt connection, the counter EMF can be calculated from

$$E_C = V_L - I_A(R_A + R_S) \tag{4-53}$$

The net air-gap field flux $\phi_F$ is given by

$$\phi_F = K_F' I_F \pm K_G' I_A \tag{4-54}$$

The $\pm$ signs arise from the fact that the series field may aid or oppose the main shunt field, respectively. Invariably, compound motors are connected cumulative.

A comparison of the torque–speed characteristic curves of all three motors is shown in Fig. 4-15. The derivation of the torque–speed curve for the compound motor is fairly involved if we wish to obtain an expression relating torque and speed in closed form. However, if we want to obtain a set of curves for a particular machine, using a digital computer the procedure is far simpler.

From Fig. 4-14b we may write

$$E_C = K\phi\omega$$

$$= K(\phi_F + \phi_S)\omega$$

$$= (K_F' I_F + K_S' I_S)\omega$$

$$= (K_F' I_F + K_S' I_A)\omega \tag{4-55}$$

$$I_A = \left(\frac{V_L - E_C}{R_A + R_S}\right) = \frac{V_L - (K_F' I_F + K_S' I_A)\omega}{R_A + R_S}$$

$$I_A\left(\frac{1 + K_S'\omega}{R_A + R_S}\right) = \frac{V_L - K_F' I_F \omega}{R_A + R_S}$$

$$I_A = \frac{V_L - K_F' I_F \omega}{R_A + R_S + K_S'\omega} \tag{4-56}$$

Sec. 4-4   Motor Connections                                                        **113**

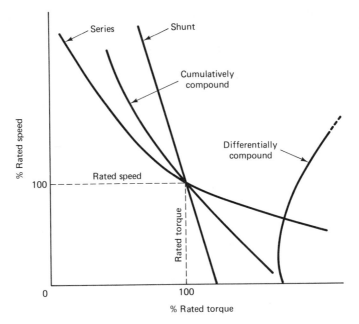

**Figure 4-15**  Comparison of motor torque versus speed.

$$P = E_C I_A \tag{4-57}$$

$$T = \frac{P}{\omega} \tag{4-58}$$

The procedure for computation would be as follows:

1. Choose a suitable $I_A$ using Eq. (4-56).
2. Using the values of $I_A$, calculate $E_C$ using Eq. (4-55).
3. Knowing $E_C$ and $I_A$, use Eqs. (4-57) and (4-58) to obtain torque.

Program 4-2 is a suggested program that will perform these calculations. To test it out, values of $K'_F = 0.345$, $K'_S = 0.0075$, $R_S = 0.345 \ \Omega$, $R_A = 0.354 \ \Omega$, and $R_F = 62.5 \ \Omega$ were used, and the results are shown in Fig. 4-16.

### PROGRAM 4-2

```
 5 ? "CLR/HOME"
10 ? "PROGRAM 4-2 COMPOUND MOTOR"
20 ? INPUT "ARMATURE RESISTANCE"; RA
30 ? INPUT "SERIES FIELD RESISTANCE"; RS
40 INPUT "SHUNT FIELD RESISTANCE"; RF
50 INPUT "LINE VOLTAGE"; VL
60 INPUT "KS PRIME"; KSP
70 INPUT "KF PRIME"; KFP
80 INPUT "RATED SPEED"; WR
```

```
 85 PRINT "TORQUE SPEED (RAD/SEC)      %RATED SPEED"
 90 W1 = WR/10
100 FOR A = 1 TO 15
110 FC = VL/RF
115 W = A * W1
120 AC = (VL − KFP * FC * W)/(RA + RS)
130 EC = (KFP * FC + KSP * AC) * W
140 P = EC * AC
150 TORQ = P/W
160 PER = A * 10
170 PRINT TORQ, W, PER
175 NEXT A
180 ? "HIT ANY KEY TO CONTINUE"
190 GET A$; IF A$ = " " THEN 190
195 GO TO 5
```

**Torque–speed relationship.**    The torque versus speed characteristic of the compound motor is intermediate between the series and shunt motors, as can be seen in Fig. 4-15, which shows comparative curves for three motors of equal rating.

The compound motor has a higher starting torque than the shunt motor, due to the presence of the series winding.   At higher torque and hence higher current, the speed of the compound motor drops off more quickly with load than does the shunt motor, but not as quickly as the series motor.

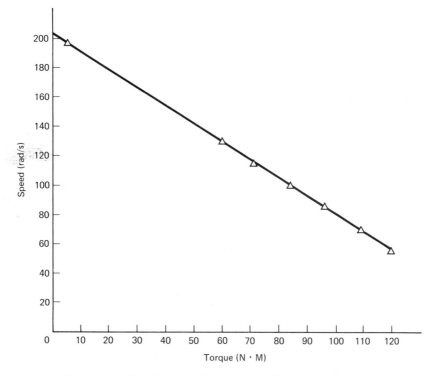

**Figure 4-16**   Calculated speed–torque curve for compound motor.

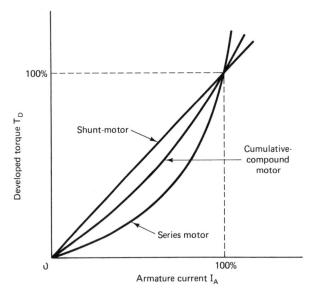

**Figure 4-17** Developed torque versus armature current for dc motors.

For the same number of shunt-field turns, the cumulative-compound motor has more net field flux than the shunt motor. Now, for the same rated output, both machines have same current and counter EMF. The counter EMF is given by

$$E_C = \begin{cases} K\phi_C\omega_C & \text{for the compound motor} \\ K\phi_S\omega_S & \text{for the shunt motor} \end{cases}$$

But

$$\phi_C\omega_C = \phi_S\omega_S$$

$$\phi_C > \phi_S$$

Therefore,

$$\omega_S > \omega_C$$

Thus the shunt motor would rotate at a slightly higher speed than the compound motor.

In summary, the developed torque versus armature current, which is proportional to loading for shunt, series, and cumulative-compound motors, is illustrated in Fig. 4-17. Each motor is assumed to develop the rated torque at rated armature current.

## 4-5 RATING AND EFFICIENCY OF DC MOTORS

In Section 4-3 the various losses in a shunt motor were discussed briefly. As the world becomes more energy conscious, it becomes more important to appreciate where energy losses occur, and a great deal of effort is being expended at present in producing energy-efficient machines.

It has previously been stated that the same dc machine can be used as either a motor or a generator. However, there is one significant difference, namely, that the machine rating as a generator is higher than it would be as a motor. The limitation in both cases is the total internal power loss that occurs in the machine. The losses in both cases being the same, the currents and the terminal voltages will also be the same. Then the terminal electrical power in both cases is the same.

This can be seen clearly in Fig. 4-18, which shows the same machine connected as a motor in Fig. 4-18a and as a generator in Fig. 4-18b. In Fig. 4-18a, an electrical input of 10 kW results in 2 kW internal losses and 8 kW appears as useful mechanical power. In Fig. 4-18b, the electrical conditions are the same, namely 2 kW internal losses but 10 kW electrical at the output terminals. Since output power determines the rating, the machine is rated as an 8-kW (10.7-hp) motor or a 10-kW generator.

Efficiency $\eta$ was defined in Eq. (3-16) as

$$\eta = \frac{P_{out}}{P_{in}} = \frac{P_{in} - P_L}{P_{in}} = 1 - \frac{P_L}{P_{in}} \qquad (3\text{-}16)$$

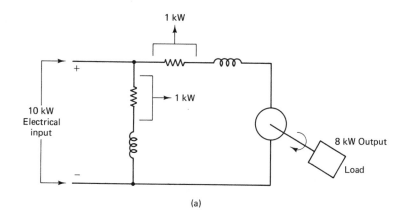

(a)

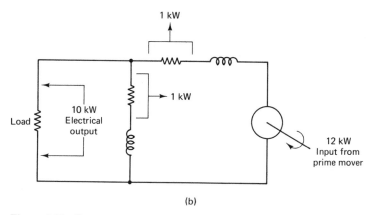

(b)

**Figure 4-18** Generator–motor rating comparison: (a) machine connected as a motor; (b) machine connected as a generator.

$P_L$ represents the total losses in a motor, including the electrical, mechanical, and electromechanical losses. It is a straightforward matter to calculate electrical losses, as they are either $I^2R$ losses in windings or $IV_B$ brush drop losses. Mechanical losses are not straightforward, as they depend on velocity, shape of components, and material. These losses include friction, windage, and brush friction losses. Electromechanical losses are eddy current and hysteresis losses. Theoretically, these are calculable, but because of the various mathematical relationships involved, they too are difficult to calculate. Very often, then, both mechanical and electromechanical losses are measured directly or predicted from data based on other machines of the same size and type. In all cases, all the input power from the line must supply all the losses plus the output power. Figure 4-19 shows the distribution of losses in a compound motor with compensating and interpole windings.

**EXAMPLE 4-13**

A 75-hp 230-V dc compound motor is connected long shunt. It runs at 1200 r/min when fully loaded. The field resistance is 82 $\Omega$, armature resistance including brushes 0.045 $\Omega$, and series-field resistance 0.03 $\Omega$. The machine has an efficiency of 86% at full load. Calculate the rotational losses.

**SOLUTION**

The output power is

$$P_{out} = 75 \text{ hp} = 75 \times 746 = 55.95 \times 10^3 \text{ W} = 55.95 \text{ kW}$$

The input power is

$$P_{in} = \frac{55.95}{0.86} = 65.1 \text{ kW}$$

Further,

$$I_L = \frac{65,100}{230} = 283 \text{ A}$$

$$I_F = \frac{230}{82} = 2.8 \text{ A}$$

Therefore,

$$I_A = 283 - 2.8 = 280.2 \text{ A}$$

Losses:

| | |
|---|---|
| Shunt field: $2.8^2 \times 82$ | $= 643$ W |
| Series field: $280.2^2 \times 0.03$ | $= 2355$ W |
| Armature: $280.2^2 \times 0.045$ | $= \underline{3533}$ W |
| Total electrical losses | $= 6531$ W |
| Total losses $= 65,100 - 55,950$ | $= 9100$ W |

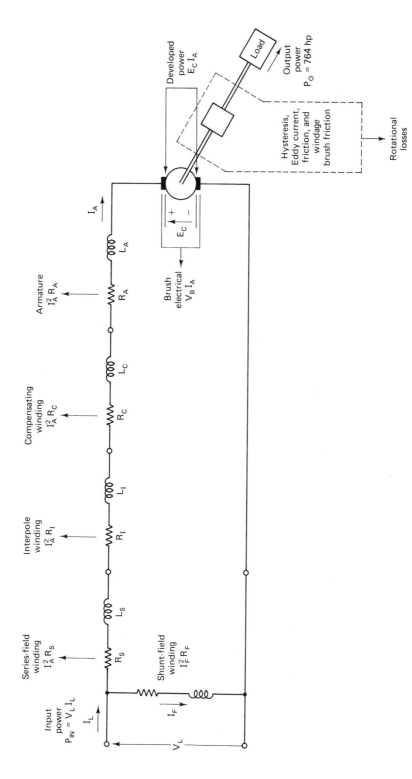

**Figure 4-19** Dc motor distribution of losses.

Therefore,

$$\text{rotational losses} = 9100 - 6531 = 2569 \text{ W} = 2.57 \text{ kW}$$

**EXAMPLE 4-14**

For the compound motor of example 4-13, calculate the constants $K_F'$ and $K_S'$.

**SOLUTION**

The full-load speed is 1200 r/min or 125.7 rad/s.

$$\text{counter EMF} = 230 - 280.2(0.045 + 0.03) = 209 \text{ V}$$

and

$$209 = (K_F' \times 2.8 + K_S' \times 280.2)125.7$$

If not given, we must assume a ratio of main field strength to series-field strength of, say,

$$\frac{K_F' I_F}{K_S' I_A} = \frac{80\%}{20\%} \quad \text{that is,} \quad \frac{K_F' \times 2.8}{K_S' \times 280.2} = 4$$

$K_F' = 400.3 K_S'$, and upon substituting and solving, we obtain $K_F' = 0.4751$ and $K_S' = 0.001187$.

## 4-6 APPLICATIONS

Dc motors are used in a great variety of industrial applications, particularly where a wide range of speeds is encountered or where precise control of speed or position is required. They are used in conveyors, pumps, machine tools, packaging machinery, cutting machines, contour-following machines, and so on. For applications involving approximately constant speed, the induction motor is more economical and rugged, while for applications involving only one speed, the synchronous motor may be the better choice.

At one time, dc applications were restricted by the availability of a suitable dc source. However, with the availability of reliable rectifiers and especially silicon-controlled rectifiers (thyristors), dc machines can easily be run from ac lines. In fact, in many instances, it is easier to obtain better control in a simpler manner from a rectified ac source than from a pure dc source.

The choice of which motor to use for a particular application may be quite straightforward or involve many factors. If the motor is to be used in a control application requiring it to have a speed that varies over a range of 15:1, at one time a dc motor would have been the obvious choice. This speed range requires a fairly sophisticated controller, involving both field and armature control. At present, there are ac variable-speed, variable-frequency controllers which may be cost competitive with dc drives.

Suppose for a moment that we have decided on using a large dc motor for a particular application. What size motor do we want? This of course depends on the characteristics of the load. A trolley car might require a very high starting torque but very little running torque, so that a series motor might be desirable. A crane motor might require a fairly high starting torque and running torque to lift a load, so that a shunt motor might be the correct choice. A shunt motor would probably be satisfactory for a fan motor that has a parabolic torque–speed characteristic. At standstill it requires no torque, but as it speeds up the torque requirements vary as the square of the speed. A shunt motor would be suitable for this application.

The rating of a motor chosen for a particular job also depends on the conditions under which it is used. Motors can be run beyond their rated values for periods ranging from seconds to minutes, depending on the application and on the size of the motor. Because of their large thermal mass, it may take a considerable amount of excess current to cause sufficient heating in a large motor to result in burnout. Thus some motors may undergo considerable overloading for several minutes or even hours. On the other hand, small motors may burn out on overload in a few minutes. The best advice in choosing a motor for a critical application would be to consult the manufacturer's representative.

A motor is rated by its power output either in kilowatts or horsepower. However, the rating is meaningful only if all conditions are known for the specified rating. For example, a shunt motor is rated at 10 hp, 2500 r/min, and 240 V, for continuous duty at 40°C ambient. In addition, it has a service factor of 1.0, class H insulation with F rating, ball bearings in a drip-proof NEMA 219 AT drip-proof frame. It is designated as suitable for use with three-pulse or six-pulse rectified three-phase 50/60-Hz supply.

It may be necessary to know all of the information above in order to make an economical choice of motor for use in a particular application. Notice that in the list of specifications there is no mention of resistance, losses, efficiency, or speed regulation, all of which are essential properties of the motor.

As stated, the motor is rated at 10 hp, 2500 r/min, 240 V. It is suitable to drive such a load, assuming that the ambient temperature does not rise above 40°C.

Internal heating is the most serious limitation on a motor's lifetime. Heating causes insulation to deteriorate, and this eventually leads to internal electrical faults, either short or open circuits. Class H insulation with F rating means that the motor is conservatively rated. Class H insulation with 40°C ambient allows for a temperature rise of 140°C, while class F only allows for rise of 115°C.

## REVIEW QUESTIONS

4-1. Explain the differences that occur in dc machines when the generated voltage is greater and less than the applied line voltage.

4-2. Explain how counter or back EMF of a motor produces its internal developed power.

4-3. Write the equation of internal developed (electromagnetic) power and describe each of its factors. How is it related to the output power of the dc motor?

4-4. Describe the electrical losses and the rotational losses in a shunt motor.

**4-5.** How do the losses due to hysteresis, eddy currents, friction, and windage depend on the speed of the motor?

**4-6.** Write the speed equations of a shunt motor as a function of both field flux and field current. Describe each factor in the two equations.

**4-7.** An understanding of load characteristics is important when considering applications of dc motors. Describe the load characteristics for a hoist traveling at uniform speed, a centrifugal pump operating at a constant head, and a household fan.

**4-8.** Plot on a torque–speed graph the operating characteristics of a shunt motor at constant torque and constant horsepower. Assume that both motor characteristics have the same values of torque at the base speed. Explain these characteristics.

**4-9.** Plot on a horsepower–speed graph, the operating characteristics of a shunt motor at constant torque and at constant horsepower. Assume that both motor characteristics have the same values of horsepower at the base speed. Explain these characteristics.

**4-10.** **(a)** Explain how a shunt motor may be operated to develop a constant torque output versus increasing speed.

   **(b)** Explain how a shunt motor may be operated to develop a constant horsepower output versus increasing speed.

**4-11.** Write out the equation of speed versus load current for a series motor and describe each of its factors. Does the equation of speed versus load current demonstrate that this characteristic is inversely related? State any assumptions that may be required.

**4-12.** List all the power losses in a compound motor; also, draw the power flow diagram from the input electrical power to the output mechanical power. How may some of these power losses be minimized?

**4-13.** Using practical data for a dc motor, plot the graph of efficiency versus output power.

**4-14.** State one advantage and one disadvantage in the application of each of the three basic types of dc motors.

**4-15.** List the important information available from the nameplate of a dc motor.

## PROBLEMS

**4-1.** A dc machine connected as a generator delivers 40 A at 220 V to a resistive load. The armature resistance of the machine is 1.38 $\Omega$. If the machine is producing rated output power, calculate:
   **(a)** Its rating as a generator.
   **(b)** Its rating as a motor if it is connected to a 220-V line.
   Neglect all other losses.

**4-2.** A 250-V shunt motor has an armature resistance of 0.32 $\Omega$ and a field resistance of 125 $\Omega$. At no load the motor draws 12 A from a 250-V supply. The no-load speed is 2500 r/min. Calculate the full-load speed if the line current is 82 A.

**4-3.** The motor of Problem 4-2 has rotational losses of 2 kW at full load. Calculate full-load:
   **(a)** Output power.
   **(b)** Output torque.
   **(c)** Efficiency.

**4-4.** The following data are provided by a motor manufacturer for a dc shunt motor.

$$\text{output power} = 25 \text{ hp}$$

$$\text{voltage} = 240 \text{ V}$$

$$\text{speed} = 1150 \text{ r/min}$$

When delivering full output, the various losses are:

$$\text{armature} = 6\% \text{ of full load}$$

$$\text{field} = 4\% \text{ of full load}$$

$$\text{rotational} = 4\% \text{ of full load}$$

Using this information, derive an equivalent model for this motor similar to Fig. 4-4.

**4-5.** Calculate the no-load speed for the motor of Problem 4-4:
  **(a)** Considering only the effect of rotational losses.
  **(b)** Neglecting rotational losses.

**4-6.** A shunt motor has a shunt-field resistance of 60 Ω, armature circuit resistance of 0.1 Ω, and a brush voltage drop of 1.7 V. Motor nameplate ratings are 500 V, 1200 r/min, and 100 hp. Full-load efficiency is 90%. At rated output, calculate:
  **(a)** The input current.
  **(b)** The developed power.
  **(c)** The rotational losses.
  **(d)** The output torque.

**4-7.** A dc shunt motor is rated 240 V, 40 hp, and 850 r/min. Field resistance is 48 Ω and armature resistance is 0.3 Ω, and the torque constant $K_T = 2.56 \text{ N} \cdot \text{m/A}$. Using Eq. (4-27b) plot a curve of speed against developed torque over the range 0 to 900 r/min and 0 to 600 N · m.

**4-8.** A dc shunt motor runs at 2510 r/min when taking an armature current of 45 A from a 500-V line. Calculate the speed at which it will run on a 240-V line when taking an armature current of 20 A. The armature resistance is 0.44 Ω. Assume that the flux per pole at 250 V is 80% of the value at 500 V.

**4-9.** A 500-V dc motor drives a 75-hp load at 1150 r/min. The shunt-field resistance is 75 Ω and the armature resistance is 0.32 Ω. The motor efficiency is 85%.
  **(a)** Find the rotational losses.
  **(b)** Calculate speed regulation.

**4-10.** A dc shunt motor has a field resistance of 192 Ω, an armature resistance of 0.38 Ω, and takes 34.4 A from a 240-V line. The rotor speed is 2500 r/min. Assuming that the flux is proportional to field current, calculate the value of a shunt-field rheostat that will increase the motor speed to 2750 r/min and draw the same armature current.

**4-11.** A 3-hp 120-V 1800-r/min dc shunt motor has an armature circuit resistance of 0.5 Ω and a shunt-field circuit resistance of 62.5 Ω. If the rated full-load armature current is 22 A, calculate:
  **(a)** The rotational losses at full load.
  **(b)** The full-load efficiency.

**4-12.** A shunt motor draws 60 A at 250 V and 1200 r/min. The armature resistance is 0.45 Ω and the field resistance is 100 Ω. Calculate:
  **(a)** The counter EMF.

**(b)** The mechanical output power, torque, and efficiency if rotational losses are 600 W.

**(c)** The no-load speed, assuming the rotational losses remain the same as in part (b).

**4-13.** The armature resistance of a 600-V series motor is 0.215 $\Omega$, the resistance of the commutating field is 0.05 $\Omega$, and the resistance of the series field is 0.08 $\Omega$. At rated voltage and current of 82 A, the speed is 600 r/min.

**(a)** Determine speed and torque when current is:

   **(i)** 95 A.

   **(ii)** 40 A.

**(b)** Determine the value of a series starting resistor that will limit input starting current so that the motor will develop 200% full-load torque on starting.

**4-14.** A 0.2-$\Omega$ diverter is connected across the series field of Problem 4-13. The line current is 82 A. Calculate the new speed, torque, and power output assuming that there are no saturation effects or rotational losses. (*Note*: The field current and armature current are not the same, so that Eq. (4-43) must be modified to

$$T_D = K'_T I_S I_A$$

where $I_S$ is the series field current and $I_A$ is the armature current.

**4-15.** A 7.5-hp 240-V long-shunt motor has a shunt-field resistance of 125 $\Omega$, armature and interpole resistance of 0.4 $\Omega$, series-field resistance of 0.05 $\Omega$, a stray-load loss of 540 W, and a full-load brush drop of 2 V. Its base speed is 1750 r/min.

**(a)** Calculate:

   **(i)** The generated counter EMF.

   **(ii)** The power input and efficiency.

   **(iii)** The speed regulation.

   **(iv)** The developed torque.

**(b)** If the five-turn series field were removed, how many turns must be added to the shunt winding to provide the same flux? It is not necessary to consider the slight voltage rise due to removing the turns.

**4-16.** A 240-V long-shunt motor takes 27 A from the line. The armature resistance is 0.4 $\Omega$, series-field resistance is 0.05 $\Omega$, and shunt-field loss is 460 W. It runs at 1750 r/min, has a stray loss of 540 W at full load and a brush voltage drop of 2 V. Calculate the efficiency and speed regulation.

**4-17.** A 50-hp 550-V shunt motor has an armature resistance, including brushes, of 0.36 $\Omega$. When operating at rated load speed, the armature takes 75 A. What resistance should be inserted in the armature circuit to obtain 20% speed reduction when the motor is developing 70% of rated torque? Assume that there is no flux change.

**4-18.** A compound motor draws 125 A from a 240-V line at 1850 r/min. The motor CEMF constant for this operating condition is 1.11 V/rad/s. The shunt-field resistance is 60 $\Omega$, and the series field resistance is 0.03 $\Omega$. Rotational losses are 1200 W. Calculate:

**(a)** The output power.

**(b)** The efficiency.

**(c)** The load torque.

**(d)** The armature resistance.

**4-19.** A 25-hp series dc motor operates at a rated speed of 600 r/min when taking 100 A full load from a 230-V supply. The resistance of the armature circuit is 0.12 $\Omega$ and that of the series winding is 0.03 $\Omega$. Calculate the speed and the torque when the line current has fallen to 50 A, assuming that the net air-gap flux is proportional to the current.

**4-20.** A dc shunt machine has an output of 25 kW at 125 V when operating as a self-excited shunt generator at 1200 r/min. The resistances of the armature and field windings are 0.1 and 25 $\Omega$, respectively. Calculate:

(a) The speed of the machine when running as a motor from a 125-V supply with an input of 25 kW. Assume that the net air-gap flux remains constant.

(b) The horsepower under these conditions if the rotational losses are 2000 W.

(c) The overall efficiency for the motor.

**4-21.** A 240-V 10-hp dc shunt motor is running at full load at a speed of 2500 r/min. It has an armature resistance of 0.38 $\Omega$ and a field resistance of 192 $\Omega$. Rotational losses are 300 W and the brush drop is 2.2 V. Calculate:

(a) The line current.

(b) The efficiency.

(c) The no-load speed.

(d) The speed regulation.

# 5

# STARTING and CONTROL
# of DC MOTORS

## 5-1 INTRODUCTION

When a stationary shunt motor is connected directly across a line, it momentarily acts like two $R$-$L$ circuits in parallel. The shunt field has a large inductance and large resistance, while the armature circuit is a small inductance and small resistance. The shunt-field current reaches its steady-state value of $V_L/R_F$ fairly rapidly.

At the same time, the armature current starts to rise; since there is no counter EMF on starting, the armature current can rise to a very high value limited only by $R_A$, the brush drop $V_B$, and the armature inductance. When the motor starts to turn, a counter EMF builds up and the armature current decreases. As the motor speeds up still further, the current drops until it produces sufficient torque to drive the load. In a small motor, say of less than 2 hp, the rotor will begin to turn quickly, since the rotor inertia is comparatively small. The large starting current will not last very long and consequently will not cause a serious heating problem in the rotor winding.

Furthermore, a small motor has a comparatively large armature resistance, so that the line current maximum value will not cause a circuit breaker to trip. On the other hand, large motors have very high armature inertia and small armature resistance, so that the inrush current can be very large and cause overheating in the rotor winding. Since torque is proportional to current and starting currents may be very high, the starting torque may be excessively large. It may be large enough to damage the coupling to the load or the load itself. A 230-V motor having a resistance of 0.14 $\Omega$ would have a starting current of 1642 A if started directly across the line.

To limit starting currents to safe levels, starting resistors are used in series with the armature. There are different types of starters but they are basically of two types, resistive and electronic. Resistive starters use variable resistors which can be switched into or out of the armature circuit to limit the current. To understand the principle, we will examine a manual three-point starter. There are many variations on this basic theme, but the operating principles are the same. A large resistor is placed in series with the armature, the motor starts, and after a certain time the resistance is decreased in steps until the desired speed or output torque is reached. The starting resistors may be switched manually or automatically in response to some predetermined criteria, such as speed, current, or time.

## 5-2 MANUAL STARTERS

The principle of a manual starter can be understood by referring to Fig. 5-1. When the switch is first closed, the field current

$$I_F = \frac{V_L}{R_F + R'_F} \quad \text{A} \tag{5-1}$$

and the armature current $I_A$ on starting is equal to $I_{ST}$,

$$I_A = I_{ST} = \frac{V_L - V_B}{R_A + 4R} \quad \text{A} \tag{5-2}$$

The motor begins to accelerate and it builds up a counter EMF

$$E_C = K_G \omega \quad \text{V} \tag{5-3}$$

As the counter EMF builds up, the armature current decreases to

$$I_A = \frac{V_L - K_G \omega - V_B}{R_A + 4R} \quad \text{A} \tag{5-4}$$

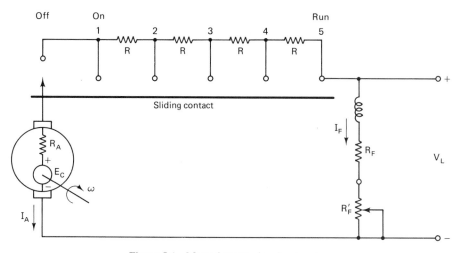

**Figure 5-1**  Manual starter for shunt motor.

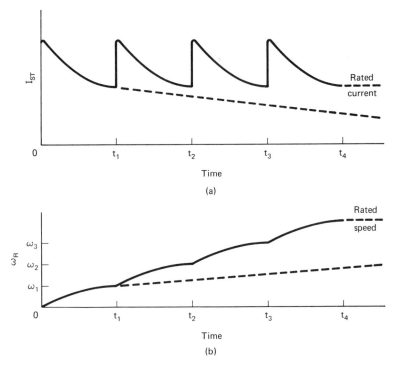

**Figure 5-2** Armature starting current and motor speed in four-step starter: (a) armature current versus time of resistor steps; (b) motor speed versus time of resistor steps.

In Fig. 5-2a the current decreases from $I_{ST}$ to the rated value in time $t_1$. At this time the speed has gone from 0 to $\omega_1$. If no further action were taken, the motor current would continue to fall along the dashed line and the speed would rise along the dashed line as shown in Fig. 5.2. In order for the speed to continue to rise toward the desired value, the series resistance is decreased. The current will then rise quickly to $I_{ST}$ and begin to decrease; at the same time the speed will increase more rapidly, as shown for the time interval $t_1-t_2$. This process is repeated until the motor reaches its rated (or desired) current and speed.

The number of resistance steps and the size of the resistor depends on the size of the motor; the larger the motor, the greater the number of steps and the larger the resistors.

### EXAMPLE 5-1

Calculate the size of a four-step starter to limit the starting current of a dc shunt motor to 150% of rated current. Assume that all four steps are equal. The motor is 75 hp, 250 V, and 850 r/min, with an armature resistance of 0.04 Ω and an efficiency of 90%. Calculate the speeds at which the resistors should be switched so that the current never falls below the rated value. Neglect brush voltage drop.

**SOLUTION**

Refer to Fig. 5-1, showing the motor and starter circuit. At full load, the motor input power is

$$P_{in} = \frac{75 \times 746}{0.9} = 62{,}167 \text{ W}$$

The line current is

$$I_L = \frac{62{,}167}{250} = 248.7 \text{ A}$$

Since field resistance is not specified, we may assume that the field current is negligible compared to the armature current and make the assumption that the rated armature current is

$$I_R = I_L = 249 \text{ A}$$

The 150% of rated current is

$$1.5 \times 249 = 373.5 \text{ A} = I_{max}$$

The current-limiting resistance on starting is then calculated from

$$I_{max} = \frac{250}{0.04 + 4R} = 1.5 \times 249$$

and

$$R = 0.1573 \ \Omega$$

In order to calculate the various speeds, we must know $K_G$, the counter EMF constant. At rated speed and current,

$$250 = 249 \times 0.04 + K_G \times 2\pi \times \frac{850}{60}$$

$$= 9.96 + 9.801 K_G$$

Therefore, $K_G = 2.6968$ V/rad/s.

If $I_A = I_R$ and there are four resistors in the starter,

$$250 = I_{max}(0.04 + 4R) + K_G \omega$$

$$= 373.5(0.04 + 4 \times 0.157) + 2.697 \omega$$

This is the starting position (i.e., $\omega = 0$). If $I_A = I_{max}$ and there are three resistors in the starter at the instant of switching,

$$250 = 373.5(0.04 + 3 \times 0.157) + 2.697 \omega$$

Therefore,

$$\omega = 21.92 \text{ rad/s} \quad \text{and} \quad E_C = 59.1 \text{ V at time } t_1$$

Similarly, for $2R$,

$$\omega = 43.67 \text{ rad/s} \qquad E_C = 117.8 \text{ V at time } t_2$$

For $1R$,

$$\omega = 65.4 \text{ rad/s} \qquad E_C = 176.4 \text{ V at time } t_3$$

For 0,

$$\omega = 87.2 \text{ rad/s} \qquad E_C = 235.2 \text{ V at time } t_4$$

Figure 5-2 shows the behavior of the motor current and speed versus time. The actual times $t_1$, $t_2$, $t_3$, and $t_4$ depend on the motor inertia and the load characteristics.

## 5-3 AUTOMATIC STARTERS

In Example 5-1, the various resistors are switched when the current reaches a predetermined value as observed on an ammeter. For some applications this is satisfactory, but in others it may be necessary to incorporate some type of automatic switching device and some type of automatic sensing device as well.

In the circuit of Fig. 5-1, note that the field is connected directly across the line. This is done to ensure that maximum field current is available to produce maximum desired torque on starting. Also, the rheostat $R'_F$ is incorporated to allow fine adjustment to the speed by varying the field current.

To develop automatic starters, we need devices that will close circuits automatically. A relay is an electromagnetic device that is actuated in one circuit and affects operations in another circuit, and a contactor may be looked upon as a large relay. The contactor consists of a magnetic coil in one circuit which closes a set of contacts in another circuit. Figure 5-3 is a schematic drawing of a contactor

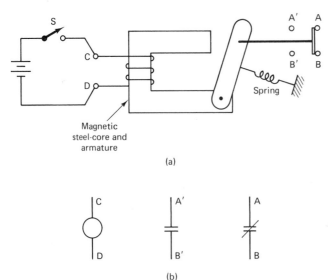

(a)

(b)

**Figure 5-3** Electrically operated contactor: (a) construction details; (b) symbol diagram.

Starting and Control of DC Motors     Chap. 5

shown in its normal or deenergized state. The circuit between terminals $A-B$ is shorted while the circuit between points $A'-B'$ is open circuited. When the switch is closed, applying a current to the coil, the pivoted arm in the magnetic circuit will close the air gap and the shorting bar will be across $A'-B'$ and $A-B$ will be open circuited. The point at which $C-D$ closes depend on the design voltage of the coil, while the physical size of the contacts depends on the load between the contacts. The symbol for a contactor is shown in Fig. 5-3b. In the symbol the contactor can operate two sets of contacts simultaneously, and in general there may be many sets of contacts. It is thus possible to control economically large amounts of power supply and without exposing operators to the hazards of high voltages and heavy currents. Sensitive relays can be used in conjunction with contactors so that it is possible to control megawatts of power in response to milliwatt signals. See Section 1-7 for a description of the magnetic tractive force of an armature of an electomagnetic contactor.

Figure 5-4 shows the schematic diagram for an automatic starter based on the counter EMF principle. As the motor speed increases, the counter EMF increases, and by taking advantage of this we can cause contactors to cut in at various different speeds.

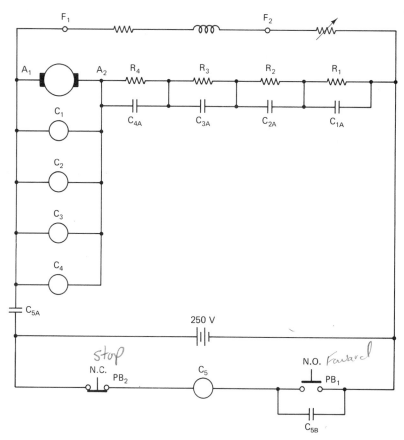

**Figure 5-4** Counter EMF-type automatic starter.

Sec. 5-3    Automatic Starters

There are two pushbuttons and five contactors in this circuit. If we use the motor of Example 5-1 then contactor $C_1$ will operate at 59.1 V, $C_2$ at 117.8 V, $C_3$ at 176.4 V, and $C_4$ at 235.2 V. To start, the normally open, spring-loaded pushbutton $PB_1$ is closed momentarily, energizing contactor $C_5$ and closing contacts $C_{5B}$ and $C_{5A}$. The closing of $C_{5A}$ places a 250-V supply across the motor armature and $R$ network and across the shunt field. This starts the motor with the four resistors in series with it. At startup, the CEMF is zero and none of the contactors pick up. As the motor speeds up it reaches 21.9 rad/s, at which time there is 59.1 V of CEMF, which causes $C_1$ to energize, which closes contacts $C_{1A}$ and short-circuits resistor $R_1$. The motor continues to accelerate, increasing its speed and CEMF until $C_{2A}$ acts to short out $R_2$, and the process continues until all four resistors are shorted.

Contactor $C_{5B}$ is used to short out pushbutton $PB_1$ so that it does not have to be held in permanently. Pushbutton $PB_2$ is normally closed, and when momentarily opened will cause contactor $C_5$ to open and disconnect the motor from the supply.

In this example, the contactors were chosen to cut in at various speeds since the contactors are effectively sensing the counter EMF, which is proportional to speed. Other automatic starter circuits can be constructed which are sensitive to current, power, time, and so on. In addition, relays may be used to sense overvoltage, undervoltage, overspeed, or any abnormal conditions that would necessitate shutting the motor down.

## 5-4 MOTOR CONTROL

The major advantage of the dc motor is that it can be operated over a wide range of speed and torque. In the preceding sections, we have seen how motors can be started using variable resistors in series with the armature. Obviously, if we leave the resistors in the circuit rather than switching them out at various points, we can change the speed of the motor. Various methods of speed control were shown in Chapter 4, where resistors were used in the armature and field circuits. Motor controllers are similar in principle to the starters discussed previously, but the resistors may be left in the circuits and therefore must have larger power ratings. By using combinations of resistors, relays, and contactors it is possible to start, stop, and reverse motors quickly and accurately. Speed, torque, or any other desired quantity can also be accurately controlled. There are many control applications using electromechanical relays and contactors and resistors, but the modern trend is toward electronic control and in many situations to computer control.

In motor controller circuitry, we actually deal with two levels of control, the signal level and the power level. At the signal level we measure the quantity of interest, such as speed, current, voltage, temperature, and so on. At the power level we are switching or varying large amounts of power by varying the current or voltage to the electrical apparatus. In the following sections, we discuss electronic controllers for dc motors and their electronic principles involved in power circuitry. We will not go into any detail regarding signal-level circuitry and only the minimum detail needed for understanding power circuitry. It should be recognized that the proper application of electronic controls to electrical machines requires an appreciation and understanding of both electronics and machines. It is becoming increasingly more important for this understanding, as the trend is continually accelerating toward the use of more and more automatic systems.

## 5-5 ELECTRONIC CONTROLLERS

We have seen that a wide range of speed–torque for dc motors can be achieved by control and adjustment of field current and armature current. Presently, this control and adjustment of motor currents is performed by electronic controllers. Dc motors provided with electronic controllers can have their inherent characteristics adjusted and their operating range with respect to their loads varied to suit the particular requirements. Motor speed, torque, current limiting for starting, motor protection, and acceleration are the most common characteristics that can be adjusted by electronic controllers.

In contrast to ac motor control, the separately excited shunt motor has excellent variable-speed characteristics and is easily adjusted by the electronic controller operating at relatively high efficiency.

Commercial electronic controllers for dc motors are supplied from ac lines alleviating the need for separate dc power sources. The ac lines may be either single-phase or three-phase.

Modern electronic controllers consist of solid-state control circuitry and electronic power devices such as controlled rectifiers and choppers. The electronic power devices acting as electronic switches control the average currents of the motor. The combination of the motor and the control equipment is called a *drive system*. The following sections describe power-switching devices and their circuits.

## 5-6 POWER-SWITCHING PRINCIPLES

Figure 5-5a shows a switch in series with a battery of voltage $E$ and a resistor of resistance $R$. If the switch is opened and closed regularly so that it is closed for time $t_1$ and open for time $t_2$, the voltage that appears across the resistor is shown in Fig. 5-5b. The average voltage read by a dc voltmeter would be

$$V_R = \frac{t_1}{t_2} E \tag{5-5}$$

The ratio $t_1/t_2$ is called the *duty cycle* of the waveform, and by controlling it, we can vary the average voltage, hence the average current through the resistor.

### EXAMPLE 5-2

A 12-V battery is connected in series with a 2-$\Omega$ resistor and a switch as shown in Fig. 5-5a. The switch operates continuously; it is closed for 2 ms and opened for 3 ms.

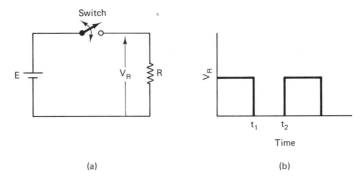

Figure 5-5  Manual-switching circuit: (a) circuit diagram; (b) voltage waveform.

(a) Draw the waveforms of voltage across the resistor and the current through it.

(b) Determine the average current in the resistor.

**SOLUTION**

(a) Figure 5-6a illustrates the voltage waveform across the resistor. At any instant of time, the current is equal to the voltage across the resistor divided by the resistance. During the ON time, the current is then 12/2 = 6 A, and 0 during the OFF time. Figure 5-6b illustrates the current waveform through the resistor.

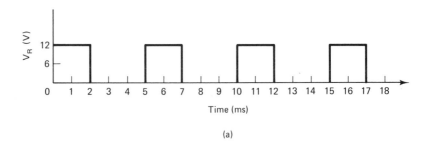

(a)

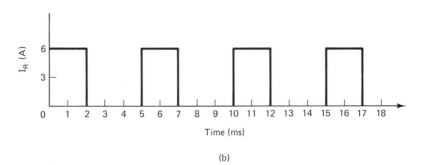

(b)

Figure 5-6  Waveforms for Example 5-2: (a) voltage waveform across resistor; (b) current waveform through resistor.

(b) The average current in the resistor is

$$I_{R_{av}} = \frac{12}{2} \times \frac{2}{2 + 3} = 2.4 \text{ A}$$

## 5-7 POWER-SWITCHING DEVICES

A mechanical switch operating under the conditions of the example would have to be very special since it is switching 2.4 A and operating at the rate of 200 operations per second. Although this is possible for mechanical switches, it is close to their limit, and it is for this reason that electronic switching devices are invariably used. The power-handling component in these switches is either a power transistor or thyristor (silicon-controlled rectifier). There are a great variety of both devices and circuits, but they all act as switches, as indicated in the previous sections. Since this textbook is not concerned primarily with electronics, but with machine principles, we will omit any details of these switches, but consider them to be four-terminal "black boxes," as indicated in Fig. 5-7. Terminals $P_1$ and $P_2$ are the power terminals, while $C_1$ and $C_2$ are the contact terminals. If a voltage of the correct magnitude and polarity exists across $C_1$ and $C_2$, a short circuit, in effect, exists between $P_1$ and $P_2$ and current $I_R$ will flow in resistor $R$. If the voltage across $C_1$ and $C_2$ is not of sufficient amplitude or correct polarity, the switch is open between $P_1$ and $P_2$, hence no current flows. A simple example of such a circuit is shown in Fig. 5-8 using a transistor operated in the "saturated" mode. A saturated transistor is one in which there is such a large current in the base circuit that the transistor conducts so well that it acts like a short circuit between the emitter and collector. If the control voltage $E_c$ is large enough, a base current $I_B$ flows which turns on the transistor, and when $E_c$ is zero, no base current flows and the emitter–collector circuit is open. A typical set of waveforms is shown in Fig. 5-9. In this configuration, the switch is controlled by the varying voltage of the signal generator $E_c$, which supplies a train of pulses to the base. When the base current is sufficiently positive, the current rises to a high enough value to turn

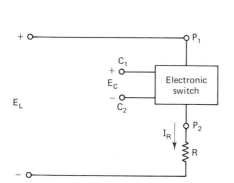

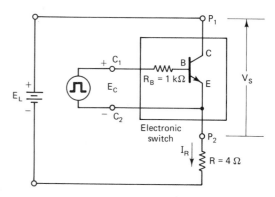

**Figure 5-7** Four-terminal black-box representation of electronic switch.

**Figure 5-8** Transistor operated in saturated mode as electronic switch.

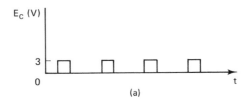

(a)

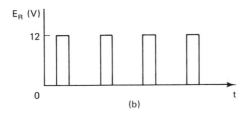

(b)

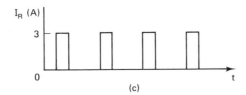

(c)

**Figure 5-9** Waveforms for transistor electronic switch: (a) control source voltage; (b) resistor load voltage; (c) resistor load current.

the transistor on completely, thus shorting $P_1$ and $P_2$ and applying full line voltage to the 4-$\Omega$ resistor. (This explanation neglects the voltage drop across the transistor itself, which in any event is very small, of the order of 0.7 V.)

## 5-8 SWITCHING IN INDUCTIVE CIRCUITS

What happens if a switch is suddenly opened in an inductive circuit? We know from circuit theory that the induced voltage is equal to the rate of change of current times the inductance. Then

$$E_I = \text{rate of change of current} \times \text{inductance}$$

or

$$E_I = \frac{\Delta I \times L}{\Delta t} \tag{5-6}$$

where 
$$E_I = \text{voltage induced in inductance, V}$$
$$\Delta I/\Delta t = \text{rate of change of current, A/s}$$
$$L = \text{inductance of coil, H}$$

If we interrupt the current quickly, $\Delta I/\Delta t$ becomes a very large number and $E_I$ also becomes large. In fact, $E_I$ may be large enough to cause voltage breakdown of the coil insulation.

## EXAMPLE 5-3

A 24-V battery is connected in series with a switch, a 5-H inductor, and a 4-Ω load. The circuit is allowed to reach steady-state conditions and the switch is then suddenly opened. If the switch opening time is 10 ms, find the voltage induced in the inductor.

## SOLUTION

The steady-state value of the current is 24/4 = 6 A. The rate of change of current is then

$$\frac{6}{10 \times 10^{-3}} = 600 \text{ A/s}$$

The induced voltage is then

$$5 \times 600 = 3000 \text{ V}$$

This voltage would appear as a spark or arc across the switch. It is obviously very much greater than might be expected from a 24-V source.

To prevent the generation of such a large voltage, it is necessary to avoid interrupting the current so abruptly. One method of doing this is to add a diode in the circuit as shown in Fig. 5-10a. Now when the switch is opened, the current that is in the inductor attempts to keep flowing and it can do so through the path provided by the diode, as indicated in the equivalent circuit of Fig. 5-10b. In any switched inductive circuit, such as a motor, some provision must be made for current to flow after the switch is opened.

The current waveform for the circuit of Fig. 5-10b is shown in Fig. 5-11. The current is seen to have an exponential rise and then an exponential decay.

An interesting observation may be made regarding switching into an inductive circuit in which the ON time and OFF time are equal, that is, a circuit with a 50% duty cycle. We can select a circuit in which the circuit time constant is less than the ON time as an example.

## EXAMPLE 5-4

In the circuit of Fig. 5-10 the switch is closed for 0.2 s and opened for 0.2 s. If $R = 2 \Omega$ and $L = 0.2$ H, find the current at the end of 0.4 s assuming that the circuit is initially "dead." The supply voltage $E$ is 15 V.

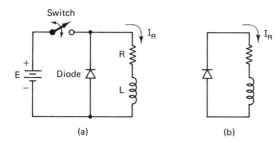

(a)                    (b)

**Figure 5-10** *L-R* circuit with shunting diode: (a) circuit diagram; (b) equivalent circuit after switch is opened.

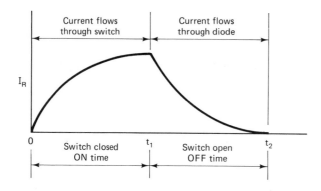

Current flows through switch | Current flows through diode

$I_R$

0

Switch closed ON time | $t_1$ | Switch open OFF time | $t_2$

**Figure 5-11** Inductor current wave-form.

**SOLUTION**

The circuit time constant is $\tau = L/R = 0.2/2 = 0.1$ s. At the end of 0.2 s, $t = 0.2$ s. Therefore, $t/\tau = 0.1/0.2 = 0.5$ and $t = 0.5\tau$. When $t = 0.5\tau$, $i(t) = 0.39I_{ss}$ and the steady-state value of current is $E/R = 15/2 = 7.5$ A. Therefore, at the end of 0.2 s, the current is

$$0.39 \times 7.5 = 2.93 \text{ A}$$

During the OFF time of the current cycle, the current starts at the value of 2.93 A and decays exponentially for 0.2 s. Again $\tau = 0.1$ s and at the end of 0.2 s, $t = 0.5\tau$. At $0.5\tau$, the current falls to 0.606 of its initial value. But the initial value is 2.93 A, so that the final value is

$$I = 0.606 \times 2.93 = 1.78 \text{ A}$$

The result, which may be surprising, is that the current has not dropped to zero. If the switch is now closed for several cycles, the current will start at 1.78 A and increase to a larger value than 2.93 A and decay to a larger value than 1.78 A, and the process will be repeated. The final result will be an average value with a ripple superimposed on it, as shown in Fig. 5-12. The average steady-state value can easily be calculated if we consider the

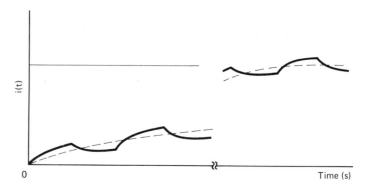

$i(t)$

0

Time (s)

**Figure 5-12** Current waveform for Example 5-5.

Starting and Control of DC Motors   Chap. 5

circuit of Fig. 5-10a. Across the diode, the voltage is either $E$ volts (switch closed) or zero (switch open, diode conducting). Since the duty cycle of the switch is 50%, the average voltage must be $E/2$. If the average or dc component of voltage is $E/2$, the average or dc component of current must be $E/2R$, since there is no dc voltage drop across the inductor.

If we replace the inductor of Fig. 5-10 by a dc motor, essentially the same phenomena occur as outlined above. However, because of the motor EMF, the switch looks at an $R-L$ circuit in series with a dc voltage. At a particular speed, the counter electromotive force is a constant proportional to speed as indicated in Section 4-3. Since motor speed is constant,

$$E_C = K_G\omega = \text{a constant}$$

In Fig. 5-13 the voltage $E_D$ will be a chopped version of $E_L$. It is equal to $E_L$ when the switch is closed, and zero when the switch is opened. The average

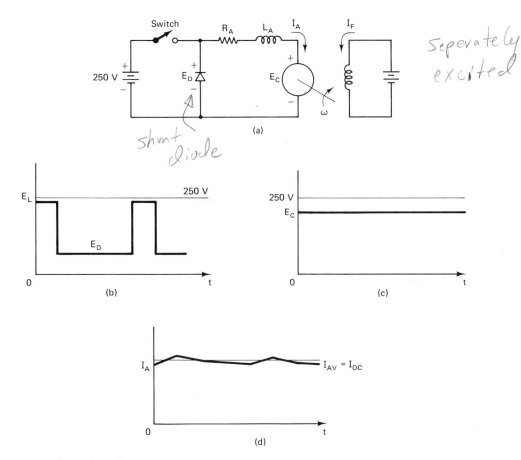

**Figure 5-13** Dc motor supplied from 250-V dc switched source: (a) circuit diagram; (b) diode voltage waveform; (c) counter EMF waveform; (d) armature current waveform.

value of $E_D$ is equal to the average counter EMF minus the (average armature current $\times R_A$) drop.

To summarize, we may say that when the motor has reached its steady-state value, a current flows which has a dc or average value with a ripple superimposed on it. The dc value can be calculated using Eq. (5-5), and the ripple content depends on the switching rate and the electrical time constant. In addition, we have not examined the transient behavior of the circuit, but this depends on the mechanical time constant, which in turn depends on the machine inertia and friction. The transient behavior of electrical machines is the subject of more advanced study and will not be considered further.

Because of the ripple in the armature current, the motor will have higher losses when energized from an electronic source unless some filtering is added to the circuit. At present there are dc motors which are specifically rated to be run from electronic sources, either directly from a dc line or from multiphase, rectified ac lines.

The actual current waveforms obtained in a dc motor operated from a switched source is more complicated than that obtained for a simple $R-L$ circuit due to saturation in the magnetic circuit. In addition, there are periodic variations in the coil inductance as the rotor moves, which further complicates accurate calculations. However, we can make some simplifying assumptions which can give us some feeling for circuit performance.

Figure 5-13(a) shows a dc motor supplied from a 250-V dc switched source. The duty cycle is 20% and the input power is 12 kW at 1250 r/min. The waveforms measured at various points are shown in Fig. 5-13b–d. It is assumed that there is a very large armature inductance, so that $I_A$ is practically pure dc. The speed is constant at 1250 r/min and therefore the counter EMF $E_C$ is constant if we neglect the effects of armature reaction as we have been doing. The $R_A I_A$ drop is small compared to $E_C$ but it is not quite zero.

If the duty cycle is 0.2, the average value across the diode

$$E_D = 0.2 \times 250 = 50 \text{ V}$$

The average input current is

$$\frac{12,000}{250} = 48 \text{ A}$$

The average power supplied to the diode is the same on either side of the switch, and this is 12 kW.

The average current into the motor is

$$250 \times I_{dc} = 50 \times I_A$$

Therefore,

$$I_A = 5I_{dc} = 5 \times 48 = 240 \text{ A}$$

$$\text{and average voltage drop} = 240 \times 0.05 = 12 \text{ V}$$

Therefore,

$$\text{average counter EMF} = 50 - 12 = 38 \text{ V}$$

$$\text{and average output power} = 38 \times 240$$

$$= 9.12 \text{ kW (12.2 hp)}$$

### EXAMPLE 5-5

If for the motor of Fig. 5-13 just described, we increase the duty cycle to 0.4 and the motor speed to 1600 r/min, estimate the power output.

### SOLUTION

Counter EMF is proportional to speed, so that the new counter EMF is

$$E_C = \frac{1600}{1250} \, 38 = 48.6 \text{ V}$$

The average applied voltage across the diode is

$$E_D = 0.4 \times 250 = 100 \text{ V}$$

Therefore,

$$\text{average current} = \frac{100 - 48.6}{0.05} = 1028 \text{ A}$$

Therefore,

$$P_{\text{out}} = 48.6 \times 1028 = 49,961 \text{ W} = 50 \text{ kW (67 hp)}$$

From this example, we can see that by varying the duty cycle electronically, we can vary the speed or current in a motor. These quantities are varied by changing the duty cycle, not by inserting a series resistor. Control comes from the electronic switch, hence there is very little power loss in the switching and the efficiency of the whole system can be very high. Using resistive control for speed wastes power and provides only limited speed range.

A second advantage of electronic control is that the output may be controlled by an electronic signal rather than by using a rheostat, which must be moved mechanically. This allows the controller to be used as part of a closed electrical control loop in which speed or torque, for example, must be held constant. Because of this property, dc motors can be used in precision applications where speed must be tightly controlled to accuracies of 0.01%, such as in papermill drives and positioned-control systems.

By using a variable duty cycle, the electronic switch may also be used as an electronic current-limiting starter. A motor started directly across the line would draw a very large current, limited only by the armature resistance since there is no counter EMF. The electronic switch can limit the starting current to a reasonable average value and allow the motor to accelerate gradually.

In the preceding sections, we have examined the principles involved in energizing dc motors from dc sources using electronic switching methods. However, except in some special applications, large amounts of power are available only from ac sources. Alternating current, of course, is readily converted into direct current using rectifiers and by using a combination of rectifiers and thyristors; thus it is possible to drive large dc motors directly from ac lines. The thyristor or silicon-controlled rectifier (SCR) is really an electronic switch that can operate over part of an ac cycle. By utilizing this switching capability synchronized to the ac line, it is possible to control very accurately the current, torque, speed, or power of a dc motor.

Figure 5-14 illustrates a full-wave diode bridge supplying a resistive load. If the anode terminal of a diode is positive with respect to the cathode, the diode will conduct current, acting as an ideal short circuit. On the other hand, if the anode is negative with respect to the cathode, it will not conduct and will act as an open circuit.

In the bridge circuit shown, when point $A$ is positive with respect to point $B$, $D_1$ and $D_4$ conduct, but $D_2$ and $D_3$ are reverse biased, hence cannot conduct. When point $B$ is positive with respect to $A$, the situation is reversed. $D_3$ and $D_2$ can now conduct but $D_1$ and $D_4$ are unable to conduct because they are reverse biased. By tracing through the circuit, we can see that the load is always subjected to the same polarity of applied voltage; hence it has a unidirectional current flow

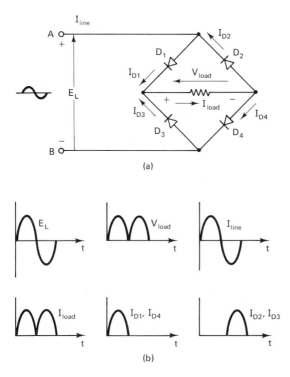

(a)

(b)

**Figure 5-14** Full-wave diode bridge supplying resisitive load: (a) circuit diagram; (b) waveforms.

through it.  Since no filtering has been included, this unidirectional voltage produces a varying current which in effect is a dc voltage with some ac variations on top of it.  We can also say that the current is dc plus a ripple component.  In the case of a rectified sine wave, the average value is 0.637 of the maximum value. The peak-to-peak current variation is 100% of the current, so that the ripple content is extremely high and the hysteresis and eddy currents that would flow would produce very high losses in the motor.

To improve the motor efficiency and reduce harmonics, an inductor is often used as part of the rectifier drive.  Because there is only a small dc drop in the inductor (due to its own resistance), the dc value of the rectified wave remains the same, but the ac component is suppressed, allowing the motor to run more efficiently and cooler.

Full-wave single-phase bridges are satisfactory for motors up to about 5-hp ratings.  However, above this size, line currents become excessive and it is necessary to operate the motors on three-phase supplies.

## 5-10 FOUR-QUADRANT OPERATION OF DC MACHINES

A dc machine that is controlled by electronic switching apparatus can be run as a motor or generator in either direction of rotation.  Because of this property, the motor-switching apparatus can utilize regenerative braking and recover some of the energy stored in the rotor.  The motor can be braked smoothly to zero in one direction and then accelerated smoothly to full speed in the reverse direction.  In order to be able to act as a motor or a generator and operate in either direction, the dc controller must be capable of four-quadrant operation.

In Fig. 5-15a we have a dc motor driving in the clockwise direction.  If we wish to brake the motor, we can increase the field current, causing the internal generated voltage to rise.  This will cause the armature current to reverse direction and in effect make it act like a generator, and since there is no energy being supplied to it to act like a generator, it will slow down as it dissipates the stored energy of the rotor and load.

On a dc line, this method of braking will allow only some energy to be returned to the line, since as the motor slows down the counter EMF will soon not be sufficient to provide any braking force.

Regenerative braking will not bring a motor to a complete stop, so that a mechanical brake would be necessary in addition to the regenerative system.  Regeneration naturally occurs in certain systems when a load tends to overhaul the motor.  An electric locomotive descending a steep hill has all the energy of the train trying to go faster than the engine.  The energy can be returned to the line and provide braking at the same time.

It is possible to force the motor quickly to zero and then remove the applied voltage.  This method of forcing the motor to zero speed by applying a reverse voltage is called *plugging*.  Since the line voltage and counter EMF are in series during the plugging interval, very large currents can flow and means must be included in the circuit to limit these large currents.

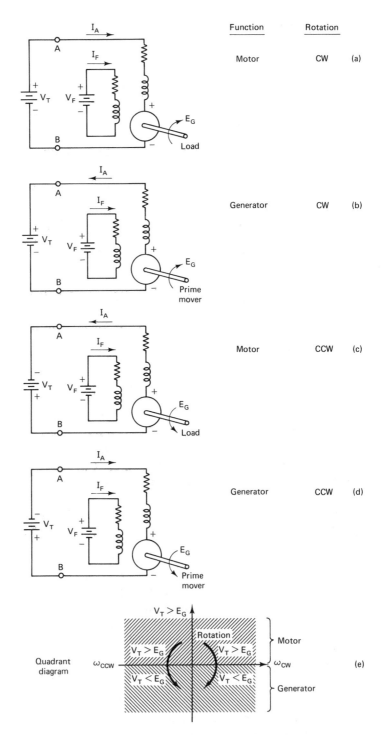

**Figure 5-15**  Four modes of operating a dc machine using a power supply: (a)$V_T$ $> E_G = E_C$; (b)$E_G > V_T$; (c) $V_T > E_G = E_C$; (d) $E_G > V_T$; (e) Quadrant diagram.

The problem of regeneration for a dc motor energized from an ac source is actually less complicated than the motor supplied from a dc source since the line automatically reverses itself each half-cycle. Furthermore, the dc counter EMF is also larger than some part of the ac wave, so that regeneration can be used over most of the cycle by careful design of the timing circuit to ensure that regeneration is taking place. Furthermore, the ac-driven system can readily be programmed to allow the motor to reverse direction, so that it is possible to design a motor drive that will be able to reverse its direction of rotation quickly and smoothly. Furthermore, it is also possible to recover some of the stored energy in the rotor which would have been dissipated in braking.

## REVIEW QUESTIONS

**5-1.** Describe the startup current in the field winding of a shunt motor based on its electrical characteristics. Draw the current versus time diagram.

**5-2.** Describe the startup current in the armature of a shunt motor based on its electrical characteristics. Draw the current versus time diagrams.

**5-3.** Clearly state the reasons for requiring the maximum value of armature current of a shunt motor to be restricted to one and a half to four times its rated value.

**5-4.** Motor starting currents are limited to safe values by means of starter resistors. Describe how the ohmic values and power ratings of these resistors are determined.

**5-5.** Draw the circuit diagram of a four-resistor manual starter for a shunt motor.

**5-6.** Sketch the waveforms of armature current and rotational speed for the shunt motor using a four-resistor manual starter. It is assumed that at each of the four steps the maximum armature currents are the same.

**5-7.** What are the factors that determine the number of resistance steps and the power ratings of the starter resistors?

**5-8.** What is an automatic starter? State the benefits obtained from this type of starter.

**5-9.** Describe the electromagnetic devices used to close and open electrical circuits of automatic starters. Draw their schematic diagrams.

**5-10.** Explain the operating principles of a counter electromotive force type of automatic starter. Draw its schematic diagram.

**5-11.** Describe the various control functions that are possible with dc motors.

**5-12.** Compare and contrast motor starters and motor controllers.

**5-13.** Electronic controllers have many advantages over electromechanical controllers. List several of the advantages.

**5-14.** Describe the circuitry and devices used in modern electronic controllers.

**5-15.** What is a drive system?

**5-16.** Why is there a need for power switching? Describe the principles of power switching using the simple dc source and mechanical switch connected to a resistive load. Draw the necessary waveforms.

**5-17.** State the two major types of power switching devices. Describe the operating characteristics of these two devices when used in the power switching mode.

**5-18.** Draw the block diagram of a four-terminal black-box representation of an electronic switch. Clearly indicate the functions of the two pairs of terminals.

**5-19.** Describe the operating principles of the four-terminal black-box representation of the power electronic switch. Draw the necessary waveforms.

**5-20.** What is the duty cycle of a switching device? What effects occur on the switched currents when the duty cycle is varied?

**5-21.** Clearly describe the problem associated with the opening of a highly inductive circuit that has been connected to a dc voltage source. How can the voltage in the inductor be calculated?

**5-22.** Explain the detrimental effects of ripple in the armature current of a shunt motor supplied by a switching voltage source.

**5-23.** Explain how the power electronic switch connected to the armature of a shunt motor may be used as a current-limiting starter. The field winding is separately excited from a fixed-voltage source.

**5-24.** Draw the full-wave diode bridge supplying a resistive load, and sketch the associated waveforms of voltages and currents.

**5-25.** The complete operating characteristics of dc machines can be illustrated by a four-quadrant diagram. Draw this diagram and the associated circuit diagrams of the machine armature connected to a voltage source. The field winding is excited by a constant-voltage source.

**5-26.** Explain the following motor control terms: dynamic braking, regenerative braking, and plugging.

**5-27.** If your school has a programmable logic controller (PLC), develop a program to start up and run a shunt motor; draw the relay ladder diagram and list the program. Connect the PLC to the motor; load the program into it and see if it works.

## PROBLEMS

**5-1.** A 20-hp 220-V 540-r/min shunt motor has an armature and a shunt-field resistance of 0.24 $\Omega$ and 157 $\Omega$, respectively. The full-load line current is 85 A. If the starter resistance is 1.7 $\Omega$, what line current does the motor draw at the instant of starting?

**5-2.** Draw the circuit diagram of a shunt motor with a shunting diode connected to a switching power supply. The field winding is excited by a constant current. Explain the operation and control of this motor under these conditions.

**5-3.** Draw the waveforms of the diode voltage, line, and counter EMF voltages and the armature current of Problem 5-2. Assume that the duty cycle is 0.2 or 20%.

**5-4.** If the duty cycle is increased to 0.4 in the circuit of Problems 5-2 and 5-3, explain the changes in the waveforms that occur and the related effects on the operation of the dc motor.

# 6

# ELECTRIC POWER GENERATION

## 6-1 INTRODUCTION

The three-phase synchronous generator is used exclusively to generate bulk power. The voltage levels at which this power is generated is typically in the range 13.8 to 28 kV. As we will see, this voltage is limited by practical considerations, such as the number of conductors that can be placed into the stator slots. Conductors must have a sufficient or adequate cross section to carry the current.

It is possible to transmit this power directly, but this results in unacceptable power losses and voltage drops, even if reasonably small distances are involved. It is therefore absolutely essential to distribute power at as high a voltage level as is practically and economically feasible.

Transmission voltages vary routinely in the many of hundreds of thousands of volts, up to the 765-kV level. This is made possible by the power transformer. It transforms the generated voltage to a level at which power transmission becomes feasible for distances well in excess of over hundreds of miles. At the receiving end, this high voltage is transformed back to moderately high voltage levels, after which it is distributed to "local" consumers and users. We discuss the power system briefly in the next section.

## 6-2 ELECTRIC POWER SYSTEMS

As we mentioned in the introduction, when power is transmitted at the generated voltage level, we do not "get too far." Let us investigate this further. Assume

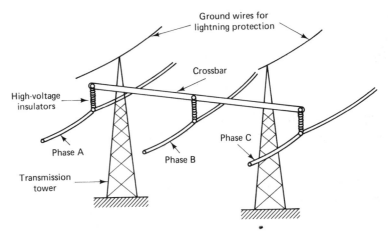

**Figure 6-1** Overhead transmission lines.

that we have a 500-MW 22.5-kV three-phase synchronous generator. To deliver this rated power at an assumed unity power factor ($\cos \theta = 1.0$) results in a line current according to the expression for three-phase power ($P = \sqrt{3}\ V_{\text{L-L}}I_L \cos \theta$) of

$$I_L = \frac{P}{\sqrt{3}\ V_{\text{L-L}} \cos \theta} = \frac{500{,}000 \text{ kW}}{\sqrt{3} \times 22{,}500 \times 1} = 12.83 \text{ kA}$$

These three currents will be carried in three identical bare overhead aluminum cable-steel-reinforced (ACSR) conductors, as shown in principle in Fig. 6-1.

Assume that the conductors have a cross-sectional area of 16 cm² ($=0.0016$ m²), which approximately amounts to a circular conductor having a diameter of 1.8 in. Let the conductor material be copper (resistivity $\rho = 0.175 \times 10^{-7}$ $\Omega$-m). From physics we have the equation

$$R = \frac{\rho l}{A} \quad \Omega \tag{6-1}$$

where $l$ is the conductor length in meters. Equation (6-1) expresses the resistance $R$ of the conductor in terms of its physical dimensions $l$ and $A$, and its material property $\rho$. Let us further assume that we want to transmit this power over a distance of 20 miles (32.2 km). Substituting our values now in Eq. (6-1) results in a conductor resistance (per line) of

$$R = 32{,}200 \times 0.175 \times \frac{10^{-7}}{0.0016} = 0.352 \ \Omega$$

We can now compute the ohmic loss per conductor as

$$P = I^2R = (12{,}830)^2 \times 0.352 = 57.94 \text{ MW}$$

which amounts to

$$\frac{57.94 \times 10^6}{32{,}200} = 1780 \text{ W/m per conductor}$$

Thus 57.94 MW/conductor gives a total loss of $3 \times 57.94 = 173.8$ MW for our three-phase system. This amounts to a power loss of

$$\frac{173.8}{500} \times 100\% = 34.8\%$$

of the generator output. In other words, more than one-third of the generated power is lost. If we look at the corresponding ohmic voltage drop in the line, we have

$$V = IR = 12.83 \text{ kA} \times 0.352 = 4.52 \text{ kV per phase}$$

This is a rather large voltage drop. It becomes much worse if we consider that we are transmitting ac power. For a typical line, its reactance is normally much larger than its resistance value. Typical reactance-to-resistance ratios for overhead lines vary from approximately 3 (at low voltages) to 20 or more (at ultrahigh voltages). This then illustrates clearly that transmitting power under these conditions is absurd. The reader can easily verify that if the voltage were stepped-up to, say, 500 kV, the corresponding power loss drops to negligible proportions. The line current in that case is "only" 577 A.

To increase the voltage level of our generator, we need a power transformer that steps-up the voltage. Power transformers are always designed for a single frequency; for our purposes it is 60 Hz. Power transformers by definition are those with a rating of 250 kVA and up. For large power transformers we typically think in terms of thousands of kVA and more, up to 1000 MVA. Of course, at the end of a transmission line we need another transformer to step-down the voltage. As we will discuss later, generators operate generally in parallel, and various generating stations supply the power grid. Referring to Fig. 6-2, we see an overview of a typical electric power system. As we can see, there are really four separate aspects to such a system:

1. Generation
2. Transmission
3. Distribution
4. Utilization

In this book we confine ourselves to the generation of power and more specifically to the utilization aspect. That is, we want electric power to drive our machines to do a specific task. To this end we need to know what machines are available and what they can do, so that we can make an educated choice as to the type of machine we choose. However, it is informative to see how we arrive from one end to the other in a power system. Observe the various voltage levels and the role of the transformers.

As mentioned, power transformers are those with ratings over 250 kVA. Transformers with ratings of 250 kVA or less are commonly known as *distribution transformers*. They are usually located in the various power distribution stations at the end of the transmission system which further distribute the power to localized centers.

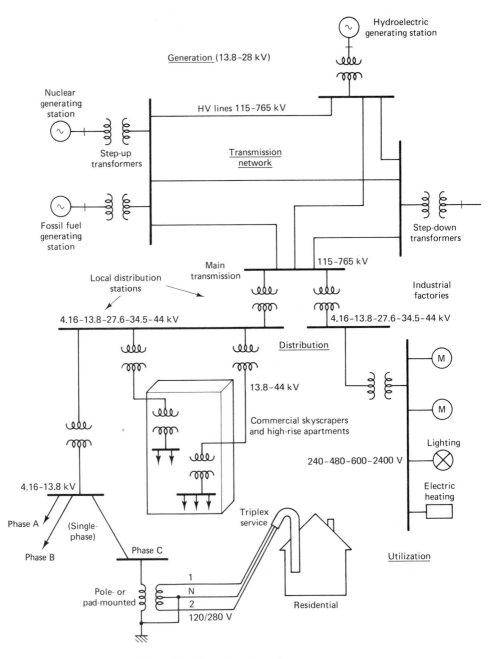

**Figure 6-2** Overview of a typical power system.

The combined specifications of voltage, current, frequency, and power for which a transformer is designed is termed its *rating*. The rated voltage is limited by the maximum allowable flux density in the core, which is less than the value at which the core material saturates. The rated current is limited to a value so as to keep the losses reasonable and not cause excessive heating of the transformer. These ratings are stamped on metallic nameplates mounted on the transformers. The schematic diagram is also shown.

We see therefore that the figure which defines the power-handling capacity of a transformer must refer to the apparent power $S$, measured in volt-amperes (VA, or the larger unit, kVA), not the active power $P$, measured in watts (or kW). The nameplate rating of a transformer tells us what it can do when operating at its maximum capacity. It does not tell us what it is doing at any given time or in a specific job application.

In studying ac machines, starting with the synchronous generator, our objective will be to analyze a machine in sufficient detail to develop an equivalent circuit, which for practical purposes describes that machine. Having such a circuit, we can then use it to predict the machine's performance without too many detailed calculations.

As before, the equivalent circuit will have an induced voltage depending on the air-gap flux. It will also have reactances and resistances as part of the ac parameters at a particular frequency. We will deal with ac quantities now, not dc values. This inevitably leads to phasor diagrams representing the various voltage and current relationships, as a visual aid in understanding machine performances.

## 6-3 CONSTRUCTION OF THREE-PHASE SYNCHRONOUS GENERATORS

A synchronous generator could be built by replacing the commutator in a dc machine by slip rings. The internal ac-generated voltage would then appear across the stationary brushes riding on the rotating slip rings. For three-phase generated voltages we would need three slip rings, which of course must be connected to appropriate points on the armature winding.

In practice, this construction is never used for commercial synchronous generators. Some exceptions do occur, however, for small specialty generators. In practice, a more dramatic change is employed, which makes the synchronous machine appear physically different from dc machines. Figure 6-3 shows the synchronous generator schematically. Unlike the dc generator, the synchronous generator must be driven at a constant speed. The reason is that the frequency of the generated voltage, hence that of the electrical network it supplies, is directly related to that speed. Thus the mechanical speed of the generator must be synchronized with the electrical frequency—from which the name *synchronous machine*.

The principle involved is simply this: generator action depends entirely on the relative motion of the conductors with respect to the field lines. This suggests that it is possible to construct an ac generator in which the role of the stationary

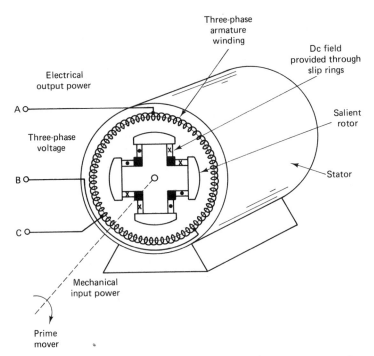

**Figure 6-3** Schematic representation of a four-pole synchronous generator showing reversal of power circuit as compared to the dc machine.

and rotating members is interchanged as compared to the dc machine. That is exactly what is done in practical generators and as shown in Fig. 6-3. The winding in which the voltages are induced, the armature winding, is placed on the stator. The field circuit is placed on the rotor. This arrangement is preferable for several reasons:

1. The armature winding being generally rated for high voltages and currents is much larger and more complex than the field winding. Therefore, it can be better secured in the stationary member, from both a mechanical and an electrical point of view.
2. The armature winding is easier to cool when stationary than when rotating. Because the stator core is larger, it makes it possible to provide better forced-air circulation by being able to provide more air ducts. Forced cooling of the armature with circulating coolants (e.g., water), necessary in large generators, is practical only in this type of construction. For extremely large generators, 500 to 1000 MVA, enclosed hydrogen-gas cooling is employed in addition to water-cooled stators. The water flows through the conductors and of course is separated from the hydrogen cooling system, which has a heat-exchange system.
3. The field coils carry relatively small currents compared to the armature circuit. Their rotating electrical connections are thus smaller. No polarity reversals are necessary and slip rings are usually employed. Also, slip rings through

which the field receives current are inherently less prone to failure than are segmented commutators.

4. No commutator action is required, making the high-power armature connections easier to make on a stationary member.

This explains why synchronous generators are different from dc machines. Not only that, but synchronous generators are built in much larger sizes than dc generators. This is possible because the serious limiting factor of commutation on dc machines is not present on ac machines. Also, to meet the ever-growing demand of electric power, designers have kept pace by producing ever-larger sizes.

As an example to give some typical values, a certain 430,000-kVA 18.2-kV three-phase 1800-r/min 60-Hz synchronous generator has a rated armature current of 13,640 A. The field circuit will take 1780 A when working from a 500-V dc source. This shows how impractical it would be to conduct 13,640 A at 18,200 V (line to line) through slip rings and a brush arrangement. The field-circuit current, on the other hand, although substantial, has a working voltage of only 500 V and needs only two slip rings. Insulating difficulties between slip rings and shaft are thus greatly reduced. Furthermore, the high-voltage armature conductors are more easily insulated on the stationary member. Last but not least, when generator sizes exceed 200 MVA, forced cooling of the armature conductors with liquid coolants is practical only with stationary windings.

These are the reasons for the inverse arrangement in construction as compared to dc machines. We will now take a closer look at some of the details of generator construction.

### Stators

As is typical in all electrical apparatus, the stator of generators consists of good magnetic iron, laminated to minimize eddy current losses. By "good magnetic iron" we mean that both the permeability and the resistivity of the material are high. Silicon iron meets this criterion. Figure 6-4 shows the stator slotted to receive the armature winding in much the same way as is done in dc generators. The number of slots is generally such that a symmetrical three-phase winding can be used. This is possible when the number of slots, divided by the number of

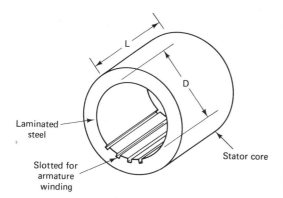

Laminated steel

Slotted for armature winding

Stator core

**Figure 6-4** Slotted and laminated stator core.

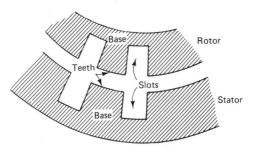

**Figure 6-5** For equal slot depth, teeth on stator become wider and stronger at the base.

poles, times the number of phases, is an integer, that is,

$$\frac{\text{slots} \times \text{phases}}{\text{poles}} \equiv \text{integer}$$

We will discuss the armature winding in more detail shortly. In slow-speed large-diameter machines such as hydroelectric generators having many poles, the axial length of the stator core is relatively short. In high-speed machines, such as those driven by steam turbines, only two or four poles are generally used. The axial length of the core is many times the diameter in those machines. The reader may wonder why is this so. This ties in with the space the poles take up. The more poles that have to be fitted in, the larger the diameter. For example, a 60-Hz 225-r/min hydro unit will have 32 poles. But why the shorter length, then? This can be explained in terms of Eq. (2-13). This equation showed us that the generated voltage is related to the radius $R$ of the armature and the length $L$ of the conductor in field. The conductor length more or less determines the axial length of the machine. Therefore, we see that $\pi R^2 \times L$ is dimensionally a volume. For a specific machine power output, this volume does not change too much. It implies that for a certain rating (volume) making the diameter of the machine larger (needed to fit all these poles in), the length of the machine must be made correspondingly shorter.

As the machine rating goes up, more armature copper is required. To accommodate the larger conductors, deeper slots are required. With deeper slots, the outer diameter is approached and the teeth get wider (see Fig. 6-5). This makes for mechanically stronger teeth on the stator. In addition to the advantages listed for inverting the power circuit, we have an added bonus here. Because if slots on a rotating armature are made deeper, we approach the center of that armature and the teeth become narrower and mechanically weaker, as shown in Fig. 6-5.

About 55% of stator circumference must be provided for teeth to safely carry the necessary magnetic flux without exceeding maximum flux density. The leaves about 45% of the circumference for slots to be filled with conductors and insulation. More coils mean a greater number of slots. On the other hand, fewer but wider slots mean fewer turns of heavier conductors. In the first case, a higher voltage with lower current rating may be in order. In the second, we may aim for a higher current rating but lower voltage.

This leads next to rating. The generator output is limited by its magnetic capability. Like all magnetic devices, they are subject to iron saturation. As magnetic flux density increases, a point is reached where any increase in excitation current results in little change in magnetic flux. This normally happens at the point where the generator voltage is at its maximum. Practical operation takes place somewhere below this point. Machine rating is normally given in kVA (kilovolt-amperes). The volt-ampere rating, generally lower than magnetic capability, is that load where the design temperature has been reached. Exceeding this rating means that the machine gets too hot and the insulation deteriorates faster. Rating determines overall size, which as we stated above, relates to machine dimensions, a given design type, cooling method, and so on.

## Rotors

The rotating member of a synchronous generator is generally constructed in two ways; (1) with salient or projected poles, or (2) with a round rotor or cylindrical rotor. Figure 6-6 indicates these rotor types. The salient rotor machine has dc current in the rotor field coils that provides the MMF $(NI)_{field}$ to set up the magnetic field shown. Because the pole faces are tapered, flux path $l_2$ contains a wider air gap than $l_1$. As we know from our study in Chapter 1, a larger air gap means more reluctance. As a result, the flux density of path $l_2$ is less than that of $l_1$.

As mentioned, the pole shoe is shaped to make the resultant flux density in the air-gap sinusoidal. The advantage of this is then a generated voltage that is also shaped sinusoidally. Considering the cylindrical rotor in Fig. 6-6b, the dc winding is placed in rotor slots. The air gap is uniform, but we see that path $l_1$ encloses more conductors carrying current than path $l_2$. We can therefore expect greater flux densities along path $l_1$ because of greater MMF ($NI$ product). Again, a sinusoidal flux density distribution is obtained of the type shown.

The salient-pole construction is generally restricted to low-speed generators, such as those driven by waterwheels. Because of their low speeds, they require a large number of poles. This can quickly be verified by Eq. (6-2), assuming a fixed frequency. For example, we need 48 poles for a 150-r/min 60-Hz generator. The salient-pole structure is generally simpler and more economical to manufacture than a corresponding cylindrical rotor type. Besides, if the number of poles becomes large, the cylindrical rotor construction becomes impossible to construct in practical machines. Therefore, cylindrical rotors are used exclusively for synchronous generators driven by steam turbines. The generator is then generally known as a turboalternator or turbogenerator. They are usually two- or four-pole, although six-pole generators are found in rare circumstances. Because they have few poles, they are used for high-speed applications. Steam turbines are inherently efficient when operating at relatively high speeds, 1800 and 3600 r/min being common for 60 Hz. Since the rotor is compact, they readily withstand the centrifugal forces developed in large generators at those speeds. As Fig. 6-6 shows, the field winding is placed in slots that run parallel to the rotor axis. It is the physical location of the slots that creates the magnetic poles, since there are no constructed field poles to be seen.

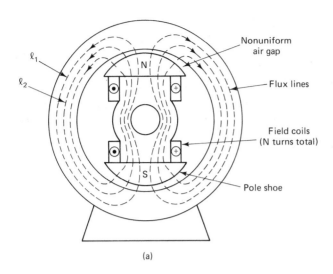

Nonuniform
air gap

$\ell_1$

$\ell_2$

Flux lines

Field coils
(N turns total)

Pole shoe

(a)

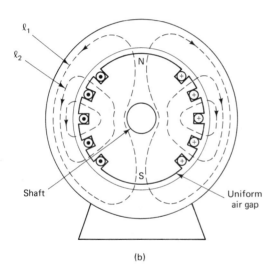

$\ell_1$

$\ell_2$

Shaft

Uniform
air gap

(b)

**Figure 6-6** Two-pole synchronous generators: (a) salient pole; (b) cylindrical rotor construction.

As mentioned earlier, the stator is laminated since it carries an ac magnetic flux. The salient-pole rotor has normally laminated poles, but the cylindrical rotor is often not laminated. Cylindrical rotors are made of a single steel forging. The slots are machined out and the primary function of the rotor teeth is to hold the field coils in against the centrifugal forces. Heavy wedges of nonmetallic steel are forced into the grooves in the teeth to close the slot.

It may be informative to the reader to refer to Fig. 6-7. It shows a typical hydrogenerator, which is normally vertical in construction. On the other hand, Fig. 6-8 shows a steam turbine generator, which is horizontally constructed.

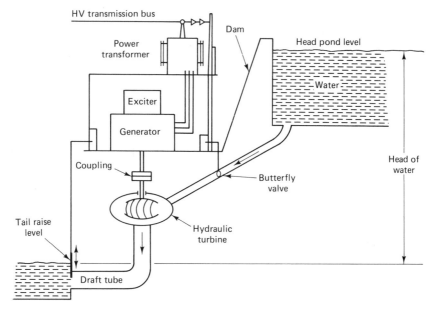

**Figure 6-7** Typical hydrogenerating plant. Hydrogenerator is vertically constructed.

## 6-4 THREE-PHASE GENERATED VOLTAGES AND FREQUENCY

Rotor speed and frequency of the generated voltage are directly related. Let us now determine how. Consider the elementary two-pole ac generator of Fig. 6-9. To simplify things, there is only one coil shown, made up of two conductors in series, *a* and *a'*. When we have a single coil like this, we talk of a concentrated

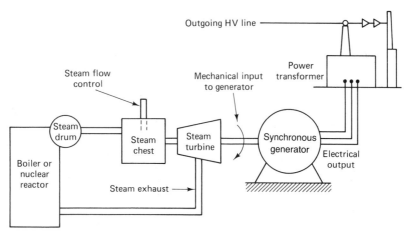

**Figure 6-8** Simplified electric generating plant using horizontal construction of steam turbine generator.

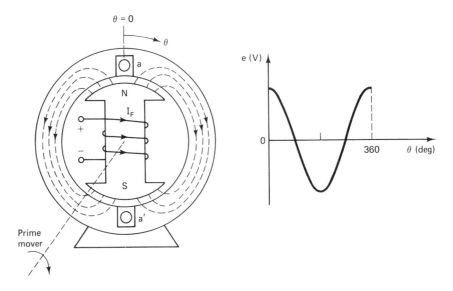

**Figure 6-9** Schematic diagram of single-phase two-pole ac generator and the generated voltage in coil $a-a'$.

winding. In real generators we have many coils in each of the three phases, which are distributed in slots over the entire stator periphery. In that case, we talk of distributed windings. Naturally, each coil could have many turns, although we only have one turn per coil here. Direct current is supplied to the field winding on the rotor through brushes and slip rings. The small air gap minimizes the magnetizing current required for a given flux density. By shaping the air gap properly, by making it nonuniform (i.e., having a variable reluctance under the pole faces), the generated voltage waveform can be made to approximate a sinusoid. Assume that the rotor is turned clockwise. This means that the rotor field flux moves with respect to the armature coil, a necessary condition for generating voltage. In fact, we have the same situation as Fig. 2-5a predicts, except that here the field moves.

We can therefore expect the generated voltage as a function of angular rotor position (or time) to be the same as that shown in Fig. 2-6. Thus turning the rotor one complete turn starting from the position shown generates one voltage waveform. Rotating the rotor one turn in 1 s gives us a frequency of 1 cycle per second [1 hertz (Hz)]. Turning the rotor around two times in 1 s yields 2 cycles, and we say that the frequency is 2 Hz. The number of cycles per second, or frequency, is thus directly proportional to the speed of the rotor. If the speed is 60 revolutions per minute, the frequency $f$ equals 1 hertz, or $f = 1$ Hz. For a frequency of $f = 60$ Hz, the rotor must turn at 3600 rotations per minute, or 3600 r/min.

Thus for $n$ revolutions of the rotor per minute, the rotor turns at a speed of $n/60$ revolutions per second. Because of the continuous nature of magnetic field lines, magnetic poles always occur in pairs. If the rotor has more than two poles, say $p$ poles, each revolution of the rotor induces $p/2$ cycles of voltage in the stator coil. Then the frequency of the induced voltage as a function of rotor speed is

$$f = \frac{p}{2} \frac{n}{60} = \frac{pn}{120} \quad \text{Hz} \tag{6-2}$$

where  $f$ = generated frequency, Hz
  $n$ = rotor speed, r/min
  $p$ = number of poles on the rotor

For the generator to produce a sinusoidal voltage of a given frequency, the rotor must be turning at a speed compatible with the frequency. The rotor is said to be turning at synchronous speed if its speed corresponds to $n$ in Eq. (6-2).

**EXAMPLE 6-1**

A four-pole ac generator operates at 1800 r/min.

  (a) What frequency does it generate?
  (b) What must the speed be if the frequency is 50 Hz?

**SOLUTION**

(a) $f = \dfrac{pn}{120} = \dfrac{4 \times 1800}{120} = 60$ Hz

(b) $n = \dfrac{120f}{p} = \dfrac{120 \times 50}{4} = 1500$ r/min

Table 6-1 gives the number of poles and speeds for the frequency (60 Hz) found on this continent and the frequency (50 Hz) most common in Europe and the rest of the world.

For three-phase synchronous generators we must have three separate armature windings, or three such coils, as shown in Fig. 6-9. The two additional coils must be placed on the generator stator in such a way that the three coils are all 120 electrical degrees apart. The machine is then called a three-phase syn-

**TABLE 6-1**  POLES AND CORRESPONDING
SPEEDS

| Poles | Speed (r/min) | |
| --- | --- | --- |
| | $f = 60$ Hz | $f = 50$ Hz |
| 2 | 3600 | 3000 |
| 4 | 1800 | 1500 |
| 6 | 1200 | 1000 |
| 8 | 900 | 750 |
| 10 | 720 | 600 |
| 12 | 600 | 500 |
| 24 | 300 | 250 |
| 48 | 150 | 125 |

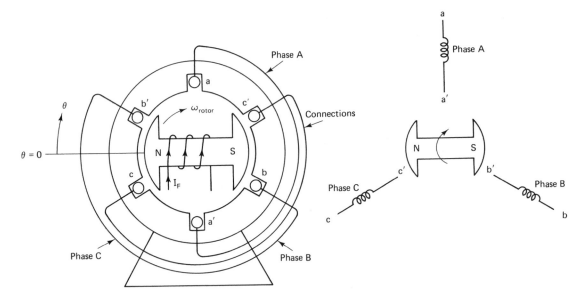

**Figure 6-10** Simple two-pole three-phase ac generator. Armature winding concentrated in one slot per phase per pole.

chronous generator. Figure 6-10 shows the machine of Fig. 6-9 with two additional coils added. Each coil is displaced by 120° to each of the other two coils. Figure 6-11 shows the voltages produced. Curves $e_A$, $e_B$, and $e_C$ show the instantaneous values of the phase voltages, while the effective or rms values $E_A$, $E_B$, and $E_C$ are used in drawing the phasor diagram. The subscripts $A$, $B$, and $C$ refer to the phase sequence of the voltages (i.e., the order in which they are generated).

The principles described for the simple ac generator in Fig. 6-9 also apply to the three separate single-phase voltages induced in the three separate coils of the generator in Fig. 6-10. These three coils on the stator are usually connected either wye (Y) or delta (Δ), to produce a three-phase voltage source as illustrated in Fig.

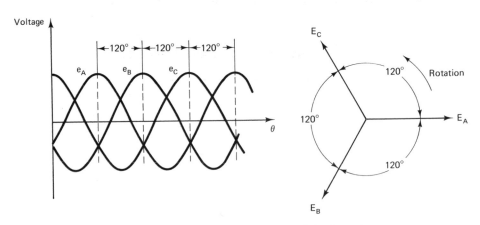

**Figure 6-11** Voltage waveforms and phasor diagram of the three-phase generator in Fig. 6-10.

Electric Power Generation   Chap. 6

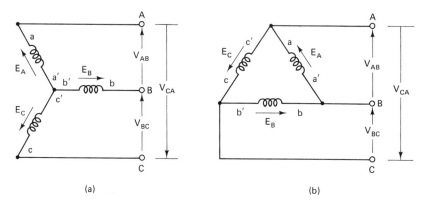

(a)                                                      (b)

**Figure 6-12** Three coils form a three-phase voltage source: (a) Y-connected; (b)
Δ-connected.

6-12.  The voltage induced in each stator coil is known as a phase voltage $E$ and
the voltage appearing between any of two line conductors is known as the line
voltage $V_{L-L}$ or the terminal voltage $V_t$.  Each of the phase voltages $E_A$, $E_B$, and
$E_C$ is completely specified by its magnitude, frequency, and phase.  The magnitude
of each phase voltage is

$$E_{max} = B_m l \omega r \quad \text{V} \tag{6-3}$$

where    $B_m$ = maximum flux density produced by the rotor field winding, Wb
        $l$ = length of both coil sides in the magnetic field, m
       $\omega$ = angular velocity of the rotor, rad/s ($2\pi \times$ frequency)
       $r$ = distance of coil sides to center of machine, m

The reader will recognize Eq. (6-3) to be identical to Eq. (2-12).  If the rotor is
driven by its prime mover at a constant speed, the voltage can be adjusted by
changing the field current.

### EXAMPLE 6-2

The generator shown in Fig. 6-10 generates a three-phase voltage of phase
sequence $A$-$B$-$C$ as illustrated in Fig. 6-11.  Assuming the following data:
$B_m = 1.2$ T, armature length 0.5 m, rotor is turned at 1500 r/min, and inside
diameter of stator core is 0.4 m.

(a) Determine the magnitude of the induced voltage per phase.
(b) Determine the expression of generated voltages in the time domain.
(c) Express the voltage as phasors.

### SOLUTION

(a) Equation (6-3) gives the magnitude of the generated voltage.  Thus

$$E_{max} = B_m l \omega r$$

$$= 1.2 \times (0.5 + 0.5) \times \frac{2\pi \times 1500}{60} \times \frac{0.4}{2} = 37.7 \text{ V}$$

Sec. 6-4    **Three-Phase Generated Voltages and Frequency**                      **161**

(b) The generated voltages are sinusoidal. Taking phase $A$ as the reference, we see that phase $B$ lags phase $A$ by 120 electrical degrees, and phase $C$ lags phase $A$ by 240 electrical degrees. From Eq. (6-2),

$$f = \frac{2 \times 1500}{60} = 25 \text{ Hz}$$

which is the frequency of the generated voltage; therefore,

$$\omega = 2\pi f = 50\pi = 157 \text{ rad/s}$$

Taking phase $A$ as the reference voltage gives

$$e_A = E_{max} \sin \omega t \qquad = 37.7 \sin 157t \qquad \text{V}$$

$$e_B = E_{max} \sin (\omega t - 120°) = 37.7 \sin (157t - 120°) \qquad \text{V}$$

$$e_C = E_{max} \sin (\omega t - 240°) = 37.7 \sin (157t - 240°) \qquad \text{V}$$

(c) Representing the voltages as phasors, we need their rms values;

$$E_{rms} = \frac{E_{max}}{\sqrt{2}} = 26.7 \text{ V}$$

Then

$$\mathbf{E}_A = 26.7 \ \underline{/0°} \text{ V} \qquad \mathbf{E}_B = 26.7 \ \underline{/-120°} \text{ V} \qquad \mathbf{E}_C = 26.7 \ \underline{/-240°} \text{ V}$$

### EXAMPLE 6-3

For the generator in Example 6-2, calculate the line voltages if the armature winding is:

(a) Y-connected.
(b) Δ-connected.

### SOLUTION

(a) From circuit theory or by application of Kirchhoff's voltage law applied to Fig. 6-12, we have

$$\mathbf{V}_{AB} = \mathbf{E}_A - \mathbf{E}_B \qquad \mathbf{V}_{BC} = \mathbf{E}_B - \mathbf{E}_C \qquad \mathbf{V}_{CA} = \mathbf{E}_C - \mathbf{E}_A$$

We can represent this graphically (see Fig. 6-13). Note that the phasors $\mathbf{V}_{AB}$, $\mathbf{V}_{BC}$, and $\mathbf{V}_{CA}$ again form a set of three-phase voltages. Their magnitudes are readily shown to be larger than their phase voltages by a factor of $\sqrt{3}$. Thus,

$$\mathbf{V}_{AB} = 26.7 \ \sqrt{3} \ \underline{/30°} = 46.2 \ \underline{/30°} \text{ V}$$

$$\mathbf{V}_{BC} = 46.2 \ \underline{/-90°} \text{ V} \qquad \mathbf{V}_{CA} = 46.2 \ \underline{/150°} \text{ V}$$

The phase relation between the corresponding line voltages is seen from the construction. Note that this set of voltages is shifted in phase by +30 electrical degrees from the set of phase voltages.

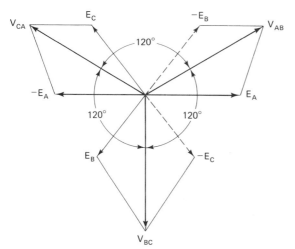

**Figure 6-13** Phasor diagram refers to Y-connection in Example 6-3.

(b) By inspection of Fig. 6-12b, we see that the line voltages are identical to the phase voltages. Thus $\mathbf{V}_{AB} = 26.7 \ \underline{/0°}$ V, $\mathbf{V}_{BC} = 26.7 \ \underline{/-120°}$ V, and $\mathbf{V}_{CA} = 26.7 \ \underline{/-240°}$ V, which form a three-phase voltage set.

As will be evident as we proceed, for a specific machine rating, connecting the armature in wye results in larger terminal voltages. The line current, however, equals the phase current. For a delta-connected armature winding, the terminal voltages equal the phase voltages, but the line currents will be larger by the factor $\sqrt{3}$. In total, we see later that power delivered by the generator is the same in both instances.

## 6-5 EXCITATION SYSTEMS

Ac generators have to be excited with dc current; they cannot be self-exciting. The excitation current is generally supplied by a small dc generator called an *exciter*. The exciter can be mounted on the end of the synchronous generator shaft as shown in Fig. 6-14, or can be driven by a separate motor. The excitation is fed to the rotor field winding via brushes and slip rings.

Another arrangement is to use a solid-state power supply (rectifier) energized

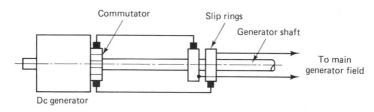

**Figure 6-14** Dc generator on synchronous generator shaft supplying main generator field excitation.

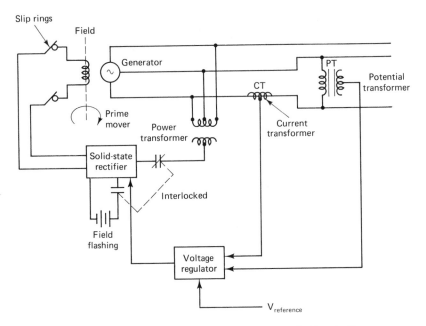

**Figure 6-15** Modern solid-state (static) excitation system (self-excited).

by the generator terminal voltage, as shown in Fig. 6-15. Still another method is the *brushless exciter*, where the excitation current is obtained from a separate ac winding placed on a separate rotor but connected to the main rotor. The ac voltage is then rectified by a rectifier placed on the rotor. This system is indicated schematically in Fig. 6-16. Pilot exciter, main exciter, and silicon diodes are located on the main turbine drive shaft. The pilot exciter generates a small ac signal, which is rectified in the regulator. The system is self-exciting and needs no outside power source. Brushless excitation systems can be build up to 7000 kW to control a 1000-MVA turbine generator. Absence of brushes and commutator should be noted, making the unit more reliable. Typical dc voltage levels for the field circuit are multiples of 125 V, up to 500 V.

## 6-6 AUTOMATIC VOLTAGE CONTROL OF GENERATORS

The loading on a synchronous generator is usually subjected to continual change with a corresponding variation in output voltage. We will see in Chapter 7 that the voltage regulation is normally large. This calls for corrective action. But how? The obvious way is to set the field current manually to keep the generator line voltage within reasonable limits.

Naturally, when the load changes are unpredictable, this method does not look too attractive. In practice, voltage control is achieved by means of automatic control of the field excitation current on individual generators. This calls for a feedback loop. Figure 6-17 shows such a system in principle. The generator voltage $V$ is sensed and compared with a reference voltage $V_{ref}$, which provides a

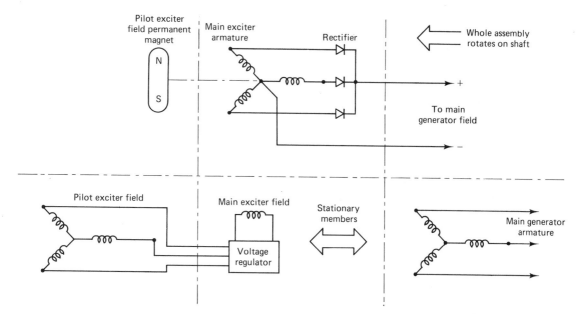

**Figure 6-16**  Brushless exciter system.

measure of the desired value of the generator voltage. When the line voltage differs from the desired value, an error signal $\Delta V$ is obtained which is

$$\Delta V = V - V_{ref} \tag{6-4}$$

This signal is amplified in the power amplifier and sent to the actuating control in the field circuit supply. Action is taken by the actuating control such that the error signal $\Delta V$ is reduced to zero. Thus, if the load is partially reduced, it calls for a decrease in field current in order to reduce $V$ (assuming a lagging power factor load).

For a generator supplying multiple users, the demand on the generator can vary by the minute or more likely in a fraction of a minute. The question then

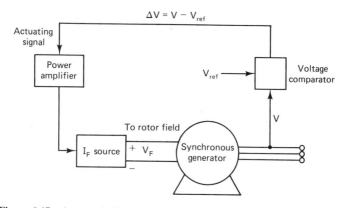

**Figure 6-17**  Automatic field excitation control for synchronous generators.

becomes: How sensitive should our control be? Clearly, it should not be so sensitive as to respond to every minor change in voltage. In other words, we should allow the voltage $V$ to change within an acceptable range, bound by an upper and lower limit. Usually, this limit amounts to a few percent of the line voltage. The lower voltage limit is more critical. This ties in with the many devices that are voltage sensitive, such as heating elements, light bulbs, and particularly, induction motors. The starting torque of induction motors is proportional to the voltage squared, as we will see in Chapter 10.

## 6-7 POWER TRANSFER FROM A SYNCHRONOUS GENERATOR

In electromechanical energy-conversion devices (and the synchronous generator is no exception) the expression of the air-gap or developed power is important. It should be stressed that we are converting the mechanical input power from the prime mover into electrical power. It was shown in Chapter 4 that once the developed torque $T$ is calculated, the developed power $P_d$ can be found from

$$P_d = \omega T \quad \text{W} \tag{6-5}$$

where $\omega$ is the mechanical speed of rotation in radians per second.

It is often useful, however, to have an equation that expresses this power in terms of machine parameters such as voltage and phase angle. A better insight regarding machine performance may then be obtained.

To do this, refer to Fig. 6-18. It shows the synchronous machine represented by a reactance $X_s$ and the corresponding phasor diagram for this circuit. $E_G$ is the internal generated voltage and $V_t$ the terminal voltage, both on a per phase basis. We have neglected the armature resistance. This is acceptable, since the

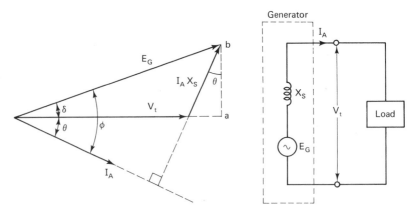

**Figure 6-18** Phasor diagram for synchronous generator with lagging power factor. Armature resistance is assumed negligible.

effect of armature resistance is very small in practical generators. We can write the expression for the developed power in one of two ways, namely,

$$P_d = 3V_t I_A \cos \theta \quad \text{W} \tag{6-6}$$

or

$$P_d = 3E_G I_A \cos \phi \quad \text{W} \tag{6-7}$$

where $\phi = \delta + \theta$ electrical degrees. Equation (6-7) follows from Fig. 6-18 upon noting that the projection of $E_G$ on the $I_A$ phasor is identical with the projection of $V_t$ on $I_A$. Similarly, we can see that the quantity $ab$ in Fig. 6-18 can also be expressed in two ways, namely,

$$ab = E_G \sin \delta = I_A X_S \cos \theta \tag{6-8}$$

From Eq. (6-8) it follows that

$$I_A \cos \theta = \frac{E_G}{X_S} \sin \delta \tag{6-9}$$

If we now substitute the result of Eq. (6-9) into Eq. (6-6), the desired result will be obtained, namely,

$$P_d = 3 \frac{V_t E_G}{X_S} \sin \delta \quad \text{W} \tag{6-10}$$

Equation (6-10) applies only under the assumption that $R_A$ is negligible. The factor of 3 refers to the three phases. Because this power is seen to depend on the angle $\delta$, it is called the power or torque angle. Thus Eq. (6-10) gives the developed power in a synchronous generator where the armature resistance is negligible.

As Eq. (6-10) indicates, no power is generated when the power angle $\delta$ is zero. Furthermore, the equation reveals that the developed power is a sinusoidal function of the power angle. Its maximum occurs when $\sin \delta = 1.0$ or $\delta = 90°$. Figure 6-19 graphically displays the variation of $P_d$ versus positive and negative values of $\delta$. Positive angles refer to generator action. Negative angles refer to motor action. We discuss the power angle in more detail in Chapters 7 and 12.

Our concern in this chapter is the synchronous generator as part of an overall

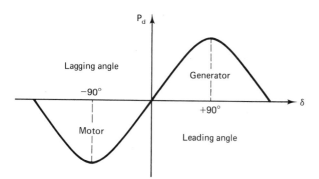

Figure 6-19  Graphical representation of Eq. (7-37), $P_d$ versus $\delta$.

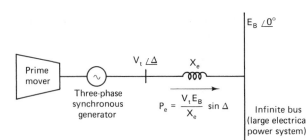

$E_B \angle 0°$

Prime mover — Three-phase synchronous generator

$V_t \angle \Delta$

$X_e$

$P_e = \dfrac{V_t E_B}{X_e} \sin \Delta$

Infinite bus (large electrical power system)

**Figure 6-20** Line diagram of generating station connected to power grid.

power grid. Figure 6-20 illustrates this. In this figure $X_e$ represents the external reactance between the generator and the system's bus. It includes such factors as the impedance of the step-up transformer and the transmission-line impedance between the generating station and the point where the connection to the system is made. This point may be a substation located at some distance from the generating station. Following a similar analysis as above, it can be shown that

$$P_e = \frac{V_t E_B}{X_e} \sin \Delta \qquad (6\text{-}11)$$

where   $P_e$ = generated power delivered to the power system
   $X_e$ = reactance beyond the generation station to the grid
   $E_B$ = power system voltage, taken as the reference voltage
   $\Delta$ = phase angle between generator voltage $V_t$ and the system bus voltage $E_B$

In essence, Eq. (6-11) tells us that the generated voltage at the generator terminals $V_t$ must lead that of the electrical system voltage $E_B$ by a phase angle $\Delta$. Then the generator will deliver the power $P_e$ to the system.

### EXAMPLE 6-4   *"alternator*

A three-phase synchronous generator operates onto a grid bus of 13.8 kV. The external reactance is 8 $\Omega$/phase. The magnitude of the generated terminal voltage is 20% higher than the bus voltage. When the machine delivers 12 MW to the grid, determine the external power factor angle $\Delta$.

### SOLUTION

The bus phase voltage is

$$E_B = \frac{13.8}{\sqrt{3}} = 7967 \text{ V/phase}$$

Since the terminal voltage is 20% larger than this,

$$V_t = 7967 \times 1.2 = 9560 \text{ V/phase}$$

and

$$\text{delivered power} = \frac{12}{3} = 4 \text{ MW/phase}$$

From Eq. (6-11) we obtain

$$4 = \frac{7.967 \times 9.56}{8} \sin \Delta$$

Therefore, $\sin \Delta = 0.42$ and $\Delta = 24.8°$.

## 6-8 PARALLEL OPERATION OF SYNCHRONOUS GENERATORS

Synchronous generators essentially all operate in parallel as part of a large power grid or power distribution system. Such a system operates 24 hours a day. Their common frequency is kept very constant ($60 \pm 0.05$ Hz) and each individual generator, even a very large 1000-MW generator, is usually relatively small compared to the overall system capacity. For instance, Ontario Hydro has, in all of Ontario, approximately 80 generating stations, containing almost 500 generators with an installed generating capacity exceeding 22,000 MW. Before an individual generator is connected to such a grid, or put on line as it is called, it must first be synchronized. This applies to all those synchronous generators operating within a station and from generating station to generating station. Each generating station must be synchronized with the rest of the power system.

Synchronization means that our generator is put in such a condition electrically so that it can look into the running system by closing the circuit breaker. Figure 6-21 illustrates the parallel operation in principle.

What conditions should be fulfilled in order to "throw" the circuit breaker to permit synchronization of our generator? To do so we must ensure that:

1. The generated frequency must match that of the power grid; therefore, the generator must be driven at synchronous speed.

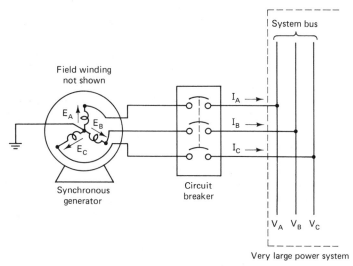

**Figure 6-21** Paralleling synchronous generator to a power grid.

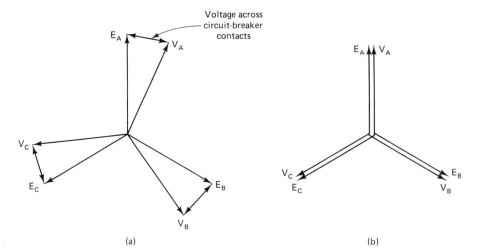

**Figure 6-22** Synchronization of generator: (a) incorrect phase relation; (b) proper moment for closing circuit breaker.

2. The field excitation must be set so as to match the system bus voltage $V$.
3. The generator voltage phase sequence must be identical to that of the system bus voltage $V$.

When these three conditions are met and zero voltage exists across the circuit-breaker terminals, the circuit breaker is closed and synchronization takes place smoothly.

When the frequency of the incoming generator is not equal to that of the power system, the phase relation between corresponding voltages will vary in accordance with the difference frequency. This condition is depicted in Fig. 6-22a. The difference voltage between the $E$ and $V$ phasors represents the voltage difference across the corresponding circuit-breaker contacts when the breaker is in the open position.

The circuit breaker can be closed only when the two sets of voltages are momentarily in phase, provided that three conditions are fulfilled: (1) equal voltage magnitudes, (2) correct frequency, and (3) proper phase sequence, as shown in Fig. 6-22b. The precise condition of synchronization may be indicated by a device known as a synchroscope. This is an instrument which in addition to indicating the instant of precise synchronism also indicates whether the incoming machine is running too fast or too slow.

With temporary wiring or in laboratory setups, an arrangement of lamps can be used to assist in synchronization. When using lamps, two possible arrangements are possible, known as the bright-lamp or dark-lamp method. These designations apply to the state of the lamps when the circuit breaker is closed.

Let us look at the dark-lamp method. Figure 6-23 shows the method. As can be appreciated, the lamps must be able to withstand double the normal voltage. This means that enough lamps may have to be placed in series if an instrument

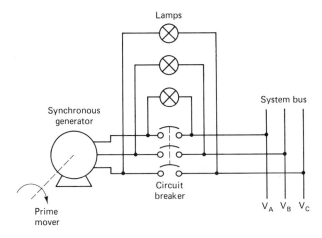

Lamps

Synchronous
generator

System bus

Circuit
breaker

Prime
mover

$V_A$ $V_B$ $V_C$

**Figure 6-23** Connections for synchro-
nizing lamps using the dark-lamp
method.

transformer is not used. A voltmeter is used to monitor the voltage level. The lamps will indicate when the phase and frequency are correct.

The procedure in synchronization is as follows. The machine to be synchronized is brought up to the proper speed by means of the prime mover and its voltage adjusted to match that of the system bus. The circuit breaker in Fig. 6-23 is still open at this point, but the lamps will flicker when speed is near synchronization. By adjusting the prime-mover speed, the cyclical change between lamps light and dark can be made very slow. When all three lamps are dark, the condition shown in Fig. 6-22b is reached and the circuit breaker closed.

Immediately after synchronization of an incoming machine with the system, the generated voltage of this machine is equal but in opposition to the system voltage. Under these conditions no resultant voltage exist between the two and there will be no current flowing from the incoming generator to the system.

### Sharing of Loads

The reader may well ask at this stage: Just how does the machine behave after synchronization? What do we control to ensure that the incoming generator delivers power to the system and thereby takes up its share of the load?

As we will see, and this point is of significance, unlike dc generators, we cannot shift part of the load onto this machine by merely adjusting the field current. As a matter of fact, this would only change the power factor between the generator current and its voltage. Let us investigate, then, what we should do. First, let us expand the idea of, say, increasing the field current of the synchronous generator. If we control the field current, we must remember that the system voltage $V$ will essentially remain constant in magnitude and phase. The voltage is determined by the combined strength of the many generators working in parallel forming the power grid. One added generator is normally insignificant in the overall picture. This leads to a term that power engineers use—"infinite bus." The voltage is unaffected by any changes of current into or out of the system.

As we control the field current, no change can occur in the real power or torque delivered by the prime mover. For our assumed situation of increasing the field current, what happens is that the magnitude of the generated voltage $E_G$ increases because the rotor field is increased. This causes $|E_G| > |V|$. This condition is referred to as an *overexcitation*. Similarly, making the field current $I_f$ smaller than what it was originally, we cause $|E_G| < |V|$. This is *underexcitation*.

Obviously, we will have a voltage difference

$$\Delta V = E - V \qquad \text{V/phase} \qquad (6\text{-}12)$$

This will cause a circulating current to flow between the machine and the system, of value

$$I = \frac{\Delta V}{|\mathbf{Z}_S|} \qquad \text{A/phase}$$

where $\mathbf{Z}_S = (R_A + jX_S)$, the synchronous impedance of the stator winding in ohms per phase.

The reactance $X_S$ for a typical generator is much larger than its effective armature resistance. It follows that this circulating current will lag the voltage $\Delta V$ that induced it by nearly 90°. We can represent these relations by the phasor diagram of Fig. 6-24. It is seen that no transfer of load takes place between the generator and the system when the field current is controlled. It merely changes the magnitude of the armature current.

To refer back to our original question, to make the incoming machine take on part of the load up to its rated capability, it is necessary to increase the input to its prime mover. This is accomplished by increasing the water flow in a hydro plant or steam flow in a thermal plant while keeping the speed constant. When increasing the prime-mover input, its speed tends to increase. This is only a momentary increase in speed. The rotor position moves ahead of its no-load position, giving it a larger angular displacement ($\delta$) of the rotor with respect to the

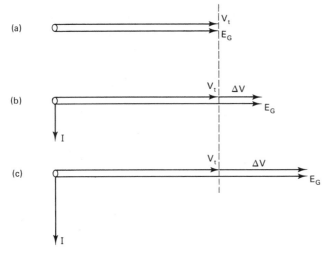

**Figure 6-24** Result of increasing the field excitation current in a synchronous generator just synchronized to the system. Field current in (b) and (c) increased over that in (a). No real power is delivered to the system.

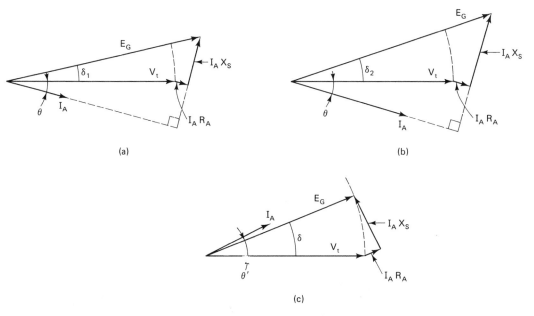

**Figure 6-25** Sharing of generator load by increasing prime-mover speed: (a) and (b) field current is slightly increased to maintain power factor; (c) field current not adjusted. Note change in power factor angle, which is now leading.

resultant field. This is shown in Fig. 6-25, illustrating that $\mathbf{E}_G$ is leading $\mathbf{V}_t$ by a larger angle $\delta$ when the load on a generator is increased.

Note also that the real component of the armature current in phase with the terminal voltage is increased. More real power is thus delivered to the system. This leads to important conclusions. In a synchronous generator power flows out of the machine when the power angle $\delta$ is positive, and $\mathbf{E}_G$ leads $\mathbf{V}_t$. When studying synchronous motors in Chapter 12, we will see the opposite conditions occur.

The reader should observe a fundamental difference here with the dc generator. In dc generators, as we have seen, the terminal voltage is always less than the generated EMF. In synchronous generators this need not be so. In the next section we discuss how the load on the synchronous generator determines the power factor angle $\delta$ and the field current determines whether $E_G$ is larger or smaller than $V_t$. For instance, overexcitation causes $E_G > V_t$. Assuming the power delivered by the machine to remain constant, we have seen that this condition leads to an adjustment of the power factor at which the machine is operating—namely, it starts to lead. In actual practice, changes in prime-mover speed require some change in field current to prevent a slight change in terminal voltage. Otherwise, the power factor at which the machine operates will change. This is evident from Fig. 6-25c.

It is seen that a balance must exist between the electrical output power at the generator and the input mechanical power to the prime mover (turbine). This balance requires that the speed of the turbine be kept constant by means of the turbine governor control.

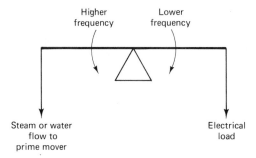

Higher
frequency

Lower
frequency

Steam or water
flow to
prime mover

Electrical
load

**Figure 6-26** Mechanical analog demonstrating balance between prime-mover input and electrical output load to maintain constant frequency.

## 6-9 AUTOMATIC FREQUENCY CONTROL

A sudden demand in load on a synchronous generator will be "felt" by the prime mover. From Fig. 6-26 we have some feeling for the frequency balance of the power system that we try to maintain. If the prime-mover power is not compensated, the frequency will change because of a slowing down in rotational speed. Even a frequency change of 0.5 Hz would be unacceptable in a modern power system. There are many reasons why the frequency must be controlled within close limits. Control of the frequency can be achieved in a similar way as that which we encountered in voltage control—it calls for a feedback loop. Figure 6-27 illustrates such a control loop schematically, as applied to a generator driven by a steam turbine. Control of the frequency by automatic regulation of the generator output is applied in most systems today.

As shown in Fig. 6-27, the output frequency is sampled and compared with a reference frequency. A frequency sensor-comparator will send a signal if there

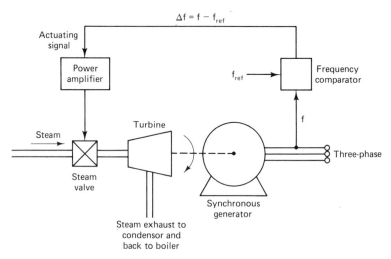

**Figure 6-27** Electronic governor.

is a frequency error signal generated,

$$\Delta f = f - f_{reference} \qquad (6\text{-}13)$$

which is a measure of the frequency deviation. This signal can be amplified and corresponding action taken to modify the steam flow to the turbine.

If the frequency drops due to a sudden load increase, the error signal in Fig. 6-27 would become negative. The actuating signal would then command a greater generator output (i.e., an increase in steam flow or an opening of the steam valve).

Again the question of responsiveness has to be considered. It will be noted that thermal systems are inherently sluggish because of their thermal inertia. Frequency control is therefore accomplished over a period of time; that is, the average frequency in a day is kept to 60.00 Hz.

## REVIEW QUESTIONS

**6-1.** How does the construction of a practical three-phase synchronous generator differ from that of a dc machine?

**6-2.** State at least three reasons for the inverse construction of the synchronous generator as compared to a dc machine.

**6-3.** Why are low-speed generators relatively short and high-speed generators relatively long in axial length?

**6-4.** What determines the frequency of the generated voltage in a synchronous generator? How are these quantities related?

**6-5.** State two ways in which the synchronous generator can be constructed.

**6-6.** For salient-pole generator construction, describe how the air-gap flux is made to approximate a sinusoidal distribution.

**6-7.** Repeat Question 6-6 for cylindrical rotor construction.

**6-8.** Which rotor construction is normally used for high-speed turboalternators? Why?

**6-9.** Which rotor construction is used for low-speed hydrogenerators? Why?

**6-10.** What determines the waveshape of the generated voltage in a synchronous generator?

**6-11.** State at least three methods of providing the required rotor excitation power. Describe one of them with the aid of a circuit diagram.

**6-12.** Upon what factors does the magnitude of the generated voltage depend in a synchronous generator?

**6-13.** State the conditions to be fulfilled before a synchronous generator can be paralleled to a power grid. What instrumentation is needed to accomplish this?

**6-14.** What adjustments must be made to a synchronous generator just paralleled to a system bus to ensure it delivers power to the system?

**6-15.** What would be the reason for operating generators in parallel?

**6-16.** By what means is the generated frequency kept constant in practice?

**6-17.** Machine ratings are specified in kVA rather than in kW. How do you explain this? What determines the rating?

## PROBLEMS

**6-1.** What speed must an eight-pole synchronous generator rotate at in order to generate 60-Hz voltage?

**6-2.** A dc motor is used as a prime mover for a synchronous generator in order to obtain a variable-frequency supply. If the speed range of the motor is 820 to 1960 r/min and the generator has four poles, what is the obtainable frequency range of the output voltage?

**6-3.** A three-phase load of 10 $\Omega$/phase may be connected by means of switches in wye or delta. If connected to a 220-V three-phase synchronous generator, calculate:
(**a**) The power dissipated in wye.
(**b**) The power dissipated in delta.

**6-4.** A 250-kVA 1260-V Y-connected synchronous generator has its armature winding reconnected in delta ($\Delta$). Determine the machine rating when $\Delta$-connected (i.e., kVA, $I_L$, and $V_{L\text{-}L}$).

**6-5.** A 1000-kVA 2200-V 60-Hz three-phase synchronous generator is Y-connected. Calculate the full-load line current.

**6-6.** If the generator in Problem 6-5 delivers a load of 720 kW at a power factor of 0.80, calculate the line current.

**6-7.** The open-circuit voltage of a 60-Hz synchronous generator is 4600 V at a field current of 8 A. Calculate the open-circuit voltage at 50 Hz if the field current is 6 A. Neglect saturation.

**6-8.** If the synchronous generator of Problem 6-7 is used to generate 50 Hz, what will be the line voltage if the armature is Y-connected?

**6-9.** A three-phase synchronous generator operates onto a 13.8-kV power grid. The synchronous reactance is 5 $\Omega$/phase. It is delivering 12 MW and 6 MVAR to the system. Determine:
(**a**) The power angle $\delta$.
(**b**) The phase angle $\theta$.
(**c**) The generated EMF $E_G$.

**6-10.** Two identical three-phase Y-connected synchronous generators share equally a load of 3200 kW at 24 kV and 0.8 lagging power factor. Each machine has a 4.8-$\Omega$/phase synchronous reactance and negligible armature resistance. One of the generators carries 50 A at a lagging power factor. What is the current supplied by the other machine?

# 7

# THREE-PHASE SYNCHRONOUS GENERATORS

## 7-1. INTRODUCTION

A dc generator has to provide a voltage that remains constant in direction and magnitude and that was accomplished by using a commutator. This enabled us to obtain a dc voltage from an alternating EMF. The commutator simply acted as a mechanical rectifier. It makes it possible via the brush rigging to make electrical connections to the rotating power circuit. After all, the generator load is stationary.

The various dc machine equivalent circuits were presented. They show the rotating armature as the power circuit, which is a dc voltage source depending on the air-gap flux and speed of rotation.

In ac machines, to which the remainder of the book is devoted, the fluxes are not constant; in fact, they vary or move about continually. We have already seen in Chapter 6 that the field winding of a synchronous generator is placed on the rotating member, called the *rotor*. The armature winding as with most ac machinery is placed on the stationary member, called the *stator*. With dc machines we have an inverse situation, in that the power circuit is now stationary.

Large-scale power generation (i.e., the conversion of mechanical energy to electrical form) is done by synchronous generators. Electrical power utilities generate three-phase power exclusively. Therefore, only three-phase synchronous generators will be studied in this chapter. Construction details of the generator were discussed in Chapter 6; we concentrate now on its specific performance characteristics. The three-phase concepts necessary in understanding three-phase machines will be explained as we go along. Readers who wish to familiarize themselves with three-phase power measurements will find Appendix D helpful.


## 7-2 MAGNITUDE OF THE GENERATED VOLTAGE

Up to now we have presented the reader with some constructional details of the synchronous generator (see Chapter 6). We have tried to avoid unnecessary complications or skipped important practical considerations altogether. We now appreciate how this machine differs from the dc generator. At this point we proceed to consider what the generated voltage is. We already encountered an expression for the voltage, as given by Eq. (6-3). That expression is not used in practice, because the value of $r$ is not readily known. Therefore, we want an alternative expression.

As expected, the average induced voltage in each phase winding as the rotor field sweeps by is governed by Faraday's law. It is in formula form as follows:

$$E_{avg} = N\frac{\Delta\phi}{\Delta t} \quad V \tag{7-1}$$

where  $E_{avg}$ = average generated voltage in the winding, V
$\Delta\phi$ = change of flux in a given time, Wb
$\Delta t$ = time in which the flux change takes place, s
$N$ = number of turns on the winding

Figure 7-1 illustrates such a coil of the armature while a field pole moves past. It should be noted that it is the relative motion between the field and the coil that counts, so that if the coil is moving from left to right with the poles stationary, identical results are obtained.

With reference to Fig. 7-1, the flux change between the two positions shown is $\phi_m$ (webers). This implies a pole movement equal to one-half of the pole pitch in which one-fourth of a voltage cycle is generated. Since one cycle occurs in $1/f$ seconds, the elapsed time for this part of the waveform awill be $1/(4f)$ seconds.

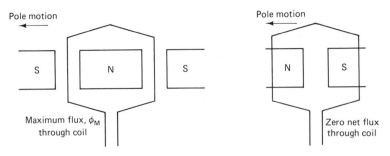

**Figure 7-1**  Flux change through a coil as poles move past the coil from right to left.

Therefore, substituting this value of elapsed time in Eq. (7-1), it becomes

$$E_{avg} = N \frac{\phi_m}{1/(4f)} = 4fN\phi_m \quad V \qquad (7\text{-}2)$$

Equation (7-2) is a general expression applicable to all generators regardless of the pole-flux distribution. Of course, if the flux distribution is sinusoidal, the generated voltage will be sinusoidal. If so, we know from circuit theory that the so-called effective or rms value of the voltage $E$ is 1.11 times the average value. Thus

$$E = 4.44fN\phi_m \quad V \qquad (7\text{-}3)$$

where $E$ = generated rms voltage in each phase winding, V

$f$ = frequency of the generated voltage, Hz

$\phi_m$ = maximum flux per pole, Wb

Practically, the flux distribution may not be a perfect sine wave. In that case, the effective voltage is slightly modified to account for the so-called harmonics. Recall that the form factor 1.11, being the ratio between effective and average voltage, applies only if the waveform is sinusoidal. However, we will not consider non-sinusoidal waveforms.

## EXAMPLE 7-1

Determine the effective voltage generated in one of the phases of a synchronous generator. The following data are given: $f = 60$ Hz, turns per phase $N = 230$, maximum flux per pole $\phi_m = 0.04$ Wb.

### SOLUTION

$$E = 4.44 \, fN\phi_m$$
$$= 4.44 \times 60 \times 230 \times 0.04 = 2450 \text{ V/phase}$$

For a three-phase generator Y-connected, this results in an open-circuit line-to-line voltage

$$V_{L\text{-}L} = 2450 \sqrt{3} = 4244 \text{ V}$$

## 7-3 ARMATURE WINDING

There are many possible ways of winding the armature of synchronous generators. The majority of them come in two types, (1) single-layer windings, and (2) double-layer windings. The material presented here deals with the principles involved rather than all the practices of armature construction. A few types will be discussed in order to explain these fundamentals.

As we have seen in Fig. 6-10, a three-phase winding results by adding two more sets of armature coils to that of the generator shown in Fig. 6-9. They are displaced 120 and 240 electrical degrees from the first coil (phase) to produce a system of three voltages equal in magnitude and displaced from each other by 120°

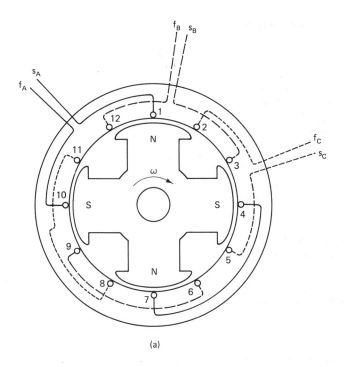

(a)

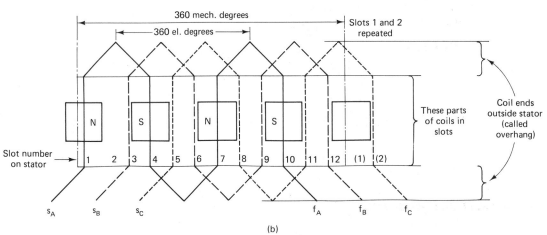

(b)

**Figure 7-2**  Three-phase synchronous generator: (a) three-phase single-layer armature winding distributed in one slot per pole per phase; (b) developed view of armature winding.

(see Fig. 6-11).  The machine is then called a three-phase generator.  Figure 7-2 illustrates a three-phase four-pole synchronous generator with a developed view of the armature winding.  We see there is only one coil side in each slot, making it a single-layer winding.  If the windings of the three phases start at $s_a$, $s_b$, and $s_c$, and finish at $f_a$, $f_b$, and $f_c$, they may be joined in two ways.  These of course are the delta ($\Delta$) and wye (Y) connections, as illustrated in Fig. 6-12. For a given number of turns per phase, the Y connection gives a higher terminal voltage than the $\Delta$ connection, but a correspondingly smaller output current. The Y-connection is sometimes called a star-connection.

Some observations can be made regarding the three-phase armature winding in Fig. 7-2. The separation between the phase windings is 120 electrical degrees, or 60 mechanical degrees. This we conclude from simple reasoning that a full EMF cycle will be generated when the four-pole rotor turns 180 mechanical degrees. A full EMF cycle represents 360 electrical degrees. Extending this to a $p$-pole generator ($p$ must always be a positive even integer), we have the following important relationship between mechanical rotor angles $\alpha_{mech}$ and electrical angles $\alpha_{el}$, namely,

$$\alpha_{el} = \frac{p}{2}\,\alpha_{mech} \tag{7-4}$$

## EXAMPLE 7-2

A three-phase synchronous generator has 12 poles. What is the mechanical angle corresponding to 180 electrical degrees?

### SOLUTION

The mechanical angle between a north and a south pole is

$$\alpha_{mech} = \frac{360 \text{ mech. degrees}}{12 \text{ poles}} = 30°$$

This corresponds to 180 electrical degrees. Using Eq. (7-4) as a check, we see that

$$\alpha_{el} = \frac{p}{2}\,\alpha_{mech} = \frac{12}{2} \times 30° = 180°, \text{ as before}$$

The generated EMFs in each coil side aid each other in each phase to add up to the total phase voltage. This is readily verified by applying the right-hand rule to the coil sides making up a phase winding.

Connecting the phases in Y as indicated in Fig. 7-3 has the added advantage that the neutral generator node can be brought out. This means that not only line voltages are available but the line-to-neutral (the phase) voltages as well. Furthermore, the generator neutral is normally grounded. It enables grounding of apparatus and devices at the consumer end of electrical energy, a very desirable feature from a safety point of view.

For the direction of rotation of the rotor indicated in Fig. 7-2a (clockwise), the resulting phase sequence of the three-phase supply is $ABC$. This means that the maximum voltage is generated in phase $A$, followed by phase $B$, and then phase $C$. Reversing the direction of rotation would result in an $ACB$ sequence,

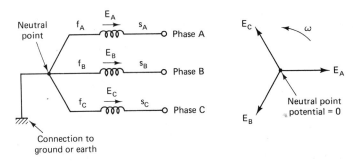

**Figure 7-3** Y-connected armature winding of Fig. 7-2; phase sequence $ABC$.

or *negative phase sequence*. The former ($ABC$) is often referred to as the *positive phase sequence*.

In summary, we have obtained a symmetrical set of three-phase EMFs,

$$\mathbf{E}_A = E_A \angle 0° \text{ V}$$

$$\mathbf{E}_B = E_B \angle -120° \text{ V} \tag{7-5}$$

$$\mathbf{E}_C = E_C \angle -240° \text{ V}$$

in which the $A$ phase is taken as the reference phasor. The boldface type for the EMFs $\mathbf{E}_A$, $\mathbf{E}_B$, and $\mathbf{E}_C$ indicates phasor quantities; $E_A$, $E_B$, and $E_C$, in italic type, are magnitudes of those phasors making an angle with the reference axis as indicated.

When the three phase windings $s_a$–$f_a$, $s_b$–$f_b$, and $s_c$–$f_c$ are interconnected as shown in Fig. 7-3, it is easy to verify that for a symmetrical three-phase EMF set we have the important relation

$$\mathbf{E}_A + \mathbf{E}_B + \mathbf{E}_C = 0 \tag{7-6}$$

**EXAMPLE 7-3**

Using the three-phase set of voltages in Eq. (7-5), prove Eq. (7-6).

**SOLUTION**

Since a vector addition is involved, we must write the equations in complex form, that is,

$$\mathbf{E}_A = E_A \angle 0° \qquad = E_A \cos 0° + jE_A \sin 0° = E_A + j0$$

$$\mathbf{E}_B = E_B \angle -120° \qquad = E_B \cos(-120°) + jE_B \sin(-120°)$$

$$= -0.5E_B + j0.866E_B$$

and

$$\mathbf{E}_C = E_C \angle -240° \qquad = E_C \cos(-240°) + jE_C \sin(-240°)$$

$$= -0.5E_C - j0.866E_C$$

Adding the complex numbers, realizing that the magnitudes $E_A$, $E_B$, and $E_C$ are the same, there results

$$(E + j0) + (-0.5E + j0.866E) + (-0.5E - j0.866E) = 0$$

where $E = E_A = E_B = E_C$.

### Double-Layer Winding

The armature winding of Fig. 7-2 has only one coil side per pole per phase. In effect, each phase winding has only two coils. As stated before, when each coil side is located in a single slot of the stator per phase, we speak of a concentrated winding. If the slots were not excessively wide, each conductor in a given slot would generate the same voltage. Each phase voltage would then be equal to the product of the voltage per conductor and the total number of conductors per phase.

In reality, this would not be a very effective way of using the stator core, because of the variation in flux density in the iron and the localization of heating effects in the slot regions. In other words, the result would be a flat-topped waveform instead of sinusoidal, which gives rise to harmonics. To overcome this problem, practical generators have the winding distributed in several slots per pole per phase. Figure 7-4 shows a section of an armature winding and the general type of coil used in practice. The winding is distributed in two slots per pole per phase.

Note that there are now 24 slots in the stator core as opposed to 12 slots shown in Fig. 7-2. There are two coil sides per slot and each coil shown has more than one turn. All coils are of the same shape. This means they can be preshaped prior to assembly. The shape of a typical coil is shown in Fig. 7-5. One side is placed on top of another belonging to the same phase. All coils so placed on the stator fit snugly together, and those parts of the coils not located in slots are generally referred to as the winding *overhang*. The overhang is outside the stator core and not in the magnetic field. Therefore, since no voltages are induced in those sections, they are kept as short as practically possible to reduce the amount of copper used. Naturally, we can not avoid this overhang altogether; after all, the two coil sides must be connected together to form a coil.

Another advantage of distributing the winding over several slots per pole is that it automatically improves the generated voltage waveform. This may not be obvious at this point, so we will proceed to show this.

### Distribution Factor

When a winding such as the one we are discussing is made up of a number of coils placed in separate slots, the EMFs generated in the various coils per phase are not in phase. In the concentrated winding each coil in a phase winding has the same relative position with respect to the poles. The total voltage per phase is the sum of the individual coil voltages and is given by Eq. (7-3). In a distributed winding the terminal EMF is less than if the winding had been concentrated. The factor by which the EMF of a concentrated winding must be multiplied to give the EMF

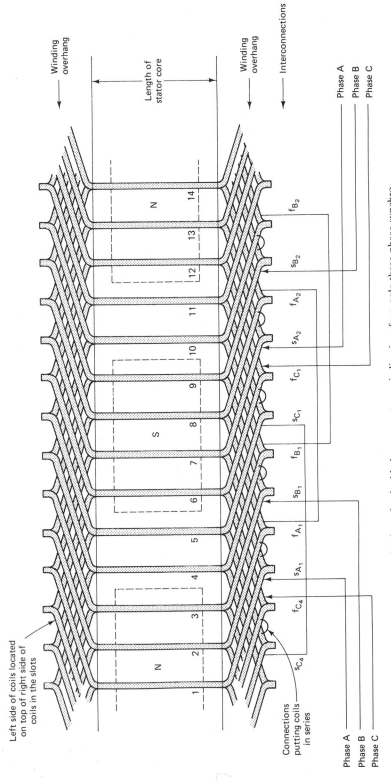

**Figure 7-4** Section of a double-layer armature winding in a four-pole three-phase synchronous generator. Winding is distributed in two slots per pole per phase.

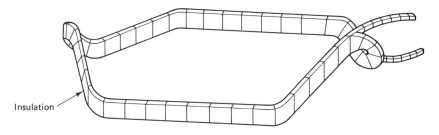

**Figure 7-5** Multiturn coil of a double-layer winding.

of a distributed winding of the same number of turns is called the *distribution factor* $k_d$ for the winding. This factor is always less than unity. Let us assume that there are $n = 4$ slots per phase per pole. The spacing between slots, in electrical degrees, is then

$$\psi = \frac{180 \text{ electrical degrees}}{n \times m} \tag{7-7}$$

where $m = 3$ is the number of phases.

Clearly, the EMF induced in slot 2 will lag the EMF induced in slot 1 by $\psi$ = 15 electrical degrees: the EMF induced in slot 3 will be $2\psi$ degrees behind in phase; and so on. Representing these EMFs by phasors $\mathbf{E}_1$, $\mathbf{E}_2$, $\mathbf{E}_3$, and $\mathbf{E}_4$, respectively, one would obtain the phasor diagram in Fig. 7-6. The total stator EMF per phase $\mathbf{E}$, is then the vector sum of all phasors, namely.

$$\mathbf{E} = \mathbf{E}_1 + \mathbf{E}_2 + \mathbf{E}_3 + \mathbf{E}_4$$

It is apparent that because of the displacement angle $\psi$, the total stator EMF $E$ is less than the arithmetic sum of the coil EMFs by the factor

$$k_d = \frac{\text{vector sum}}{\text{arithmetic sum}} = \frac{\mathbf{E}_1 + \mathbf{E}_2 + \mathbf{E}_3 + \mathbf{E}_4}{4 \times E_{\text{coil}}}$$

It can be shown that

$$k_d = \frac{\sin\left(\frac{1}{2}n\psi\right)}{n \sin(\psi/2)} \tag{7-8}$$

with the parameters as defined before.

Going back to our question of how a distributed winding improves the generated voltage waveform, consider Fig. 7-7. For simplicity we take three coils per coil group; that is, there are three coils in series per phase for every pair of poles. This means that there are 20 electrical degrees between successive coils.

To dramatize the point we are trying to make, it is assumed that each coil has a generated EMF waveform which is a nonsinusoidal curve, such as shown. Adding the three component EMFs point by point gives a resultant waveform $e_t$. Observe that the resultant waveform much better approximates a sinusoidal waveform than do the component waves. In fact, the resultant waveform will approach a perfect sinusoidal waveform as the number of component waves is increased.

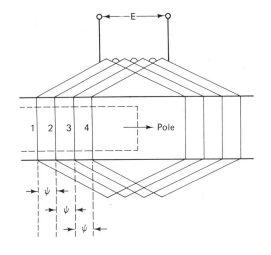

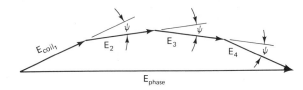

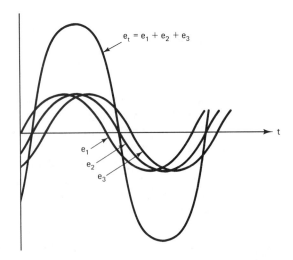

**Figure 7-6** Phasor diagram of induced voltage in a group of four coils in a distributed winding.

Note that the maximum of this resultant waveform is not three times the maximum of the coil EMF, due to the distribution factor.

### Fractional-pitch Coils

In a full-pitch coil the generated voltages in both coil sides are in phase. In practice, the distance between the two coil sides is often made less than full pitch. The whole idea, as we have seen earlier, is to generate sinusoidal voltages, thereby

**Figure 7-7** Total EMF $e_t$, of three non-sinusoidal EMFs becomes nearly sinusoidal.

eliminating harmonics. This is accomplished more easily using fractional-pitch windings than full-pitch windings. In fractional-pitch coils, the coil sides are out of phase, the number of EMFs to be added is doubled, and the waveform is improved. From a practical point of view, in machines with six or fewer poles it is often impractical to place a full-pitch coil into a machine without damaging the coil, because excessive deformation of the coil form is required to fit in the winding. Fractional-pitch coils generally have shorter end connections than do full-pitch coils; therefore, there is some saving in copper and weight.

The disadvantage of using fractional-pitch coils is that there is a slight reduction in generated EMF compared to that of full-pitch coils. It is customary to express this reduction in generated voltage by a *pitch factor*, $k_p$. Its value may be calculated by the equation

$$k_p = \sin \frac{p°}{2} \tag{7-9}$$

where $p°$ is the span of the coil in electrical degrees.

## EXAMPLE 7-4

Assume that the generator of Example 7-1 is a 12-pole machine and has a three-phase stator winding placed in 144 slots; the coil span is 10 slots. Calculate:

(a) The distribution factor $k_d$.
(b) The pitch factor $k_p$.
(c) The winding factor $k_w = k_d \times k_p$.
(d) The resulting terminal voltage on open circuit if the winding is Y-connected.

## SOLUTION

(a) The number of slots per pole per phase is

$$n = \frac{\text{slots}}{\text{phases} \times \text{poles}} = \frac{144}{3 \times 12} = 4$$

The electrical angle between slots is

$$\psi = \frac{180}{n \times m} = \frac{180}{4 \times 3} = 15°$$

Then from Eq. (7-8),

$$k_d = \frac{\sin\left(\frac{1}{2} n\psi\right)}{n \sin\left(\psi/2\right)} = \frac{\sin\left(\frac{1}{2} \times 4 \times 15\right)}{4 \times \sin\left(15/2\right)} = 0.958$$

(b) The pole pitch is $144/12 = 12$ slots, the coil span in electrical degrees is

$$p = \frac{\text{coil pitch}}{\text{pole pitch}} \times 180° = 150 \text{ electrical degrees}$$

Then, according to Eq. (7-9),

$$k_p = \sin \frac{150°}{2} = 0.966$$

(c) The winding factor is

$$k_w = k_d \times k_p = 0.958 \times 0.966 = 0.925$$

(d) The phase voltage is reduced by the winding factor, or

$$E = k_w \times 2450 = 2266 \text{ V/phase}$$

which results in a line voltage on open circuit of

$$V_{L-L} = 2266 \sqrt{3} = 3925 \text{ V}$$

The minor disadvantage of the distributed winding with a short-pitch coil is a somewhat reduced voltage. However, the many advantages that are gained far outweigh this disadvantage.

## 7-4 VOLTAGE REGULATION

The principles of the three-phase synchronous generator have been discussed and we have seen how a three-phase voltage source is obtained. Before the generator can supply an electrical load, it must fulfill the following requirements:

1. It must operate at the correct or synchronous speed as it determines the frequency, since the number of poles is fixed for any given machine.
2. The dc rotor field excitation must be present.
3. It must have the correct output voltage, which can be set by adjusting the field excitation current.

When a load is then placed on the generator, the terminal voltage will be affected despite the fact that the dc field excitation remains constant. The way in which the terminal voltage changes depends on the character of the load (i.e., upon its power factor). Resistive and inductive loads will cause the terminal voltage to drop by as much as 25 to 50% below the no-load value, whereas a capacitive load will tend to raise it above the no-load value.

In Chapter 3 when dc generator operation was discussed, it was shown that two factors are responsible for the change in voltage. These were the armature resistance voltage drop and the change in flux due to armature reaction (see Section 3-6). As we may expect, these two factors are again present in an ac generator, but there is a third factor as well: the armature reactance voltage drop. This voltage drop is caused by the inductance $L_A$ of the armature winding, which is considerable. It asserts itself as a reactance $X_A$,

$$X_A = 2\pi f L_A = \omega L_A \quad \Omega \tag{7-10}$$

where $L_A$ = armature winding inductance per phase, H
$\omega = 2\pi f$ = radial frequency, rad/s

The reactance of the armature then gives rise to an additional voltage drop. As we will see, it must be accounted for vectorially to arrive at the generated voltage. As will be shown later, the change in flux due to the armature reaction effect is generally treated as a voltage drop. It, too, must be taken into account vectorially. We discuss these voltage drops in more detail in the following sections. Before we proceed, let us refer to Fig. 7-8, which depicts typical generator terminal characteristics.

The terminal voltage is presented as a function of the load current at various power factors, the field excitation being held constant. The voltage regulation of an ac generator is defined as the rise in voltage when the load is reduced from full-rated value to zero, speed and field current remaining constant. It is normally expressed as a percentage of full-load voltage; thus

$$\text{voltage regulation} = \frac{V_{NL} - V_{FL}}{V_{FL}} \times 100\% \qquad (7\text{-}11)$$

where $V_{NL}$ = no-load terminal voltage, which equals the generated EMF, V
$V_{FL}$ = full-load terminal voltage, V

The subtraction in Eq. (7-11) is algebraic, not vectorial. Figure 7-8 shows us that the percent regulation varies considerably depending on the power factor of the load. For a leading load power factor it even becomes negative, which tells us that the terminal voltage increases upon loading.

### EXAMPLE 7-5

When the load is removed from an ac generator its terminal voltage rises from 480 V at full load to 660 V at no load. Calculate the voltage regulation.

**SOLUTION**

From Eq. (7-11),

$$\text{voltage regulation} = \frac{660 - 480}{480} \times 100\% = 37.5\%$$

In general, the voltage change is considerable when the load is removed and it is for this reason that automatic voltage regulators are used in conjunction with synchronous generators. This inherent voltage change cannot be compensated for as with dc generators by the use of compounding.

Furthermore, the voltage variations, particularly at low lagging power factors are considerably larger than those displayed by dc machines, due to the armature reaction and armature reactance effects. Also, generators generally feed comparatively long transmission lines consisting of wires and transformers, whose impedances introduce additional voltage drops. These combined factors act simultaneously to cause large voltage fluctuations with changing loads, which are

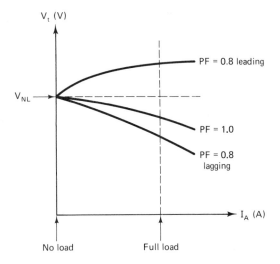

**Figure 7-8** Terminal voltage versus armature current at different load power factors.

intolerable in power distribution systems. Therefore, specially designed regulators are employed to act on the dc field excitation, so that a drop in voltage is accompanied by an increase of the flux. This control is discussed in principle in later sections.

## Generator Voltage Drops

When a load is placed on a generator, for instance, a three-phase motor is connected to it, current will flow in the generator armature windings. The armature winding will have an armature resistance $R_A$ in ohms per phase. The result is an armature voltage drop

$$V_A = I_A R_A \quad \text{V/phase} \tag{7-12}$$

where  $I_A$ = phase current, A
$R_A$ = phase winding resistance, $\Omega$

As the three phases are completely symmetrical, the voltage drops will be the same in all phases. This also applies to the currents. It is for this reason that all values are referred to on a per phase basis, with the understanding that all phases behave the same way.

In addition to the resistive voltage drop, there is a reactive voltage drop, due to the armature winding inductance. Its value is

$$V_X = I_A X_A \quad \text{V/phase} \tag{7-13}$$

where $X_A$ is the armature winding reactance per phase as determined by Eq. (7-10).

Knowing the magnitudes of these voltage drops alone is not sufficient to determine the generated voltage. The voltages have to be added vectorially to the terminal voltage; thus it is essential that for a given load current the power factor is known. Figure 7-9 readily demonstrates the dependence of $E_G$ on the power factor. For unity and lagging power factors $E_G$ is larger than $V_t$. When the power factor is leading, the opposite occurs, namely, $V_t$ is larger than $E_G$. Note, however, that $E_G$ leads $V_t$ in phase for all power factors. We discuss this property in a later section.

We now proceed to consider the third factor responsible for a voltage change upon loading the generator. This is the armature-reaction-effect, which plays a significant role in the ultimate value of the generated voltage.

## Armature Reaction

If the rotor field in the generator is excited and no current flows in the armature winding (no-load condition), the flux paths for a two-pole machine are as shown in Fig. 6-6. When the armature winding is carrying current, it will set up its own field. When both fields are present simultaneously, as would occur under normal loading conditions, they will react with each other and a single resultant flux pattern will exist.

This is not unlike armature reaction effects, as we discussed for dc machines. We therefore expect the generated voltage in synchronous generators to be affected

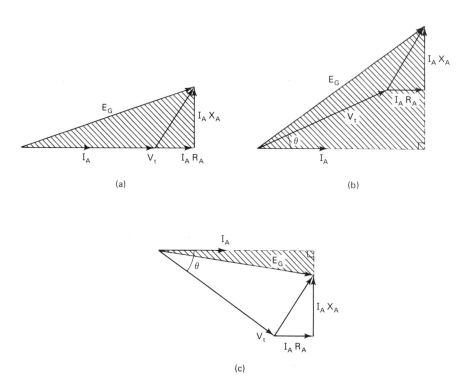

**Figure 7-9** Effect of power factor on generated voltage considering armature resistance and reactance only for (a) unity PF; (b) PF lagging; (c) PF leading.

by armature reaction as well. It is, but the nature of the resultant field is now dependent on the phase relation between armature current and voltage. In other words, the power factor is instrumental in establishing the resultant field pattern.

Let us refer to Fig. 7-10, which shows a portion of the field structure rotating clockwise. Assume that the coils on the armature are full pitch. When the adjacent pole centers are opposite the coil sides, as in Fig. 7-10a, maximum voltage is induced in the coil. The direction of the induced voltage is readily confirmed by the right-hand rule. If the current in the coil is in phase with the induced voltage (i.e., the power factor is unity), a flux is set up around the coil indicated by the arrows. The direction of this armature coil flux is seen to aid the flux of the north pole and oppose the flux of the south pole within the area of the coil span. The amounts added and subtracted just about balance except that the degree of saturation in the pole shoe tips is slightly different. This is the result of the nonlinear behavior of the magnetic iron. The net effect is a slight reduction of flux cutting the armature coil.

For a zero-power-factor lagging current, the current in the armature coil will reach its maximum when the rotor has moved through 90 electrical degrees. This means that the south pole will center on the coil axis, as shown in Fig. 7-10b. The

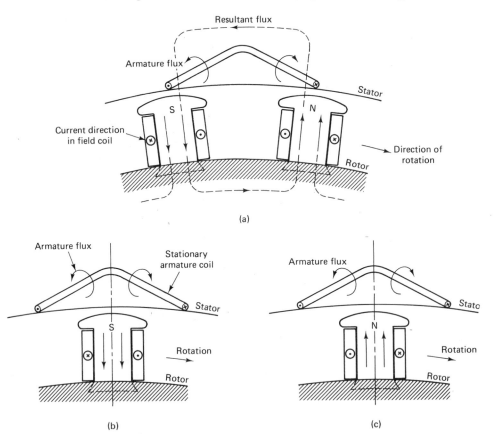

(a)

(b)                                              (c)

**Figure 7-10**  Armature reaction fluxes: (a) unity PF; (b) PF = 0 lagging; (c) PF = 0 leading.

Three-Phase Synchronous Generators   Chap. 7

armature coil flux is now directly opposed to the main pole flux and acts to demagnetize the poles.

For a zero-power-factor leading current, the armature coil current will have reached its maximum point 90 electrical degrees earlier. In this situation, the armature coil will center on the north-pole axis, as shown in Fig. 7-10c. Here the armature flux is completely aiding the main pole flux and acts to magnetize the poles. In most practical situations the armature current will be neither in phase nor leading or lagging the voltage by 90 electrical degrees, but will be somewhere in between.

The action of the armature field to produce a change in magnitude and distribution of the main pole fluxes is called *armature reaction*. It means the net air gap flux is modified, which translates into a lower or higher generated voltage $E_G$; similarly, for the generator terminal voltage $V_t$. We can now appreciate the particular generator characteristics of Fig. 7-8.

In total, there are three factors affecting the generated voltage:

1. The armature resistance voltage drop
2. The armature inductance voltage drop
3. The armature reaction effect

We will see that the armature reaction produces an effect similar to that which would be obtained by adding additional reactance to the armature circuit. It can therefore be treated as an additional reactive voltage drop.

At this point in our discussion, let us represent the various quantities in a phasor diagram, to obtain a qualitative idea of what is happening. Figure 7-11a applies to the zero-power-factor lagging load current (PF = 0, i.e., the load is purely inductive). Notice that the main pole flux $\phi_F$ is responsible for the no-load generated voltage $E_G$, the latter being taken as the reference phasor. The EMF $E_G$ lags $\phi_F$ by 90 electrical degrees, since an induced voltage lags behind the flux that produces it by 90 electrical degrees.

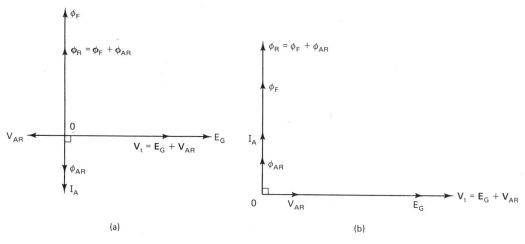

**Figure 7-11**  Phasor diagrams for zero-power-factor load current: (a) lagging; (b) leading.

With the generator supplying a zero-power-factor lagging load current $I_A$, the resulting armature flux $\phi_{AR}$ (in phase with the current $I_A$) develops a voltage $V_{AR}$ in the armature winding. Again $V_{AR}$ lags the flux $\phi_{AR}$ that induces it by 90 electrical degrees. It is now apparent what is happening. The armature reaction flux $\phi_{AR}$ reacts with the main field pole flux. Since they are oppositely directed (180° out of phase) we can obtain the resultant flux $\phi_R$ simply by subtracting $\phi_{AR}$ from $\phi_F$. In general, these two fluxes are never directly opposite and they should be added vectorially. This we indicate by

$$\boldsymbol{\phi}_R = \boldsymbol{\phi}_F + \boldsymbol{\phi}_{AR} \qquad \text{Wb} \qquad\qquad (7\text{-}14)$$

where $\qquad \phi_R$ = resultant air-gap field, Wb
$\qquad\qquad \phi_F$ = main field pole flux, Wb
$\qquad\qquad \phi_{AR}$ = armature reaction flux, Wb

Boldface type indicates vector quantities; that is, they have magnitude and direction.

Similar reasoning is now extended to the voltages generated. The armature reaction effect is responsible for the voltage $V_{AR}$. It must be added to $E_G$ vectorially to obtain the terminal voltage. Thus

$$\mathbf{V}_t = \mathbf{E}_G + \mathbf{V}_{AR} \qquad \text{V} \qquad\qquad (7\text{-}15)$$

where $\qquad V_t$ = generator terminal voltage, V
$\qquad\qquad E_G$ = generated voltage, V
$\qquad\qquad V_{AR}$ = armature reaction voltage, V

As we can see, the two voltages $E_G$ and $V_{AR}$ are also 180 electrical degrees apart, so that the vectorial addition amounts to subtracting $V_{AR}$ from $E_G$ to arrive at $V_t$. In most practical situations they too are not 180 electrical degrees apart.

It is now an easy matter to extend this discussion to the zero-power-factor leading load current (i.e., purely capacitive load, again PF = 0). This results in the phasor diagram of Fig. 7-11b. Applying Eq. (7-14) to the fields results as expected in a net air-gap flux that is larger than the main pole field, because the armature reaction field is aiding the main pole field. Similarly, the resulting terminal voltage $V_t$ is larger than the generated voltage $E_G$ by an amount equal to $V_{AR}$. Equation (7-15) readily confirms this.

The magnitudes of $\phi_{AR}$ and $V_{AR}$ naturally depend on the load current and their specific phase angle will be determined by the power factor of the load. This is what we deal with in the next section.

## 7-5 GENERAL PHASOR DIAGRAM

When a generator is loaded, the armature winding carries a current and all three voltage drops (resistive, reactive, and armature reaction) will be present. The task remaining is how to account for them in arriving at the generated voltage, assuming that the terminal voltage and character of the load are known. To start, we assume the armature current $I_A$ to lag the terminal voltage $V_t$ by the load angle $\theta$, and $V_t$ will be taken as the reference phasor, as portrayed in Fig. 7-12a. It also shows the air-gap voltage $E_A$ which is induced by the resultant air-gap flux $\phi_R$. Since an induced voltage lags behind the flux that creates it by 90 electrical degrees, it is apparent that $\phi_R$ leads $E_A$ by 90°. The vectorial difference between $E_A$ and $V_t$ is the impedance drop $I_A Z_A$ within the armature. This drop consists of the effective resistance voltage drop $I_A R_A$ in phase with the current, and the armature reaction voltage drop $I_A X_A$, which leads the current by 90°.

As it is, Fig. 7-12a is not complete. We have seen in Fig. 7-11a that the resultant flux is due to the interaction of the armature field with the main pole field of the rotor. The armature reaction field is in phase with the armature current since it is established by it.

With reference, then, to Fig. 7-12b, we can construct the main pole field phasor by application of Eq. (7-14). The armature reaction voltage $V_{AR}$, produced by the armature reaction field, lags that flux by 90 electrical degrees and is also shown.

Finally, to arrive at the general phasor diagram of Fig. 7-12c, we have to add to Fig. 7-12b the generated voltage $E_G$ which would exist if the armature does not carry current. In short, the open-circuit terminal voltage of the machine would be $E_G$, induced in the armature by the main pole field $\phi_F$. Since $E_G$ is induced by $\phi_F$ it must lag it by 90 electrical degrees.

Because $\boldsymbol{\phi}_R = \boldsymbol{\phi}_F + \boldsymbol{\phi}_{AR}$, it must follow that $\mathbf{E}_A = \mathbf{E}_G + \mathbf{V}_{AR}$. From the construction of Fig. 7-12c it is evident that $V_{AR} = I_A X_{AR}$ and $I_A X_A$ are in phase. This implies that the armature reaction produces an effect similar to that which would be obtained by adding additional reactance in the armature circuit. The summation of the voltage drops $V_{AR}$ and $I_A X_A$ gives the total reactive drop produced in the armature circuit by the armature flux. It results in

$$I_A(X_A + X_{AR}) = I_A X_S \qquad (7\text{-}16)$$

where $\quad X_S = X_A + X_{AR}$.

This voltage drop is normally referred to as the *synchronous reactance voltage drop* and $X_S$ is called the *synchronous reactance*. It is the value of $X_S$ that can be determined from tests, as discussed in the next section.

An interesting observation with regard to Fig. 7-12c should be pointed out. The angle $\delta$ indicated between $E_G$ and $V_t$ is called the *power* or *torque angle* of the machine. It varies with load and is a measure of the air-gap power developed in the machine. The angle $\beta$ is the angular slippage of the resultant field with respect to their no-load position. As seen, the resultant field $\phi_R$ in a generator slips back by the angle $\beta$ from the no-load position as the load current increases. The reader may note that the power angle is zero in Fig. 7-11a, because a purely inductive load does not consume power.

### Calculation of Voltage Regulation from the Equivalent Circuit

To simplify calculating procedures we will reduce the complexity of the phasor diagram in Fig. 7-12c. It is clear that when dealing with the synchronous reactance $X_S$ ($= X_A + X_{AR}$), the distinction between armature reaction and armature re-

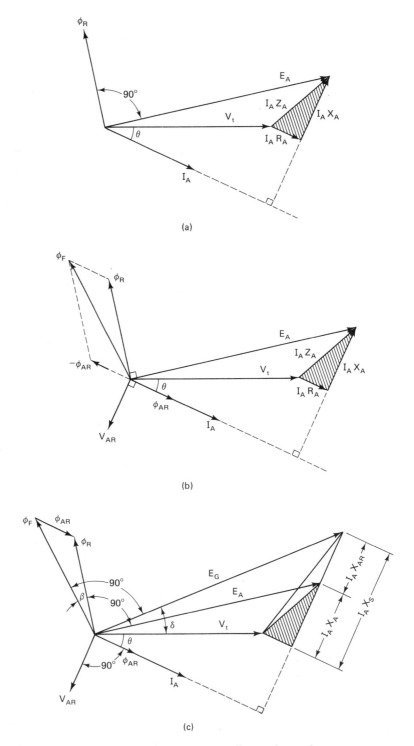

**Figure 7-12** Development of general phasor diagram for synchronous generator with lagging power factor load.

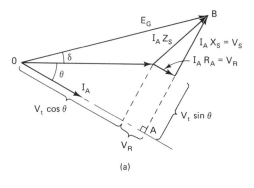

(a)

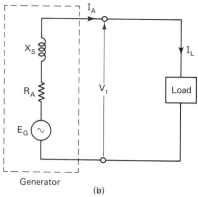

Generator

(b)

**Figure 7-13** Simplified equivalent ac circuit (per phase) for synchronous generator: (a) phasor diagram; (b) electric circuit.

actance voltage drops need not be made. If we also delete the field phasors, the diagram reduces to that shown in Fig. 7-13a. With reference to this figure, using $V_t$ as the reference phasor, $E_G$ can be solved for, since

$$\mathbf{E}_G = \mathbf{V}_t + \mathbf{I}_A \mathbf{Z}_S \qquad \text{V} \qquad (7\text{-}17)$$

where 
$E_G$ = generated EMF per phase, V
$V_t$ = terminal or load voltage per phase, V
$I_A$ = armature phase current, A
$Z_S$ = armature phase impedance = $R_A + jX_S$, Ω

Note that all parameters are phase quantities.

The reader may recall from circuit theory that the phasor diagram in Fig. 7-13a is applicable to the equivalent circuit represented in Fig. 7-13b. This suggests therefore that the synchronous generator on a per phase basis can be represented by an R-L circuit. The resistance represents the armature winding resistance $R_A$, and the reactance represents the synchronous reactance $X_S$, both phase quantities. The power factor angle θ in the circuit is dictated by the load connnected. The power consumed by the load is

$$P_L = V_t I_A \cos \theta \qquad \text{W/phase} \qquad (7\text{-}18)$$

The machine is a three-phase synchronous generator; thus

$$P_L = 3V_t I_A \cos \theta \qquad \text{W} \qquad (7\text{-}19)$$

Sec. 7-5    General Phasor Diagram

In general, we specify power in terms of line quantities, that is, line current $I_L$ and line-to-line voltage $V_{\text{L-L}}$. To convert Eq. (7-19) in those terms, we have to know how the generator winding is connected.

For Y-connected armature windings, which is normally done to bring the grounded neutral out, $I_A = I_L$ but $V_t = V_{\text{L-L}}/\sqrt{3}$. Making these substitutions in Eq. (7-19) gives

$$P_L = 3 \frac{V_{\text{L-L}}}{\sqrt{3}} I_L \cos \theta = \sqrt{3}\, V_{\text{L-L}} I_L \cos \theta \qquad (7\text{-}20)$$

which is the general expression for the three-phase power delivered to the balanced load in Fig. 7-13 and is independent of the armature winding connection.

To determine the voltage regulation we need the generated voltage $E_G$. This can be determined by using Eq. (7-17) or alternatively, solving for $E_G$ directly from the phasor diagram in Fig. 7-13a. To use the first method, write $I_A$ and $Z_S$ in component form as

$$\mathbf{I}_A = I_A (\cos \theta - j \sin \theta) \qquad (7\text{-}21)$$

and

$$Z_S = R_A + j X_S \qquad (7\text{-}22)$$

Thus, for inductive loads, which are characterized by a lagging power factor, we obtain by substituting Eqs. (7-21) and (7-22) into Eq. (7-17),

$$\mathbf{E}_G = V_t + I_A (\cos \theta - j \sin \theta)(R_A + j X_S) \qquad (7\text{-}23)$$

in which $V_t$ is taken as the reference phasor (i.e., $\mathbf{V}_t = V_t \angle 0°$.

Naturally, to solve Eq. (7-17) for $E_G$, the equivalent-circuit parameters for the generator as well as the load details must be known. Section 7-6 will tell us what tests are necessary to determine $R_A$ and $X_S$.

The second method uses the property that the triangle $OABO$ in Fig. 7-13a forms a right-sided triangle. We see that side $OA = V_t \cos \theta + V_R$ and $AB = V_t \sin \theta + V_S$. $E_G = OB$, the hypotenuse of this triangle. Thus

$$E_G = [(V_t \cos \theta + V_R)^2 + (V_t \sin \theta + V_S)^2]^{1/2} \qquad (7\text{-}24)$$

The reader familiar with complex algebra will realize that Eq. (7-24) follows from Eq. (7-23), by rearranging the former equation (i.e., select $I_A$ as the reference phasor) and determining the absolute value of $E_G$.

However, as is apparent, and this will apply to the remainder of the devices to be discussed in this book, a phasor diagram is an extremely useful tool for solving machine problems. It provides a visual aid of how various quantities relate to each other. The influence of specific parameters on others can easily be appreciated. Furthermore, unknown quantities are relatively easily obtained by using simple trigonometric relations to the diagram construction.

To complete our discussion, we will assume a capacitive load on the generator. This results in an armature current leading the terminal voltage. Figure 7-14 applies to this situation. Again the terminal voltage is taken as the reference phasor. In

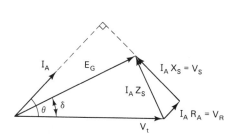

**Figure 7-14** Phasor diagram for a synchronous generator on a per phase basis supplying a capacitive load.

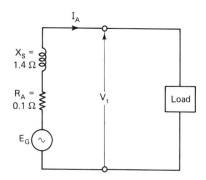

**Figure 7-15** Equivalent circuit diagram for Example 7-6.

*Capacitive conditions*

this case there results

*Vectorial*
$$\mathbf{E}_G = \mathbf{V}_t + I_A(\cos\theta + j\sin\theta)(R_A + jX_S) \qquad (7\text{-}25)$$

or

*Algebraic*
$$E_G = [(V_t\cos\theta + V_R)^2 + (V_t\sin\theta - V_S)^2]^{1/2} \qquad (7\text{-}26)$$

When the load is <u>purely resistive</u> the power factor is unity, $\cos\theta = 1.0$, and $\sin\theta = 0.0$, which reduces the expressions to

$$\mathbf{E}_G = \mathbf{V}_t + I_A(R_A + jX_S) \qquad (7\text{-}27)$$

or

$$E_G = [(V_t + V_R)^2 + (V_S)^2]^{1/2} \qquad (7\text{-}28)$$

In practical generators the value of $R_A$ is normally much smaller than $X_S$, particularly for the larger machines. This implies that the quantity $V_R$ is normally much smaller than $V_S$, and $V_R$ can often be neglected to simplify a problem. Typical synchronous generator data are given in Appendix E.


### EXAMPLE 7-6

A 250-kVA 660-V 60-Hz three-phase synchronous generator is Y-connected. The armature resistance is 0.10 $\Omega$/phase and the synchronous reactance is 1.40 $\Omega$/phase (see Fig. 7-15). Determine the voltage regulation for a load having a power factor 0.866 lagging.

### SOLUTION

As indicated in the text, all calculations will be made on a per phase basis because the voltage change is the same as that obtained between line terminals. At rated load,

$$I_A = \frac{\text{kVA} \times 1000}{\sqrt{3} \times V_{\text{L-L}}} = \frac{250,000}{\sqrt{3} \times 660} = 219\ \text{A}$$

which is the armature phase current.

$$V_t = \frac{V_{\text{L-L}}}{\sqrt{3}} = 381 \text{ V/phase}$$

Then

$$I_A R_A = 219 \times 0.10 = 21.9 \text{ V}$$

$$I_A X_S = 219 \times 1.40 = 306.6 \text{ V}$$

At 0.866 lagging power factor (i.e., $\cos \theta = 0.866$ and $\sin \theta = 0.5$) there results from Eq. (7-24),

$$E_G = [(381 \times 0.866 + 21.9)^2 + (381 \times 0.5 + 306.6)^2]^{1/2} = 609 \text{ V}$$

$$\text{voltage regulation} = \frac{609 - 381}{381} \times 100 = 59.8\%$$

## 7-6 SYNCHRONOUS IMPEDANCE AND REACTANCE

In this section we want to discuss how the equivalent-circuit parameters are obtained. From Example 7-6 it can be seen that for any given load the voltage regulation can be determined, provided that we have the machine model. Voltage regulation, of course, is of prime concern, since the line voltage must be kept constant within reasonable limits when the load on the generator changes.

To determine the voltage regulation on a large machine experimentally is not practically feasible. It would require exactly full-load conditions while maintaining the terminal voltage, frequency, and predetermined power factor of the load.

In actual practice the regulation is calculated from data obtained from a series of relatively simple tests in order to arrive at the equivalent-circuit parameters. The quantities to be determined are (1) the synchronous reactance $X_S$, and (2) the armature resistance $R_A$, both on a per phase basis.

Several methods are in existence to determine the synchronous reactance; only one will be discussed, the synchronous impedance method. To obtain the necessary data, three simple tests are required:

1. The open-circuit test
2. The short-circuit test
3. The armature resistance test

### Open-Circuit Test

This test is performed at synchronous speed with the armature circuit open-circuited as shown in Fig. 7-16a. Provisions must be made to adjust the field current $I_f$ so that when starting at zero, it may be raised until the line-to-line output voltage is somewhat above rated EMF. Recording $I_f$ and $V_{\text{oc}}$ (open-circuit terminal voltage) for a sufficient number of steps, the open-circuit saturation curve may be plotted

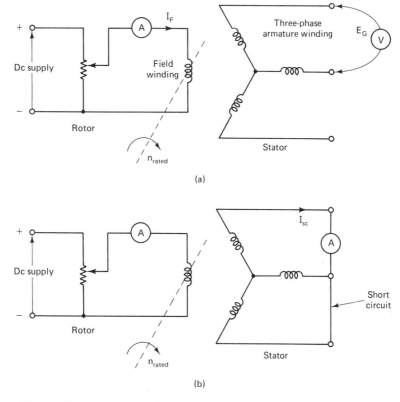

**Figure 7-16** Equivalent circuits for synchronous generator under test: (a) open-circuit test; (b) short-circuit rest (Y-connected armature winding is assumed).

(see Fig. 7-17). In plotting this curve, the open-circuit line-to-line voltage $V_{oc}$ is divided by $\sqrt{3}$ to represent the per phase value, which is then plotted as a function of $I_f$.

### Short-Circuit Test

In reference to Fig. 7-16b, the generator terminals are short circuited with an ammeter in one of the lines. The field current is reduced to zero and the generator operated at rated speed. The field current is then gradually increased to a safe maximum armature current obtainable, possibly twice rated value. Care must be taken not to allow this armature current to remain at high values for a long period.

During this test the values of $I_f$ and the short-circuit current $I_{sc}$ are recorded. Plotting $I_{sc}$ versus $I_f$ gives the short-circuit characteristic, which is also illustrated in Fig. 7-17. Referring now to Fig. 7-16b, we see that with the load terminals short circuited, the generated voltage is equal to the synchronous impedance drop. By maintaining the field current constant and removing the short circuit, the open-circuit voltage then produced is obviously $E_{G_{oc}}$. Thus with $I_f$ known, the open-circuit characteristic can be used to obtain $E_{G_{oc}}$, as illustrated in Fig. 7-17. Sim-

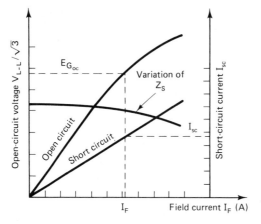

**Figure 7-17** Open-circuit and short-circuit characteristic curves for a synchronous generator, on a per phase basis.

ilarly, at this value of field current the corresponding short-circuit current that resulted can be found. The synchronous impedance is then the ratio of these two quantities:

$$Z_S = \left. \frac{E_{G_{oc}}}{I_{sc}} \right|_{I_f = \text{constant}} \qquad \Omega \qquad (7\text{-}29)$$

The value of synchronous impedance that is obtained can be seen to depend on the value of $I_f$ chosen. Therefore, we will select the value of $I_f$ that results in rated terminal voltage on the magnetization curve, which results in

$$Z_S = \frac{V_t}{I_{A\,sc}} \qquad (7\text{-}30)$$

where $I_{A\,sc}$ is the armature current from the short-circuit characteristic at the field current corresponding to rated terminal voltage of the machine on the open-circuit characteristic.

### Resistance Test

Having determined the synchronous impedance, the synchronous reactance $X_S$ can be obtained knowing the armature resistance. This, of course, can be determined from a simple resistance test. For this assume that the generator is Y-connected if it is three-phase. This assumption does not in any way affect the final result, since an identical voltage regulation is obtained if a Δ-connection is assumed. With the field circuit open, the dc resistance is measured between any two of the output terminals. Since during this test two phases are in series, the armature per phase is half the measured value. In practice, this resistance value is multiplied by a factor to arrive at the effective ac resistance $R_{A\,\text{eff}}$. This factor depends on slot size and shape, size of the armature conductors, and the particular winding construction. On practical machines it ranges between 1.2 and 1.5, depending on the size of the machine. A typical value to use in the calculations would be 1.25. Then

$$X_S = \sqrt{Z_S^2 - R_{A\,\text{eff}}^2} \qquad (7\text{-}31)$$

**EXAMPLE 7-7**

A 500-kVA 2300-V three-phase synchronous generator (see Fig. 7-18) was tested according to the test procedures outlined to determine its regulation at a full-load power factor of 0.866 lagging. The data obtained are as follows:

$$\text{dc resistance test: } V_{L\text{-}L} = 8 \text{ V}, \quad I_L = 10 \text{ A}$$
$$\text{Open-circuit test: } I_f = 25 \text{ A}, \quad V_{L\text{-}L} = 1408 \text{ V}$$
$$\text{Short-circuit test: } I_f = 25 \text{ A}, \quad I_{sc} = 126 \text{ A}$$
$$= \text{ rated full-load current}$$

Assume that the generator is Y-connected.

**SOLUTION**

$$V_{oc} = \frac{V_{L\text{-}L}}{\sqrt{3}} = 812.9 \text{ V/phase}$$

$$Z_s = \frac{V_{oc}}{I_{A\,sc}} = \frac{812.9}{126} = 6.45 \ \Omega/\text{phase}$$

$$R_{A\,\text{eff}} = 1.25 \times \frac{8 \text{ V}}{2 \times 10 \text{ A}} = 0.50 \ \Omega/\text{phase}$$

where the factor 1.25 represents the correction necessary to the dc resistance to arrive at the effective ac resistance. Since during the resistance measurement two phases are in series, the measured value must be divided by 2.

$$X_S = \sqrt{Z_S^2 - R_A^2} = \sqrt{6.45^2 - 0.5^2} = 6.43 \ \Omega$$

$$V_R = I_A R_A = 126 \times 0.5 = 63 \text{ V}$$

$$V_S = I_A X_S = 126 \times 6.43 = 810.2 \text{ V}$$

$$V_t = \frac{V_{L\text{-}L}}{\sqrt{3}} = 1328 \text{ V/phase at full load}$$

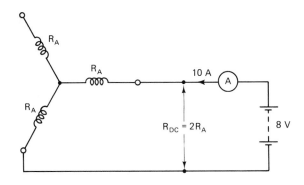

**Figure 7-18** Dc resistance measurement.

The generated voltage per phase from Eq. (7-24) is then

$$E_G = [(V_t \cos \theta + V_R)^2 + (V_t \sin \theta + V_S)^2]^{1/2}$$

$$= \{[(1328 \times 0.866) + 63]^2 + [(1328 \times 0.5) + 810.2]^2\}^{1/2}$$

$$= \sqrt{1214.8^2 + 1474.2^2} = 1910 \text{ V}$$

The voltage regulation is then

$$\frac{1910 - 1328}{1328} \times 100 = 43.8\%$$

## 7-7 LOSSES AND EFFICIENCY

To calculate the efficiency of a synchronous generator, a procedure is followed similar to that of determining the efficiency of dc generators. It will be recalled that it is necessary to establish the total losses when operating under load. For generators this includes:

1. Rotational losses such as friction and windage
2. Eddy current and hysteresis losses in the magnetic circuit
3. Copper losses in the armature winding and in the field coils
4. Load loss due to armature leakage flux causing eddy current and hysteresis losses in the armature-surrounding iron

With regard to the losses, the following comments may be made:

1. The rotational losses, which include friction and windage losses, are constant, since the speed of a synchronous generator is constant. It may be determined from a no-load test.
2. The core loss includes eddy current and hysteresis losses as a result of normal flux density changes. It can be determined by measuring the power input to an auxiliary motor used to drive the generator at no load, with and without the field excited. The difference in power measured constitutes this loss.
3. The armature and field copper losses are obtained as $I_A^2 R_A$ and $V_f I_f$. Since per phase quantities are dealt with, the armature copper loss for the generator must be multiplied by the number of phases. The field winding loss is as a result of the excitation current flowing through the resistance of the field winding.
4. Load loss or stray losses result from eddy currents in the armature conductors and increased core losses due to distorted magnetic fields. Although it is possible to separate this loss by tests, in calculating the efficiency, it may be accounted for by taking the effective armature resistance rather than the dc resistance.

After all the foregoing losses have been determined, the efficiency $\eta$ is calculated as,

$$\eta = \frac{kVA \times PF}{kVA \times PF + (total\ losses)} \times 100\% \tag{7-32}$$

where $\eta$ = efficiency, %
kVA = load on the generator (output)
PF = power factor of the load

The quantity (kVA $\times$ PF) is, of course, the real power delivered to the load (in kW) by the synchronous generator. Thus Eq. (7-32) could in general be stated as

$$\eta = \frac{P_{out}}{P_{in}} \times 100 = \frac{P_{out}}{P_{out} + P_{losses}} \times 100 \tag{7-33}$$

The input power $P_{in} = P_{out} + P_{losses}$ is the power required from the prime mover to drive the loaded generator. To calculate the load power $P_{out}$, the reader may also use the expressions given by Eqs. (7-19) and (7-20).

It should be noted that in determining the efficiency of a practical generator, the power supplied to the field circuit is not always included. As we have indicated throughout this chapter, in large machines the power supplied to the field by the excitation system is large. However, when we compare this power input with that supplied by the prime mover, we quickly realize that in the overall scheme the field power is rather small. A matter of fact, it only represents a fraction of 1%.

To give some practical figures on this, a 1000-MVA synchronous generator (which is a very large machine, even by today's standards) would require a 6- to 7-MW field circuit supply. This amounts to about 0.6 to 0.7% of the prime-mover input power. This is no small task.

Realizing that the generated voltage is of the order of 24 kV, the field supply at the 500-V level, the currents involved are enormous. This calls for forced cooling in all parts of the machine to keep to acceptable levels the heat created by the various losses. On the positive side, with forced cooling the allowable current density (current per unit area) in the copper conductors can be at least doubled. This keeps conductor sizes "reasonable." When forced cooling is implemented, the overall result is a machine of greatly reduced physical size. It is for this reason, then, that the field circuit loss is not always considered when calculating efficiencies. Its inclusion would barely alter the results.

**EXAMPLE 7-8**

A 2000-kVA 2300-V three-phase synchronous generator has a dc armature winding resistance of 0.068 $\Omega$ between terminals. The field takes 35 A from a 220-V dc supply. Friction and windage loss is 22.8 kW, and core loss including stray-load losses is 41.2 kW. Calculate the efficiency of the generator at full load and a power factor of 0.80 lagging. Assumne a Y-connected generator. The effective armature resistance may be taken as 1.25 times its dc value.

## SOLUTION

The armature current is

$$I_A = \frac{kVA \times 1000}{\sqrt{3} \times V_{L-L}} = \frac{2,000,000}{\sqrt{3} \times 2300} = 503 \text{ A}$$

and

$$R_A \text{ (per phase)} = \frac{0.068}{2} \times 1.25 = 0.0425 \text{ }\Omega$$

Losses:

| | | |
|---|---|---|
| Friction and windage | = | 22.8 kW |
| Core loss and stray loss | = | 41.2 kW |
| Armature winding: $3 \times (503)^2 \times 0.0425$ = | | 32.3 kW |
| Field winding: $(220 \times 35)/1000$ | = | 7.7 kW |
| Total | = | 104.0 kW |

The efficiency is

$$\eta = \frac{P_{out}}{P_{out} + P_{losses}} \times 100$$

$$= \frac{2000 \times 0.80}{2000 \times 0.8 + 104} \times 100 = 93.9\%$$

It can be shown that the maximum efficiency occurs at a load where the constant losses equal the variable losses. Those that are considered constant include friction and windage, core loss, and field copper loss. The armature copper losses are variable since they depend on the armature current.

For most electrical machines—and the synchronous generator is no exception—the efficiency improves rapidly from a low value to a maximum with increasing load. The point at which the efficiency reaches a maximum can be controlled at the design stage. Generally, it is not selected at 100% loading conditions. Machines seldom run at full load continuously—this applies to transformers and motors alike.

## REVIEW QUESTIONS

**7-1.** What is a distributed winding?

**7-2.** What are the advantages of a double-layer winding over a single-layer winding?

**7-3.** What are mechanical degrees? Electrical degrees? How do they relate? Give your answer in reference to a six-pole generator.

**7-4.** How can a three-phase armature winding of a synchronous generator be connected? What advantages does either connection offer?

**7-5.** What is a distribution factor? What does it affect?

**7-6.** What methods can be used to ensure that the generated voltage waveform in a synchronous generator is sinusoidal?

**7-7.** What is a fractional coil pitch? How does it influence the magnitude of the generated waveform? The waveform?

**7-8.** State the three factors responsible for the change in generator terminal voltage upon loading.

**7-9.** Under what conditions can the voltage regulation of a synchronous generator become negative?

**7-10.** What is the synchronous reactance voltage drop? What factors contribute to it?

**7-11.** Synchronous generators generally have a large voltage regulation. What causes this, and how is this compensated for in practical installations?

**7-12.** What is the power angle or torque angle in synchronous generators? Upon what factors does it depend?

**7-13.** What tests are necessary to determine the equivalent circuit model for a synchronous generator?

**7-14.** The armature circuit resistance is generally determined by applying a known dc voltage and measuring the resultant dc current. Why is an ac source not used?

**7-15.** The synchronous impedance of a synchronous generator is not a constant value over the entire operating range. Why is this so? What value would you use?

**7-16.** In determining the efficiency of a synchronous generator the field circuit input is often neglected. Explain how we justify this.

## PROLEMS

**7-1.** An eight-pole three-phase armature has a total of 504 turns. The flux per pole is 0.0218 Wb and is sinusoidal. If the frequency is 60 Hz, find the generated voltage per phase.

**7-2.** A 60-Hz three-phase hydroelectric generator has a rated speed of 120 r/min. There are 558 stator slots with two conductors per slot. The air-gap dimensions are $D = 6.1$ m and $L = 1.14$ m. The maximum flux density $B_m = 1.2$ T. Calculate the generated voltage per phase.

**7-3.** The maximum flux per pole of an eight-pole generator is 0.016 Wb, and the rated frequency is 60 Hz. There are a total of 1728 conductors in the slots. What is the open-circuit line voltage if the armature is Δ-connected and each phase winding consists of two parallel paths?

**7-4.** A six-pole synchronous generator has a total of 108 slots. Determine:
(a) The number of slots per pole.
(b) The electrical and mechanical angle between adjacent slots.
(c) The distribution factor.

**7-5.** Determine the distribution factor of a three-phase eight-pole synchronous generator winding having 120 slots on the stator.

**7-6.** If the armature winding in Problem 7-5 has a coil span of 12 slots, calculate the pitch factor.

**7-7.** The armature winding of the synchronous generator in Problems 7-5 and 7-6 has four turns per coil. Assume a double layer Y-connected armature winding. If the speed is 900 r/min and the pole flux is 0.04 Wb, calculate:
(a) The voltage per phase.
(b) The terminal voltage.

**7-8.** If the generator in Problem 7-3 has a rated current of 1250 A, what is the armature conductor current?

**7-9.** A 250-kVA 660-V 60-Hz three-phase synchronous generator is Y-connected. The effective armature resistance is 0.2 $\Omega$/phase and the synchronous reactance is 1.4 $\Omega$/phase. At full load and unity power factor, calculate the voltage regulation.

**7-10.** Calculate the voltage regulation of the generator in Problem 7-9 at full load and:
(a) The 0.866 lagging power factor.
(b) The 0.70 leading power factor.

**7-11.** A 1000-kVA 4600-V three-phase 60-Hz Y-connected ac generator has a no-load voltage of 8350 V. The generator is operated at rated volt-amperes and 0.75 power factor lagging and rated voltage. Calculate:
(a) The synchronous reactance (neglect armature resistance).
(b) The voltage regulation.
(c) The torque angle.
(d) The air-gap power.
(e) The new voltage and kVA rating if the armature winding is reconnected in delta.

**7-12.** A 25-kVA 220-V three-phase generator delivers rated kilovolt-amperes at a power factor of 0.8 lagging. The effective ac resistance between Y-connected armature-winding terminals is 0.20 $\Omega$. The synchronous reactance is 0.6 $\Omega$/phase. The field current is 9.3 A at 115 V dc. The friction and windage loss is 460 W and the magnetic (iron) core loss is 610 W. Calculate:
(a) The voltage regulation.
(b) The efficiency at full load.

**7-13.** One of the three-phase waterwheel generators at Grand Coulee Dam, Washington, is rated at 108,000 kVA, PF unity, 13.8 kV, Y-connected, 60 Hz, 120 r/min. Determine:
(a) The number of poles.
(b) The kilowatt rating.
(c) The current rating.
(d) The input at rated kW load if the efficiency is 97% (excluding field loss).
(e) The prime-mover torque applied to the shaft.

**7-14.** A three-phase star-connected generator is rated at 1600 kVA, 13,000 V. The armature effective resistance and synchronous reactance are 1.5 and 30 $\Omega$/phase, respectively. Calculate the voltage regulation and torque angle for a load of 1,280 kW at power factors of:
(a) 0.8 lagging.
(b) Unity.
(c) 0.8 leading.

**7-15.** The no-load characteristic of a three-phase 60-Hz 1600-kVA 11,000-V synchronous generator is as follows:

| Volts (line-to-line kV) | 6.45 | 9.0 | 11.0 | 12.2 | 13.4 | 14.0 | 14.5 |
|---|---|---|---|---|---|---|---|
| Field current (A) | 100 | 150 | 205 | 250 | 300 | 350 | 400 |

Electric Power Generation    Chap. 7

The field current is 186 A when the generator delivers rated current to a short circuit. Calculate the voltage regulation at 0.8 PF lagging. Consider the armature winding resistance to be negligible.

**7-16.** A 100-kVA 1100-V three-phase generator is tested to determine its regulation under various conditions of load and power factors. The data obtained are as follows:

Short-circuit test: Field current = 12.5 A dc

Line current = rated value

Open-circuit test: Field current = 12.5 A dc, the same value

as during the short-circuit test

Line voltage = 420 V

The effective resistance of the armature = 0.90 Ω between terminals. The generator is Y-connected. Determine:

**(a)** The synchronous impedance and reactance per phase.

**(b)** The voltage regulation of the generator at 0.8 lagging and 0.8 leading power factors.

**7-17.** Repeat Problem 7-16 for a Δ-connected generator.

**7-18.** If a field current of 10 A in a synchronous generator results in a current of 150 A on short-circuit and a terminal voltage of 720 V on open-circuit, determine the internal voltage drop when the generator delivers a load current of 60 A. Armature resistance may be neglected.

# 8

# SINGLE-PHASE
# TRANSFORMERS

## 8-1 INTRODUCTION

In Chapter 6 the need for high-voltage transmission was discussed if electrical power is to be provided at considerable distances from a generating station. Furthermore, at some point this high voltage must be reduced because ultimately it must supply a load. The transformer then makes it possible for various parts of a power system to operate at different voltage levels. In this chapter we discuss transformer principles and some of its constructional details.

## 8-2 TWO-WINDING TRANSFORMERS

A transformer in its simplest form consists of two stationary coils coupled by a mutual magnetic flux (see Fig. 8-1). The coils are said to be mutually coupled; the flux that links with one coil also links entirely or is a major part of the other coil.

Transformers may have cores of air, ferrite, or ferromagnetic material, depending on the frequency in the particular application. Powdered steel as an insulating medium is suitable for cores in the lower radio-frequency ranges. Air-cored transformers must be used at the higher frequencies.

In power applications, laminated steel-core transformers are used, which this text is restricted to. Transformers are efficient and reliable devices and are used principally to change voltage levels for economic reasons. Transformers are very efficient devices; because rotational losses normally associated with rotating ma-

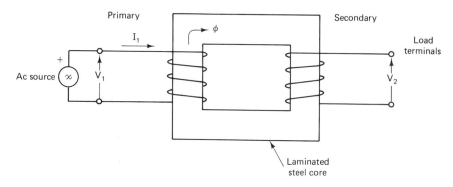

**Figure 8-1** Simple transformer.

chines are absent, relatively little power is lost when transforming power from one voltage level to another. Typical efficiencies are in the range 92 to 99%, the higher values applying to the large power transformers.

As shown in Fig. 8-1, the current flowing in the coil connected to the ac source is called the *primary winding*, or simply *primary*. It sets up a flux φ in the core which varies periodically both in magnitude and direction. The flux links the second coil, called the *secondary winding*, or simply the *secondary*. The flux is changing; therefore, it induces a voltage in the secondary by electromagnetic induction in accordance with Lenz's law. Thus we see that the primary receives its power from the source while the secondary supplies this power to the load. This action is known as *transformer action*. Before we examine this concept in more detail, let us look briefly at some transformer constructional aspects.

## 8-3 CONSTRUCTION OF TRANSFORMERS

There are basically two types of iron-core construction. They differ from each other in the way the core is constructed to accommodate the windings. We have already encountered one type, the *core type* of construction, as shown in Fig. 8-1 and repeated in Fig. 8-2a. As seen, it consists of two separate coils, one on each of two opposite legs of a rectangular core. Normally, this is not a desirable design. Its disadvantage is the large leakage fluxes associated with it, which as we will see later on results in poor voltage regulation. Therefore, to make sure that most of the flux set up by the primary will link the secondary, a construction of Fig. 8-2b is employed. This is called the *shell type* of transformer. Here we see that the two windings are wound concentrically. The higher-voltage winding is wound on top of the lower-voltage winding. The low-voltage winding is then located closer to the steel than the high-voltage winding, which is preferable from an electrical-insulating point of view. From an electrical point of view there is not much difference between the two transformer construction types. In general, the core type has a larger flux path and shorter mean length of coil turn. It has a smaller iron cross section than an identically rated shell type of transformer, so it needs more turns on the windings. The flux that may be reached in the core is not as

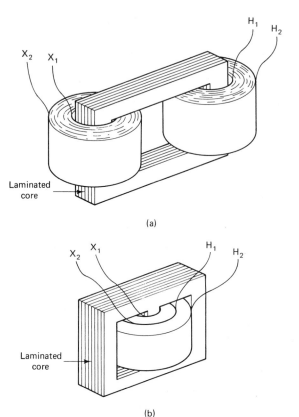

(a)

(b)

**Figure 8-2** Winding and core arrangements of transformer construction: (a) core type; (b) shell type.

high as in the shell type of transformer. Shell-type construction offers the advantage of providing better mechanical support and bracing of coils.

Cores may be built of laminations cut from alloyed sheet steel. Most laminating materials have an approximate alloy content of 3% silicon and 97% iron. The silicon content reduces the magnetizing losses, in particular, that part due to hysteresis loss. It makes the material brittle, which causes manufacturing problems, specifically, in stamping operations (punching out of laminations). Therefore, there is a practical limit on the silicon content. Most laminated material is cold rolled and often specially annealed to "orient" the "grain" or iron crystals. This provides a very high permeability and low hysteresis to flux in the direction of rolling. Transformer laminations are usually from 0.010 to 0.025 in. (up to 0.035 in.) thick for 60-Hz operation. They are coated in one of several ways on one side, for instance by a thin layer of varnish or paper, to insulate the laminations from each other.

Coils are prewound and the core design must be such that it permits placing it in the coil. Of course, a core must then be made in at least two sections, so that the coil can be placed on it.

The laminations for the core-type transformer of Fig. 8-2a may be made up of U- and I-shaped laminations as shown in Fig. 8-3a. Usually, the layers of

Single-Phase Transformers    Chap. 8

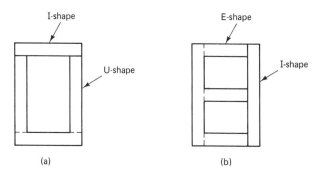

Figure 8-3 Core construction built of separate laminations.

laminations are placed so that the gaps between lamination ends of one layer are overlapped by the laminations in the next layer.

The core for the shell-type transformer of Fig. 8-2b is normally made up of E- and I-shaped laminations, illustrated in Fig. 8-3b. The position of the stampings is interchanged in alternate layers. The air gaps are then formed at three places in each layer where the E butts into the I, and are therefore bridged by the solid iron in the adjacent layers.

Many other popular types of core constructions are in existence, depending on manufacturer preferences. Any of the schemes involves a cost of material handling which is an important fraction of the total cost of a transformer. The large number of schemes used in practice indicates the thought given by engineers to solve this difficult problem. Obviously, we must restrict ourselves here to the principles involved.

As a rule, the number of butt joints are to be limited. The joints are tightly made and laminations interleaved so as to minimize the reluctance of the magnetic circuit. The whole core is built up or stacked to the proper dimension and the laminations are bolted firmly together.

As we can appreciate, core legs of square or rectangular cross sections are the result if we stack the laminations to the required core cross section. This permits coils to be fitted on them with either square, rectangular, or circular coil spools or forms.

As the transformer size increases, we soon reach a point where this construction is unsatisfactory. Therefore, in the larger transformers a stepped core arrangement is used to minimize the use of copper and reduce copper loss. Figure 8-4 illustrates what we mean by a stepped core. What is achieved in this construc-

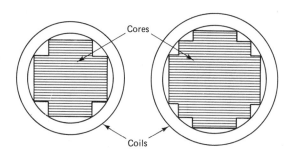

Figure 8-4 Stepped transformer cores.

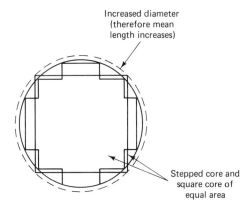

Increased diameter
(therefore mean
length increases)

Stepped core and
square core of
equal area

**Figure 8-5** Longer copper length per turn is required if equal-area square core is used.

tion is that each length of copper conductor embraces the maximum cross-sectional area of steel.

This statement is readily proven. Figure 8-5 shows the stepped core and square core of equal cross-sectional area. Note that the enclosing circle for the square core is slightly larger. This naturally implies more copper is needed. As a matter of fact, in this figure the mean length of the dashed circle is almost 10% longer than that of the drawn one. Using the stepped core in large transformers does save a considerable amount of copper, hence reduced costs. The amount of saving is partially offset by the more complex arrangement of the stepped core. But this is not the major reason to use this core construction in large transformers. Another consideration ties in with the electrical forces between current-carrying conductors, particularly under short-circuit conditions. Transformers should be able to withstand momentary short-circuit conditions without being damaged. In such instances, the mechanical stress forces resulting are proportional to the square of the current. The net effect is that the coils tend to bow outward. Rectangular coils are therefore suited only for small transformers; the practical limit seems to be about 200 kVA per coil. Thus when mechanical forces on a coil become too large for rectangular coils, round forms must be used.

In practical transformers the primary and secondary windings have two or more coils per leg. They may be arranged in series or parallel, thereby providing several possible voltages. As we already mentioned, the low-voltage winding is placed next to the core. This is because of insulating considerations. The high-voltage winding is generally on the outside. Sometimes both windings are interleaved to reduce leakage fluxes. Other problems in the construction of transformers making it necessary to clamp the laminations and impregnate the coils are the forces. These exist between laminations due to the cyclic magnetic forces and forces that exist between parallel conductors carrying current—thus the need to bolt laminations together securely. Insufficient clamping usually results in *hum*, giving rise to objectionable audible noise.

Transformers are usually air-cooled even if placed in metal cases. Larger sizes are placed in tanks with special transformer oil. The oil has a dual function; it insulates while providing cooling. Still larger sizes have tanks with corrugated sides or cooling fins or radiators to dissipate the heat to the surrounding air. Figure

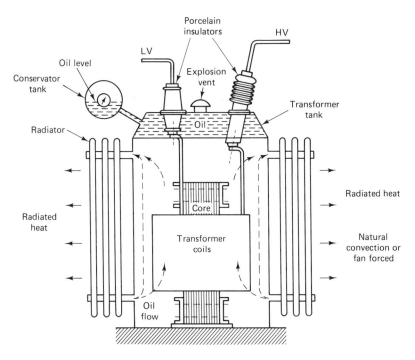

**Figure 8-6** Large oil-cooled three-phase power transformer (sectional view).

8-6 shows a typical self-cooled transformer. The oil moves around by natural convection, since warmer oil flows up. It flows down again through the radiator, which gives up this heat to the surrounding air. In larger oil-cooled units, the oil must be pumped around to maintain acceptable temperature levels. No matter what size of transformer is dealt with, they all operate on the same principle, which we deal with next.

## 8-4 TRANSFORMER PRINCIPLES

When a sinusoidal voltage $V_p$ is applied to the primary winding with the secondary winding open circuited, as shown in Fig. 8-7a, there will be no energy transfer. The impressed voltage causes a small current $I_0$ to flow in the primary winding. This no-load current has two functions:

1. It produces the magnetic flux in the core, which varies sinusoidally between zero and $\pm\phi_m$ ($V_p$ is assumed sinusoidal), where $\phi_m$ is the maximum value of the core flux.
2. It also provides a component to account for the hysteresis and eddy current power losses in the core. These combined losses are normally referred to as *core losses*.

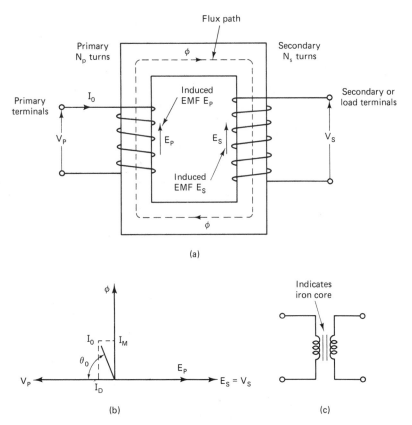

**Figure 8-7**  Transformer with open-circuited secondary: (a) schematic; (b) phasor diagram; (c) transformer symbol.

The no-load current $I_0$ is usually a few percent of the rated full-load current of the transformer (about 2 to 5%). Since at no load the primary winding acts as a large reactance due to the iron core, the no-load current will lag the primary voltage by nearly 90°. Figure 8-7b illustrates this relation, where $\theta_0$ is the no-load power factor angle. It is readily seen that the current component $I_m = I_0 \sin \theta_0$, called the *magnetizing current*, is 90° in phase behind the primary voltage $V_p$. It is this component that sets up the flux $\phi$ in the core, and $\phi$ is therefore in phase with $I_m$.

The second component, $I_c = I_0 \cos \theta_0$, is in phase with the primary voltage. It is the current component that supplies the core losses. The phasor sum of these two components represents the no-load current, or

$$\mathbf{I_0} = \mathbf{I_m} + \mathbf{I_c} \quad \text{A} \quad (8\text{-}1)$$

It should be noted that the no-load current is distorted and nonsinusoidal, as shown in Fig. 8-8. This is the result of the nonlinear behavior of the core material. Since the excitation current varies from about 1% to no more than 5% of full-load current, which is sinusoidal, the effect of magnetic saturation on our transformer analysis is neglected. When dealing with three-phase systems, the effects of the predominant (third) harmonic may have to be considered. This is examined in Chapter 9.

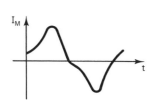

**Figure 8-8** No-load current
waveform of a transformer.

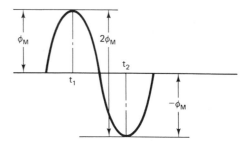

**Figure 8-9** Sinusoidal flux wave.

If we assume for now that there are no other losses in the transformer, we can further show the induced voltages in the primary winding, $E_p$, and in the secondary winding, $E_s$. Since the magnetic flux set up by the primary windings links the secondary winding, there will be an induced EMF, $E_s$, in the secondary in accordance with Faraday's law, namely $E = N\Delta\phi/\Delta t$. This same flux also links the primary itself, inducing in it an induced EMF, $E_p$.

As discussed earlier, they must lag the flux that induces them by 90°. Therefore, they are 180° out of phase with the applied primary voltage. Since no current flows in the secondary winding, $E_s = V_s$. Figure 8-7c shows the symbol for the transformer. The two lines indicated between the windings identify the presence of the iron core. The reader may note that the representation of the phasors $E_p$ and $E_s$ in Fig. 8-7b are not of the same magnitude. This indicates a turns ratio unequal to unity, as we will see shortly.

The no-load primary current $I_0$ is small, a few percent of full-load current. Thus the voltage drop in the primary winding is negligible and we can say that $V_p$ is nearly equal to $E_p$.

The primary voltage and the resulting flux are sinusoidal; thus the induced quantities $E_p$ and $E_s$ vary as a sine function. The average value of the induced voltage is given by

$$E_{avg} = \text{turns} \times \frac{\text{change in flux in a given time}}{\text{the given time}} \qquad (8\text{-}2)$$

which is Faraday's law applied to a finite time interval.

Referring to Fig. 8-9, it can be seen that the flux change in the time interval $t_1 - t_2$ is $2\phi_m$, where $\phi_m$ is the maximum value of the flux wave, in webers. The interval $t_2 - t_1$ represents the time in which this flux change occurs and equals one-half cycle or $1/(2f)$ seconds, where $f$ is the supply frequency, in hertz. It follows that

$$E_{avg} = N \frac{2\phi m}{1/(2f)} = 4fN\phi_m \qquad (8\text{-}3)$$

where $N$ is the number of turns on the winding.

From ac circuit theory, the effective or root-mean-square (rms) voltage $E$, for a sine wave, is 1.11 times the average voltage $E_{avg}$, thus

$$E = 4.44fN\phi_m \qquad (8\text{-}4)$$

Since the same flux links with the primary and secondary windings, the voltage per turn in each winding is the same, hence

$$E_p = 4.44fN_p\phi_m \tag{8-5}$$

$$E_s = 4.44fN_s\phi_m \tag{8-6}$$

where $N_p$ and $N_s$ are the number of turns on the primary and secondary, respectively.

The ratio of primary induced voltage to the secondary induced voltage is called the *transformation ratio*. Denoting this ratio by $a$, it is seen that

$$a = \frac{E_p}{E_s} = \frac{N_p}{N_s} \tag{8-7}$$

which is obtained by dividing Eq. (8-5) and (8-6).

### EXAMPLE 8-1

The 6600-V primary winding of a 60-Hz transformer has 1320 turns. Calculate:

(a) The maximum flux $\phi_m$.
(b) The number of turns on the 400-V secondary winding.

### SOLUTION

From Eq. (8-4),

(a) $\phi_m = \dfrac{6600}{4.44 \times 60 \times 1320} = 0.0188$ Wb

(b) $N_s = \dfrac{400}{4.44 \times 60 \times 0.0188} = 80$ turns

Assume that the output power of a transformer equals its input power—not a bad assumption in practice considering the high efficiencies. What we really are saying is that we are dealing with an ideal transformer (i.e., it has no losses). Thus

$$P_{in} = P_{out}$$

or

$$V_p I_p \times \text{primary PF} = V_s I_s \times \text{secondary PF}$$

where PF stands for power factor.

For the assumption stated above it means that the power factor on primary and secondary sides are equal; therefore,

$$V_p I_p = V_s I_s$$

from which we obtain

$$\frac{V_p}{V_s} = \frac{I_s}{I_p} \simeq \frac{E_p}{E_s} = a \tag{8-8}$$

Equation (8-8) shows us that as an approximation, the terminal voltage ratio equals the turns ratio. The primary and secondary currents, on the other hand, are inversely related to the turns ratio.

The turns ratio gives a measure of how much the secondary voltage is raised or lowered in relation to the primary voltage. To calculate the voltage regulation we need more information, as we will see.

**EXAMPLE 8-2**

A 100-kVA 2400/240-V transformer has 60 turns on the secondary winding. Calculate:

(a) The approximate value of primary and secondary currents.
(b) The number of primary turns.

**SOLUTION**

(a) Full-load primary current $= \dfrac{\text{kVA} \times 1000}{V_p} = \dfrac{100,000}{2400} = 41.7\text{ A}$

secondary current $= \dfrac{100,000}{240} = 417\text{A}$

(b) $a = \dfrac{2400}{240} = 10$

Therefore,

$$N_p = aN_s = 10 \times 60 = 600 \text{ turns}$$

Although the true ratio of transformation, Eq. (8-7), is constant, the ratio of the terminal voltages, $V_p/V_s$, varies somewhat depending on the load and its power factor. In practice, however, this ratio is obtained from the transformer nameplate data, which list the primary and secondary voltages under full-load conditions. For most practical purposes and as a good approximation, it is this transformation ratio which is dealt with in practical calculations. We shall adopt this method here.

When secondary voltage $V_s$ is reduced compared to the primary voltage, the transformation is said to be a *step-down transformer*; conversely, if this voltage is raised, it is called a *step-up transformer*. In a step-down transformer, the ratio of transformation $a$ is greater than unity ($a > 1.0$); for a step-up transformer it is smaller than unity ($a < 1.0$). In the event when $a = 1$, the transformer secondary voltage equals the primary voltage. This is a special kind of transformer, used in instances where electrical isolation is required between the primary and secondary circuits while maintaining the same voltage level. Therefore, this transformer is generally known as an *isolation transformer*.

As will now be apparent, it is the magnetic flux in the core that forms the connecting link between primary and secondary circuits. In the next section we see how the primary winding current adjusts itself to the secondary load current when the transformer supplies a load.

Looking into the transformer terminals from the source, an impedance is seen which by definition equals $V_p/I_p$. From Eq. (8-8) we have $V_p = aV_s$ and $I_p = I_s/a$. In terms of $V_s$ and $I_s$ the ratio of $V_p$ to $I_p$ is

$$\frac{V_p}{I_p} = \frac{aV_s}{I_s/a} = \frac{a^2 V_s}{I_s}$$

But $V_s/I_s$ is the load impedance $Z_L$; thus we can say that

$$Z_{in}(primary) = a^2 Z_L \qquad (8\text{-}9)$$

This equation tells us that when an impedance is connected to the secondary side, it appears from the source as an impedance having a magnitude which is $a^2$ times its actual value. We say that the load impedance is "reflected" or "referred to the primary." It is this property of transformers that is used in impedance-matching applications. We have more to say about transferring impedance values in Section 8-8.

## 8-5 TRANSFORMER UNDER LOAD

With the secondary of a transformer open-circuited, the primary winding is simply a high impedance taking a very low current. There is a component of the no-load current in phase with the primary impressed voltage to take care of the losses due to hysteresis and eddy currents, which must be supplied by the input circuit. This in-phase component is called the *core-loss component*. The other component of the current is in quadrature and is responsible for setting up the flux, and is called the *magnetizing current*. This was discussed in Section 8-4. Since the secondary voltage depends on the core flux $\phi_c$, it must be clear that the flux should not change appreciably if $E_s$ is to remain essentially constant under normal loading conditions.

We will now see how the core flux remains practically constant. Let us assume that a load is now applied to the secondary terminals of the transformer. A current $I_s$ will flow in the secondary circuit because the induced EMF $E_s$ will act as a voltage source. The resulting secondary current $I_s$ flows in the secondary winding and produces its own flux, called the *secondary leakage flux*, $\phi_{ls}$. This flux has such a direction that at any instant in time it opposes the main core flux

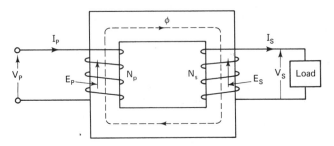

**Figure 8-10** Transformer supplying a load.

that created it in the first place. Of course, this is Lenz's law in action. If we assume that $\phi_c$ increases the current, $I_s$ must have the direction shown in Fig. 8-10 if its resulting flux is to oppose the core flux. Thus the MMF represented by $N_sI_s$ tends to reduce the core flux $\phi_c$. This means that the flux linking the primary winding reduces and consequently reduces the primary induced voltage, $E_p$. This reduction in induced voltage causes a greater difference between the impressed voltage and the counter-induced EMF, thereby allowing more current to flow in the primary. The fact that the primary current $I_p$ increases means that the two conditions stated earlier are fulfilled because:

1. The power input increases to match the power output.
2. The primary MMF increases to offset the tendency of the secondary MMF to reduce the core flux.

In general, it will be found that the transformer reacts almost instantaneously to keep the resultant core flux essentially constant. Moreover, the core flux $\phi_c$ drops very slightly between no load and full load (about 1 to 3%), a necessary condition if $E_p$ is to fall sufficiently to allow an increase in $I_p$. The no-load phasor diagram of Fig. 8-7b can now be extended so that it includes the load current. This is shown in Fig. 8-11 for a secondary current $I_s$ assumed to lag the secondary voltage by the load power factor angle $\theta$. On the primary side, $I'_p$ is the current that flows in the primary winding to balance the demagnetizing effect of $I_s$. Its MMF $N_pI'_p$ sets up a flux linking the primary only; it is called the *primary leakage flux* $\phi_{lp}$. Since the core flux $\phi_c$ remains constant, $I_0$ must be the same current that energizes the transformer at no load. The primary current $I_p$ flowing in the primary winding is therefore the vector sum of the currents $I'_p$ and $I_0$.

Since the no-load current is negligibly small compared to the primary current $I_p$ when the transformer is loaded (2 to 5%), it is correct to assume that the primary ampere-turns equal the secondary ampere-turns, because it is under this condition that core flux is essentially constant. Thus $N_pI_p = N_sI_s$, which yields Eq. (8-8). Thus we will assume that $I_0$ is negligible since it is only a small component of the full-load current.

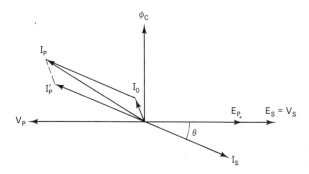

Figure 8-11   Elementary phasor diagram of transformer with load.

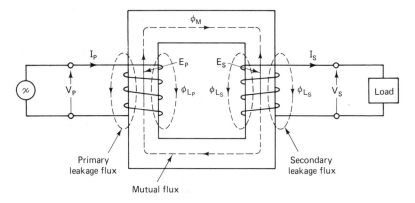

**Figure 8-12** Leakage fluxes.

## Phasor Diagrams

When a current flows in the secondary winding, the resulting ampere-turns $(N_sI_s)$ creates a separate flux, apart from the flux $\phi_c$ produced by $I_0$, that links the secondary winding only. This flux does not link with the primary winding and is therefore not a mutual flux; it is called the *secondary leakage flux*, $\phi_{ls}$. In addition, the load current that flows through the primary winding also creates a flux that links with the primary winding only; it is called the *primary leakage flux* $\phi_{lp}$. Figure 8-12 illustrates these fluxes. The secondary leakage flux gives rise to an induced voltage that is not counterbalanced by an equivalent induced voltage in the primary. Similarly, the induced voltage in the primary is not counterbalanced in the sec-

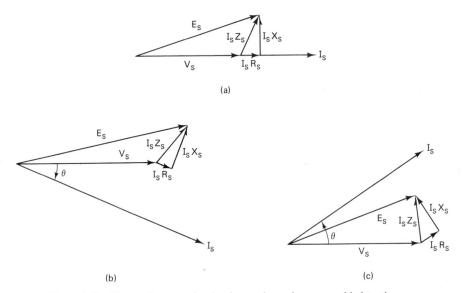

**Figure 8-13** Phasor diagrams showing how voltage drops are added to the secondary terminal voltage to arrive at the induced secondary voltage: (a) unity power factor; (b) lagging power factor; (c) leading power factor.

Single-Phase Transformers    Chap. 8

ondary winding. Consequently, these two induced voltages behave like voltage drops, generally called *leakage reactance voltage drops*. Furthermore, each winding has some resistance, which produces a resistive voltage drop.

Since the voltage drops are all directly proportional to the load current, it means that at the no-load condition there will be no voltage drop in either winding. Let us now see how the voltage drops are accounted for under load conditions.

The leakage reactance voltage induced in the secondary winding of a transformer by the changing leakage flux lags this flux by 90°. The flux and current are in phase; it then follows that the induced leakage reactance voltage lags behind the current by 90°.

Since the flux is varying sinusoidally, the induced leakage reactance voltage will also be sinusoidal. This voltage will react with the secondary induced EMF $E_s$ and tend to change the terminal voltage $V_s$. Therefore, the secondary winding not only develops an induced EMF to supply the terminal voltage $V_s$ for the load, but also has to overcome the two internal voltage drops: the ohmic voltage drop, $I_s R_s$, and the leakage reactance voltage drop, $I_s X_s$. These three voltage components $V_s$, $I_s R_s$, and $I_s X_s$ are added vectorially to yield the produced secondary EMF. Depending on the magnitude and power factor of the load (leading or lagging), the leakage reactance drop may raise or lower the terminal voltage. This may be demonstrated by the phasor diagrams in Fig. 8-13 for the three types of load.

Applying the same reasoning to the primary winding of the transformer, it should be apparent that $E_p$ is obtained by vectorially subtracting the resistance voltage drop $I_p R_p$ and the leakage reactance voltage drop $I_p X_p$ from $V_p$. This is illustrated in Fig. 8-14 for the three different types of power factors.

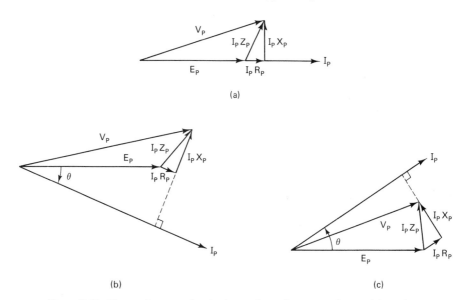

**Figure 8-14** Phasor diagrams showing how voltage drops are subtracted from the primary terminal voltage to arrive at the induced primary voltage: (a) unity power factor; (b) lagging power factor; (c) leading power factor.

## 8-6 EQUIVALENT TRANSFORMER CIRCUIT

When performing calculations on transformers it is more convenient to combine the resistance and reactance voltage drops that occur on the primary and secondary sides into a single $IR$ and $IX$ value. Obviously, when the primary side is the high-voltage side, for example, and the secondary side the low-voltage side, their respective voltage drops can not simply be added numerically. However, it is possible to refer all quantities to either the primary side or the secondary side. To do so, let us assume that the transformation ratio $a$ is larger than unity, which implies a step-down transformer. Next, combine the phasor diagrams of Figs. 8-13b and 8-14b, for example, into a single diagram (see Fig. 8-15a), but at the same time multiplying the secondary voltages by the transformation ratio $a$. Such multiplication converts the secondary values into primary terms and is equivalent to considering the transformer as having a transformation ratio of 1:1.

The next step is to rearrange the various quantities without altering the initial and final values. This is done in Fig. 8-15b. Realizing that $I_s = aI_p$, it is possible to make the following substitutions:

$$I_p R_p + a(I_s R_s) = I_p R_p + a(aI_p R_s) = I_p(R_p + a^2 R_s) \qquad \text{V}$$

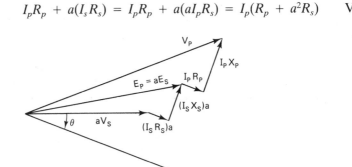

(a)

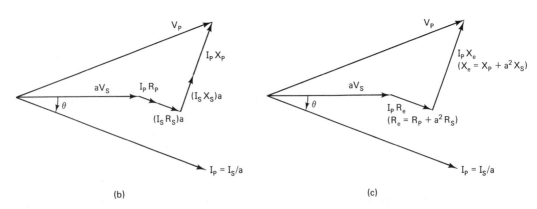

(b)

(c)

**Figure 8-15** Phasor diagram development for transformer equivalent values (referred to primary side).

and

$$I_p X_p + a(I_s X_s) = I_p X_p + a(aI_p X_s) = I_p(X_p + a^2 X_s) \quad \text{V}$$

The ohmic values in the final parenthetical terms in the two previous equations are called the equivalent resistance $R_{ep}$ and equivalent reactance $X_{ep}$ terms of the primary side. Thus in primary terms,

$$R_{ep} = R_p + a^2 R_s \quad \Omega$$
$$X_{ep} = X_p + a^2 X_s \quad \Omega \tag{8-10}$$

The final result then leads to a simplified phasor diagram of Fig. 8-15c. Note that resistive and reactive voltage drops are combined into a single phasor so that calculations are simplified. Similarly, it is possible to show that primary values can be referred to the secondary side. The equivalent resistance $R_{es}$ and the equivalent reactance $X_{es}$ referred to the secondary side becomes in secondary terms,

$$R_{es} = R_s + \frac{R_p}{a^2} \quad \Omega$$
$$X_{es} = X_s + \frac{X_p}{a^2} \quad \Omega \tag{8-11}$$

Since the resistance and reactance have properties that are 90° apart, their vector sum gives what is known as an impedance. For the transformer, therefore,

$$Z_e = \sqrt{R_e^2 + X_e^2} \tag{8-12}$$

where $Z_e$ is the equivalent impedance of the transformer. It will next be shown how the values of $R_e$ and $X_e$ are obtained from two simple tests. The corresponding impedance angle is

$$\alpha = \tan^{-1} \frac{X_e}{R_e}$$

Again, the phasor diagram of Fig. 8-15c suggests an equivalent transformer circuit diagram. Figure 8-16a shows this when the values are referred to the primary side. Figure 8-16b shows the resulting circuit with the quantities in secondary terms. The following example will show us how to obtain the equivalent values in secondary terms if they are shown in primary terms, or vice versa.


**EXAMPLE 8-3**

A 20-kVA 2400/240-V 60-Hz transformer has the following resistance and leakage reactance values: $R_p = 0.8\ \Omega$, $X_p = 3.0\ \Omega$, $R_s = 0.0084\ \Omega$, and $X_s = 0.028\ \Omega$. Calculate the equivalent transformer values:

(a) In primary terms.
(b) In secondary terms.

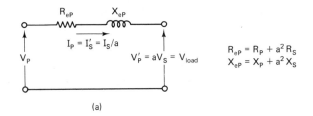

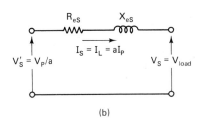

(a)

(b)

R_{eP} = R_P + a^2 R_S
X_{eP} = X_P + a^2 X_S

R_{eS} = R_S + R_P/a^2
X_{eS} = X_S + X_P/a^2

**Figure 8-16** Transformer equivalent-circuit diagram: (a) in primary terms; (b) in secondary terms.

## SOLUTION

(a) In primary terms, the transformer ratio is

$$a = \frac{2400}{240} = 10$$

From Eq. (8-10) we have

$$R_{\mathrm{ep}} = R_p + a^2 R_s = 0.8 + 100 \times 0.0084 = 1.64 \ \Omega$$

and

$$X_{\mathrm{ep}} = X_s + a^2 X_s = 3.0 + 100 \times 0.028 = 5.8 \ \Omega$$

or

$$Z_{\mathrm{ep}} = \sqrt{R_{\mathrm{ep}}^2 + X_{\mathrm{ep}}^2} \ \left/\alpha = \tan^{-1} \frac{X_{\mathrm{ep}}}{R_{\mathrm{ep}}}\right.$$

$$= \sqrt{1.64^2 + 5.8^2} \ \left/\alpha = \tan^{-1}\left(\frac{5.8}{1.64}\right)\right. = 6.03 \ \underline{/74.2^\circ} \ \Omega$$

(b) In secondary terms, from Eq. (8-11),

$$R_{\mathrm{es}} = 0.0084 + \frac{0.8}{100} = 0.0164 \ \Omega$$

$$X_{\mathrm{es}} = 0.028 + \frac{3.0}{100} = 0.058 \ \Omega$$

or

$$Z_{es} = \sqrt{0.0164^2 + 0.058^2} \underline{/\alpha = \tan^{-1}\left(\frac{0.058}{0.0164}\right)}$$

$$= 0.0603 \underline{/74.2°} \ \Omega$$

## EXAMPLE 8-4

For the transformer of Example 8-3, calculate the equivalent resistive and reactive voltage drops for a primary current of 8.3 A:

(a) In primary terms.
(b) In secondary terms.

## SOLUTION

(a) In primary terms,

$$I_p R_{ep} = 8.3 \times 1.64 = 13.61 \ V$$

and

$$I_p X_{ep} = 8.3 \times 5.8 = 48.14 \ V$$

(b) In secondary terms,

$$I_s R_{es} = 10 \times 8.3 \times 0.0164 = 1.36 \ V$$

and

$$I_s X_{es} = 10 \times 8.3 \times 0.058 = 4.81 \ V$$

Note that the transformation ratio $a = 2400/240 = 10$, which indicates that the voltage drops can be readily transformed to either side.

## Determination of Transformer Equivalent Parameters

Under the assumption made earlier, that the no-load component of the current in the primary winding can be neglected, the calculations are made simpler. This neglection is permissible in most power transformer calculations, since $I_0$ is at most but a small percentage of the load current. Even considering its effect would add very little to the accuracy of the final result. Thus the electric circuit representing the transformer merely acts as an impedance drop in series with the load. With a load, the load voltage $V_s$ is vectorially less than $V_p/a$, by the equivalent resistance and reactance voltage drops. This drop depends not only on the load current but on the load power factor as well.

## Short-Circuit Test

As the name implies, the short-circuit test on a transformer is carried out with one of the windings short circuited. This test is performed to determine experimentally the value of the equivalent impedance. Knowing the resistance value from a resistance test enables us to calculate the equivalent reactance. The windings will carry rated current without requiring that the transformer supplies a load. The power input to the transformer will be relatively small. In this way it simulates the leakage flux patterns in the primary and secondary winding because they depend on the load current. As one of the windings is short circuited, the other winding has an applied voltage which is low, usually about 5 to 15% of rated value. The equivalent circuit under these conditions is that of Fig. 8-17, where the secondary voltage $V_s = 0$.

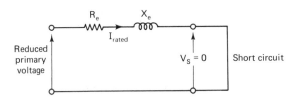

**Figure 8-17** Equivalent transformer, secondary short-circuited.

With the primary voltage greatly reduced, the core flux will be reduced to the same extent. Since the core loss is somewhat proportional to the square of the flux, it is practically zero. Thus a wattmeter used to measure the input power will register the copper losses only; the output power is zero.

Figure 8-18 illustrates the circuit diagram used to perform a short-circuit test. It is seen that the low-voltage side is short circuited while the measurements are taken on the high-voltage side. This nearly is always done for the following two reasons:

1. The rated current on the high-voltage side is lower than on the low-voltage side; therefore, dangerously high currents are avoided. It also permits commonly used laboratory instruments.
2. Since the impressed voltage on the high-voltage side is a fraction of rated voltage, an appreciable voltmeter deflection is more readily obtained on the high-voltage side.

A rheostat or variac inserted in the input side adjusts the primary current to rated value. From the input data of watts, amperes, and volts the equivalent reactance can be calculated, all in terms of the high-voltage side, namely:

$$R_{eH} = \frac{P_{sc}}{I_{sc}^2} \tag{8-13}$$

$$Z_{eH} = \frac{V_{sc}}{I_{sc}} \tag{8-14}$$

and

$$X_{eH} = \sqrt{Z_{eH}^2 - R_{eH}^2} \tag{8-15}$$

where $I_{sc}$, $V_{sc}$, and $P_{sc}$ are the short-circuit amperes, volts, and watts, respectively.

Single-Phase Transformers    Chap. 8

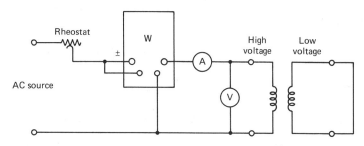

**Figure 8-18** Short-circuit test condition.

## EXAMPLE 8-5

The following data were obtained in a short-circuit test upon a 20-kVA 2400/240-V 60-Hz transformer:

$$V_{sc} = 72 \text{ V} \qquad I_{sc} = 8.33 \text{ A} \qquad P_{sc} = 268 \text{ W}$$

Instruments were placed in the high-voltage side with the low-voltage side short circuited. Obtain the equivalent transformer parameters referred to the high-voltage side.

## SOLUTION

$$R_{eH} = \frac{P_{sc}}{I_{sc}^2} = \frac{268}{8.33^2} = 3.86 \ \Omega$$

$$Z_{eH} = \frac{V_{sc}}{I_{sc}} = \frac{72}{8.33} = 8.64 \ \Omega$$

and

$$X_{eH} = \sqrt{Z_{eH}^2 - R_{eH}^2}$$

$$= \sqrt{8.64^2 - 3.86^2} = 7.73 \ \Omega$$

All measurements in Example 8-5 were taken on the high-voltage side, resulting in equivalent values referred to the high-voltage side of the transformer. However, if the transformer is used as a step-up transformer, it may be necessary to express the equivalent values in terms of the low-voltage side. This is readily obtained without having to perform an additional test. Referring to Example 8-5, the transformation ratio $a = 10$ (2400/240). Realizing that, for instance, $R_{eH} = R_p + a^2 R_s$, it is readily verified that with $a = 10$, the equivalent value transformed to the low-voltage side, in this example, is

$$R_{eL} = \frac{R_{eH}}{a^2} = \frac{R_p + a^2 R_s}{a^2} = R_s + \frac{R_p}{a^2}$$

Thus the low-voltage-side equivalent resistance becomes

$$R_{eL} = \frac{R_{eH}}{a^2} = 0.0386 \ \Omega$$

Sec. 8-6    Equivalent Transformer Circuit                    **229**

Similarly,

$$X_{eL} = \frac{X_{eH}}{a^2} = 0.0773 \ \Omega$$

and

$$Z_{eL} = \frac{Z_{eH}}{a^2} = 0.0864 \ \Omega$$

Therefore, only one short-circuit test needs to be performed. Yet by a simple calculation, the equivalent circuit is referred to either set of windings.

### Open-Circuit Test

This test is carried out with one side open circuited while the other is connected to a source whose voltage is rated value. As discussed, with no load connected, the resulting current flowing in the winding connected to the source will be extremely low. Because of this, the copper loss in which $I_0$ flows will be negligible. Therefore, this no-load current has two components: one producing the core flux and the other overcoming the hysteresis and eddy current losses. For all practical purposes, a wattmeter in the input circuit under no-load conditions will register the core loss.

Although it does not matter on which side the test is performed, it will be safer if performed on the low-voltage side. The circuit for no-load conditions is shown in Fig. 8-19. If the applied voltage is sinusoidal, it was seen earlier that the flux changes in a sine-wave fashion between $\pm \phi_m$. Because of this changing flux, two distinct components of power loss are developed in the iron core. They are known as the hysteresis and eddy current loss, as discussed earlier. The hysteresis-loss results because of the cyclic change in the magnetic state of the iron. As the flux "reverses" its direction, the magnetic particles in the iron tend to realign themselves accordingly, resulting in molecular friction, which manifests itself as a loss of energy. The hysteresis loss is directly proportional to the flux and the number of flux reversals per second, which is determined by the line frequency. Laminating the core does not reduce this loss; only the use of high-quality steel does. From theory and experimental work it has been shown that the hysteresis loss is

$$P_h = K_h f B_m^n \qquad \text{W} \tag{8-16}$$

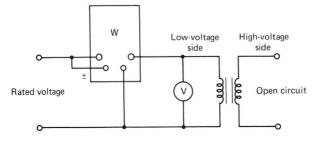

**Figure 8-19** Open-circuit test connections.

where    $K_h$ = constant depending on core volume and steel quality
        $B_m$ = maximum core flux density
        $f$ = supply frequency
        $n$ = Steinmetz constant, which varies from 1.5 for older steels to 2 for newer steels

Eddy current loss is electromagnetic in character and is caused by the local flow of currents in the iron laminations, created in exactly the same way as the ohmic losses in the transformer windings. These losses depend only on the maximum value of the flux and are therefore constant for a given transformer, independent of the load. Eddy current losses can be effectively controlled by laminating the core. It can be shown that the eddy current loss is

$$P_e = k_e f^2 B_m^2 t^2 \tag{8-17}$$

where    $k_e$ = constant depending on core volume and resistivity of the iron
        $t$ = thickness of core laminations

Since the total core loss $P_c$ is the sum of the hysteresis and eddy current loss, it follows that

$$P_c = P_h + P_e = k_h f B_m^n + k_e f^2 B_m^2 t^2 \tag{8-18}$$

It is important to note that hysteresis loss is directly proportional to the frequency, while the eddy current loss depends on the frequency squared.

## EXAMPLE 8-6

A 2400-V 60-Hz transformer has a core loss of 630 W of which one-third is eddy current loss. Determine the core loss when this transformer is connected to a 2400-V 50-Hz source. Assume the constant $n = 2$.

### SOLUTION

$P_e = 630/3 = 210$ W, hence $P_h = 630 - 210 = 420$ W. From Eq. (8-4) we have

$$V = 4.44 \, fN\phi_m$$

Therefore, since the applied voltage $V$ remains the same, a reduction in frequency (50/60 x) causes a corresponding increase in the flux (60/50 x). From Eq. (8-18) we can determine the core at 50 Hz, then, as follows:

$$P_c(50 \text{ Hz}) = k_h f B_m^2 + k_e f^2 B_m^2 t^2$$

$$= 420 \left(\frac{50}{60}\right)\left(\frac{60}{50}\right)^2 + 210\left(\frac{50}{60}\right)^2\left(\frac{60}{50}\right)^2$$

$$= 504 + 210 = 714 \text{ W}$$

This indicates an increased core loss when the transformer is operating at lower than design frequency, which would result in a higher working temperature.

Sec. 8-6    Equivalent Transformer Circuit                                    **231**

The regulation of a transformer by definition is the difference between the no-load and full-load secondary voltage expressed as a percentage of the full-load voltage. This, as we see, is similar to the definition for voltage regulation of dc and ac generators. There is an additional condition in the use of transformers that must be fulfilled—that the primary or applied voltage remains constant. Also, the power factor of the load must be stated since, as can be seen from the phasor diagrams developed for the transformer, the voltage regulation does depend on the load power factor, as with synchronous generators.

In general,

$$\text{voltage regulation} = \frac{V_{\text{no load}} - V_{\text{load}}}{V_{\text{load}}} \times 100\% \qquad (8\text{-}19)$$

The numerical values employed in the calculations depend on which winding is used as a reference for the equivalent circuit. Similar results are, of course, obtained whether all impedance values are transferred to the primary or secondary side of the transformer.

### EXAMPLE 8-7

Using the transformer data of Example 8-5, calculate the voltage regulation of the transformer for an 0.80 lagging power factor load. Assume that full-load is delivered.

### SOLUTION

At rated output, $V_p' = aV_s = 2400$ V and $I_p = 8.33$ A. The transformer equivalent values determined in Example 8-5 are

$$R_{\text{eH}} = 3.86 \; \Omega$$

$$X_{\text{eH}} = 7.73 \; \Omega$$

The resistive and reactive voltage drops in primary terms are

$$I_p R_{\text{eH}} = 8.33 \times 3.86 = 32.2 \text{ V}$$

$$I_p X_{\text{eH}} = 8.33 \times 7.73 = 64.4 \text{ V}$$

For PF = 0.80 lagging, $\cos \theta = 0.8$ and $\sin \theta = 0.6$. The vector diagram for this example is given in Fig. 8-20, with the current taken as the reference vector. As shown, the horizontal component of $V_p$ is

$$V_{p\,\text{hor}} = aV_s \cos \theta + I_p R_{\text{eH}} = 2400 \times 0.8 + 32.2 = 1952 \text{ V}$$

and the vertical component of $V_p$ is

$$V_{p\,\text{vert}} = aV_s \sin \theta + I_p X_{\text{eH}} = 2400 \times 0.6 + 64.4 = 1504 \text{ V}$$

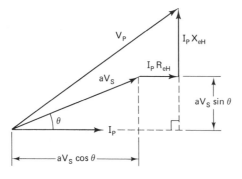

**Figure 8-20** Phasor diagram for Example 8-5.

Therefore,

$$V_p = \sqrt{1952^2 + 1504^2} = 2465 \text{ V}$$

$$\text{voltage regulation} = \frac{2465 - 2400}{2400} \times 100 = 2.71\%$$

## 8-8 TRANSFORMER EFFICIENCY

The efficiency of a transformer is high and typically in the range 95 to 98%. This implies that the transformer losses are as low as 2 to 5% of the input power. In ferrite-cored and audio transformers, the efficiencies are much lower in comparison to power transformers. Many audio transformers have efficiencies ranging from 20% to approximately 60%.

In calculating the efficiency, it is generally much better to determine the transformer losses rather than measuring the input and output powers directly. This becomes apparent when dealing with large transformers, since the power handled is far too great to apply merely for the purpose of a test. It was shown that the short-circuit test gives the copper losses at full-load conditions. Since the copper loss is proportional to the square of the current, it can therefore be calculated at any other load. The only other loss present is that in the iron core.

This core loss can be determined from the open-circuit test, as was described earlier. It consists of two parts: the hysteresis loss and the eddy current loss. The hysteresis loss is due to the reversal of the flux in the core and depends on the frequency of the supply voltage and the maximum value of the flux density setup. The eddy currents that result in the iron are a direct consequence of the induced EMF created in the iron by virtue of the changing flux in it. Its value is similarly determined by Lenz's law, and therefore is directly proportional to the rate of change of flux. The resulting eddy current loss will therefore also be proportional to the square of this eddy current. It is thus evident that both of the loss components making up the core loss are dependent on the flux and the frequency. The flux is determined by the difference of the input voltage and induced voltage at the primary, as discussed under loading of the transformer. This difference is essentially constant from no-load to full-load conditions. It may therefore be assumed that the core losses remain substantially constant throughout the operating range of the transformer. In this discussion it is assumed that the frequency of the supply voltage does not change, which, practically is true. Hence the power measured during the open-circuit test is most conveniently done if carried out on the low-voltage side and provides exactly the required core loss. Remember that the copper losses during this test are neglected since the magnetizing current is very small.

In general, the efficiency of any electrical apparatus is

$$\eta = \frac{\text{output power}}{\text{input power}} = \frac{\text{output power}}{\text{output power + losses}} \tag{8-20}$$

where $\eta$ (the Greek lowercase letter eta) is the symbol used to denote efficiency. When Eq. (8-20) is multiplied by the factor 100, the efficiency will then be in percent.

Transformer ratings are based on output kVA (MVA); therefore, this equation may also be written as

$$\eta = \frac{\text{kVA}_{\text{out}} \times \text{PF}}{\text{kVA}_{\text{out}} \times \text{PF} + \underbrace{\text{copper loss}}_{\text{variable}} + \underbrace{\text{iron loss}}_{\text{fixed}}} \tag{8-21}$$

The core loss in a transformer is practically constant for all load conditions. The copper loss varies proportional to the square of the current. Thus we can say that the copper losses are variable with loading. From the preceding discussion we have seen that the magnetic core losses are essentially independent of loading or fixed in value, since the voltage is fixed.

It is therefore a simple matter to calculate the efficiency versus load current curve. Because of this relationship, it can be shown that the efficiency of a transformer is at a maximum when the fixed losses are equal to the variable losses. In other words, the copper losses equal the iron losses. We demonstrate the performance data of a typical transformer in the following example.

**EXAMPLE 8-8**

A 10-kVA 2400/240-V 60-Hz transformer was tested with the following results:

input during short circuit test = 340 W

input during open circuit test = 168 W

Determine the efficiency of this transformer versus load power over the entire load range up to 130% overload. Use a load PF = 0.8.

## SOLUTION

Because of the repeated calculations required, we make use of a computer-aided solution. The program is written in BASIC and should run on any computer using this language. Minor modifications may be necessary to suit a specific manufacturer's model used. Using this program, the printout format and results are as given in Table 8-1. Figure 8-21 shows these results graphically.

As can be seen, specific transformer data must be typed in. This makes the program generally applicable and it can easily be used for transformers of various sizes.

**TABLE 8-1** COMPUTER PROGRAM AND RESULTS

```
10 PRINT "TRANSFORMER EFFICIENCY CURVE"
15 PRINT
20 PRINT "TYPE IN NAMEPLATE KVA"
25 INPUT KVA
30 PRINT "KVA RATING = " ; KVA
35 PRINT "TYPE IN CORE LOSS"
40 INPUT CL
45 PRINT "CORE LOSS = " ; CL
50 PRINT "TYPE IN COPPER LOSS AT FULL LOAD"
55 INPUT CUL
60 PRINT "COPPER LOSS = " ; CUL
65 PRINT "SPECIFY THE POWER FACTOR"
70 INPUT PFL
75 PRINT "PF = " ; PFL
80 PRINT
85 PRINT "KW     CU LOSS     EFFICIENCY"
90 FOR I = 1 TO 13
95 LET PCL = I/10
100 LET PCUL = PCL*PCL*CUL
105 LET LOAD = KVA*PCL*PFL
110 LET EFF = LOAD/(LOAD + (CL + PCUL)/1000)
115 PRINT LOAD; TAB 5; PCUL; TAB 14; EFF
120 NEXT I
```

Transformer Efficiency Curve
Type in Nameplate kVA
  kVA rating = 10
Type in core loss
  Core loss = 168
Type in copper loss at full load
  Copper loss = 340
Specify the power factor
  PF = 0.8

| kW | Copper loss | Efficiency (%) |
|------|------|------|
| 0.0 | 0.0 | 0.0 |
| 0.8 | 3.4 | 0.8236 |
| 1.6 | 13.6 | 0.8981 |
| 2.4 | 30.6 | 0.9236 |
| 3.2 | 54.4 | 0.9350 |
| 4.0 | 85.0 | 0.9405 |
| 4.8 | 122.4 | 0.9430 |
| 5.6 | 166.6 | 0.9436 |
| 6.4 | 217.6 | 0.9432 |
| 7.2 | 275.4 | 0.9420 |
| 8.0 | 340.0 | 0.9403 |
| 8.8 | 411.4 | 0.9382 |
| 9.6 | 489.6 | 0.9359 |
| 10.4 | 574.6 | 0.9334 |

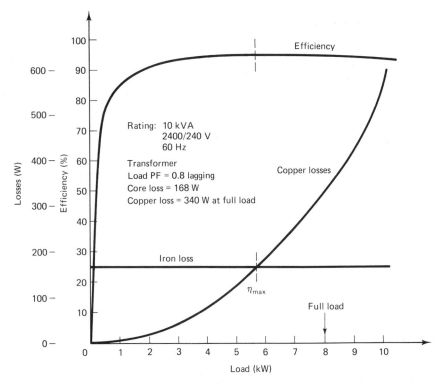

**Figure 8-21** Graphical representation of results from computer program for Example 8-8.

## REVIEW QUESTIONS

**8-1.** Explain why bulk electric power is transmitted at high voltage levels.

**8-2.** What classifies a transformer as a power transformer? As a distribution transformer?

**8-3.** A power system consists of four different engineering functions from a practical point of view. Can you identify them?

**8-4.** What is the rating of a transformer? What parameters are important, and why? What is the real basis for rating a transformer?

**8-5.** What is transformer action?

**8-6.** Distinguish between a core-type and a shell-type transformer.

**8-7.** Why is transformer sheet steel often specially annealed?

**8-8.** What is a stepped core? Where is it used, and why?

**8-9.** Why are round coil forms used in the larger transformers?

**8-10.** Why do transformers hum? How can we minimize this audible noise?

**8-11.** What purpose is served by placing transformers in an oil-filled tank?

**8-12.** With a transformer open-circuited on the secondary, a small current will flow in the primary. What is the function of this current?

**8-13.** What is the transformation ratio? How can it be determined?

**8-14.** What is the purpose of a transformer having a transformation ratio equal to unity?

**8-15.** Describe how the primary current adjusts itself when a transformer load is increased.

**8-16.** What is the voltage regulation of a transformer?

**8-17.** What is the equivalent impedance of a transformer? How is it calculated in primary terms? In secondary terms?

**8-18.** What tests are necessary to determine the equivalent impedance of a transformer?

**8-19.** What information is obtained from a short-circuit test? An open-circuit test?

**8-20.** Why is a short-circuit test performed at greatly reduced voltage? Explain why the core loss is negligible in this test.

**8-21.** Why is the core loss not affected by the load on a transformer?

## PROBLEMS

**8-1.** A single-phase transformer has 400 primary turns and 800 secondary turns. The net iron cross-sectional area of the core is 40 cm². If the primary winding is connected to a 60-Hz supply at 600 V, calculate:
(a) The maximum value of the core flux density.
(b) The secondary induced voltage.

**8-2.** Determine the rated primary and secondary currents of a 20-kVA 1200/120-V single-phase transformer. If this transformer delivers a load of 12 kW at PF = 0.8, what are the primary and secondary currents?

**8-3.** A single-phase transformer has 180 and 45 turns on its primary and secondary windings, respectively. The corresponding resistances are 0.242 and 0.076 Ω, respectively. Determine the equivalent resistance:
(a) In primary terms.
(b) In secondary terms.

**8-4.** A 60-Hz single-phase transformer has a turns ratio of 8. The resistances are 0.90 and 0.05 Ω, the reactances 5 and 0.14 Ω for the high- and low-voltage winding, respectively. Determine:
(a) The voltage to be applied to the high-voltage side to obtain a full-load current of 180 A in the low-voltage winding on short circuit.
(b) The PF under the conditions of part (a).

**8-5.** A 50-kVA 4400/220-V 60-Hz single-phase transformer takes 10.8 and 544 W at 120 V in a short-circuit test performed on the high-voltage side. Calculate the voltage to be applied to the high-voltage side on full load at a power factor 0.8 lagging when the secondary terminal voltage is 220 V.

**8-6.** In a 50-kVA 2400/240-V transformer the iron and copper losses are 680 and 760 W, respectively.
(a) Calculate the efficiency at unity power factor at:
(i) Full load.
(ii) Half load.
(b) Determine the load for maximum efficiency.

**8-7.** A short-circuit test was performed on a 10-kVA 2400/240-V transformer with the following data recorded: $V_{sc}$ = 138 V, $P_{sc}$ = 202 W, and $I_{sc}$ = 4.17 A. Calculate in primary terms:

(a) $R_e$, $Z_e$, and $X_e$.

(b) The voltage regulation when supplying full load at a power factor of 0.866 lagging.

**8-8.** A transformer that can be considered ideal has 200 turns on its primary winding and 500 turns on its secondary winding. The primary is connected to a 220-V sinusoidal supply and the secondary supplies 10 kVA to a load.

(a) Determine the load voltage, secondary current, and primary current.

(b) Find the magnitude of the load impedance as seen from the supply.

**8-9.** A 2300/208-V 500-kVA 60-Hz single-phase transformer is tested by means of the open-circuit and short-circuit tests. The test data obtained are:

$$\text{Open-circuit test: } V_1 = 208 \text{ V, } I_1 = 65 \text{ A, } P = 1800 \text{ W}$$

$$\text{Short-circuit test: } V_H = 95 \text{ V, } I_H = 217.4 \text{ A, } P = 8200 \text{ W}$$

Calculate the efficiency of the transformer at rated volt-amperes and half-rated volt-amperes.

**8-10.** For the transformer in Problem 8-9, determine the voltage regulation when supplying full load at a power factor of 0.8 lagging.

**8-11.** A 25-kVA 440/220-V 60-Hz distribution transformer has a core loss of 740 W. Maximum efficiency occurs when the load is 15 kVA. Determine the efficiency at full load.

**8-12.** The transformer in Problem 8-11 is connected to a 380-V 50-Hz supply. What are the core losses then? Assume that the hysteresis loss is 60% of the core loss and the Steinmetz constant is $n$ = 1.8.

**8-13.** A 75-kVA 6600/230-V transformer requires 310 V across the primary to circulate full-load current on short circuit. The power absorbed is 1.6 kW. Determine the voltage regulation.

**8-14.** A 5-kVA 2300/230-V 60-Hz distribution transformer was tested with the following results:

$$\text{short-circuit test input} = 112 \text{ W}$$

$$\text{open-circuit test input} = 40 \text{ W}$$

Calculate:

(a) The efficiency at rated kVA and 0.8 power factor lagging.

(b) The efficiency at half-rated kVA and unity power factor.

(c) The maximum efficiency at unity PF.

**8-15.** The magnetic core losses of a 24-kVA 2400/120-V 60-Hz single-phase transformer are 400 W. The copper losses at full load are 900 W. The transformer supplies 85 A at a power factor of 0.82 leading. Determine the efficiency at this load.

**8-16.** A single-phase load is supplied through a 33,000-V feeder whose impedance is $Z_f$ = 105 + $j360$ Ω and a 33,000/2400-V transformer whose equivalent impedance is $Z_e$ = 0.26 + $j1.08$ Ω, referred to its low-voltage side. The load is 180 kW at 0.85 leading PF and 2250 V. Determine:

(a) The voltage at the sending end of the feeder.

(b) The voltage at the primary transformer terminals.

(c) The power and reactive power input at the sending end of the feeder.

# 9

# THREE-PHASE
# TRANSFORMERS
# and CONNECTIONS

## 9-1 INTRODUCTION

Polarities of transformers identify the relative directions of induced voltages in the two windings. They are a direct result of the relative direction in which the windings are wound. For operating transformers in parallel and for various transformer bank connections, it is essential that we know these polarities.

Transformers are connected in parallel to share loads exceeding the capacity of individual transformers. Since power systems are three-phase systems, there is the obvious need for three-phase transformers to link the various parts of a power grid together. This can be accomplished with three single-phase transformers or a single three-phase transformer. Then there are special transformers, such as the autotransformer, which could be a single-winding transformer having a tap brought out. Other examples are the instrument transformers. They are used in conjunction with ammeters, voltmeters, wattmeters, and so on. In this chapter we discuss these special applications.

## 9-2 POLARITY OF TRANSFORMER WINDINGS

Transformers are often manufactured with multiple windings. This enables a selection of voltages by simply interconnecting the coils as desired. In each instance it is necessary to know the relative polarities of the terminals in order to make the proper connections. Figure 9-1 shows a multicoil transformer with two low-voltage and two high-voltage windings, designated by $X$ and $H$, respectively. As illus-

**239**

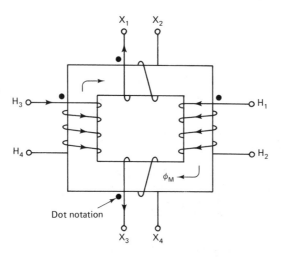

Dot notation

**Figure 9-1** Instantaneous transformer windings polarities using dot convention and conventionally used terminal markings.

trated, the instantaneous polarity is coded by a number subscript. In Fig. 9-1, the odd numbers denote the instantaneous positive polarity. They also correspond to the dot markings, indicating positive induced EMF. The instantaneous polarity of the winding terminal is dependent on the flux direction and therefore has a direct relation to the polarity of the primary winding.

Assume that the primary windings $H_1-H_2$ and $H_3-H_4$ are energized such that $H_1$ is instantaneously positive with respect to $H_2$ and $H_3$ is positive with respect to $H_4$. The current flow is as shown, which sets up a core flux in the clockwise direction. As determined by Lenz's law, EMFs are set up in the remaining low-voltage windings since they link this flux. The direction of the induced EMFs in these windings is such that if they were connected to a load, the current direction in them would be such that their created fields in turn oppose the flux.

A convenient method of determining the winding polarities of a single-phase transformer is illustrated in Fig. 9-2. The impressed voltage to one of the windings may be rated voltage, although this test could be carried out at a reduced voltage. Since the resulting excitation current, the flux, and the induced voltages are all alternating quantities, the terminal polarities are also changing constantly. At some instant, however, terminal 1 will be positive with respect to terminal 2. On an actual transformer these terminals may be so marked. To determine the relative polarity of terminals 3 and 4 at the instant terminal 1 is positive with respect to

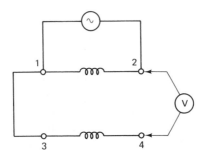

**Figure 9-2** Test for polarity.

terminal 2, a terminal of the secondary is connected to either of the primary windings. (In Fig. 9-2, terminals 1 and 3 are joined.)

A voltmeter is now connected between the free ends of the primary winding and that of the secondary. If the voltmeter reads the sum of the impressed primary voltage and the induced secondary voltage, it indicates that the joint terminals have opposite instantaneous polarities. In Fig. 9-2, terminal 3 will then be negative with respect to terminal 4. If, on the other hand, the voltmeter reads the difference of the primary and secondary voltages, it implies that points of like instantaneous polarities have been joined.

To facilitate connections of transformers without having to resort to tests, transformer manufacturers have standardized terminal markings to indicate their polarity. It is standard practice that high-voltage terminals are labeled $H_1$ and $H_2$ and that the low-voltage terminals are labeled $X_1$ and $X_2$. It is understood that like-marked terminals are to be joined when transformers are to operate in parallel.

## 9-3 AUTOTRANSFORMERS

In principle and in general construction, the autotransformer does not differ from the conventional two-winding transformer so far discussed. It does differ from it, however, in the way the primary and secondary windings are interrelated.

It will be recalled that in discussing the transformer principles of operation, it was pointed out that a counter EMF was induced in the winding, which acted as a primary to establish the excitation ampere turns. The induced voltage per turn was the same in each and every turn linking with the common flux of the transformer. Therefore, fundamentally it makes no difference in the operation whether the secondary induced voltage is obtained from a separate winding linked with the core or from a portion of the primary turns. The same voltage transformation results in the two situations. When the primary and secondary voltage are derived from the same winding, the transformer is called an autotransformer.

An ordinary two-winding transformer may also be used as an autotransformer by connecting the two windings in series and applying the impressed voltage across the two, or merely to one of the windings. It depends on whether it is desired to step the voltage down or up, respectively. This is shown in Fig. 9-3a for the step-

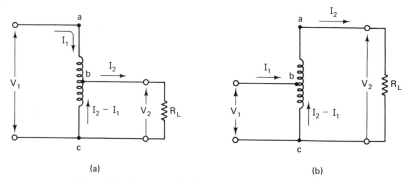

(a)  (b)

**Figure 9-3** Autotransformers: (a) step-down; (b) step-up.

down connection; the step-up connection is illustrated in Fig. 9-3b. In Fig. 9-3a the input voltage $V_1$ is connected to the complete winding (a–c) and the load $R_L$ is connected across a portion of the winding, that is, (b–c). The voltage $V_2$ is related to $V_1$ as in the conventional two-winding transformer, that is,

$$V_2 = V_1 \times \frac{N_{bc}}{N_{ac}} \tag{9-1}$$

where $N_{bc}$ and $N_{ac}$ are the number of turns on the respective windings. The ratio of voltage transformation in an autotransformer is the same as that for an ordinary transformer, namely,

$$a = \frac{N_{ac}}{N_{bc}} = \frac{V_1}{V_2} = \frac{I_2}{I_1} \tag{9-2}$$

with $a > 1$ for step-down.

Assuming a resistive load for convenience, then,

$$I_2 = \frac{V_2}{R_L}$$

Assume that the transformer is 100% efficient. The power output is

$$P = V_2 I_2 \tag{9-3}$$

Note that $I_1$ flows in the portion of winding ab, whereas the current $(I_2 - I_1)$ flows in the remaining portion bc. The resulting current flowing in the winding bc is always the arithmetic difference between $I_1$ and $I_2$, since they are always in the opposite sense. Remember that the induced voltage in the primary opposes the primary voltage. As a result, the current caused by the induced voltage flows opposite to the input current. In an autotransformer, the secondary current is this induced current, that is,

$$I_1 + (I_2 - I_1) = I_2 \tag{9-4}$$

Hence the ampere-turns due to section $bc$, where the substitutions $I_2 = aI_1$ and $N_{bc} = N_{ac}/a$ are made according to Eq. (9-2), is

$$\text{ampere-turns due to section } bc = (I_2 - I_1)N_{bc}$$

$$= \frac{(aI_1 - I_1)N_{ac}}{a} = I_1 N_{ac}\left(1 - \frac{1}{a}\right) = I_1 N_{ab}$$

$$= \text{ampere-turns due to section } ac$$

Thus the ampere-turns due to sections $bc$ and $ac$ balance each other, a characteristic of all transformer actions.

Equation (9-3) gives the power determined by the load. To see how this power is delivered, we can write the equation in a slightly modified form. By substituting Eq. (9-4) into Eq. (9-3), we obtain

$$P = V_2 I_2 = V_2[I_1 + (I_2 - I_1)] \tag{9-5}$$
$$= V_2 I_1 + V_2(I_2 - I_1) \quad \text{W}$$

This indicates that the load power consists of two parts. The first part is

$$P_c = V_2 I_1 \equiv \text{conducted power to load through } ab \qquad (9\text{-}6)$$

The second part is

$$P_{tr} = V_2(I_2 - I_1) \equiv \text{transformed power to load through } bc \qquad (9\text{-}7)$$

We will see in the following examples that most of the power to the load is directly conducted by winding $ab$. The remaining power is transferred by the common winding $bc$. To show these powers $P_c$ and $P_{tr}$ in terms of the total power $P$, we proceed as follows:

$$\frac{P_c}{P} = \frac{V_2 I_1}{V_2 I_2} = \frac{I_1}{I_2} = \frac{1}{a} \qquad (9\text{-}8)$$

and

$$\frac{P_{tr}}{P} = \frac{V_2(I_2 - I_1)}{V_2 I_2} = \frac{I_2 - I_1}{I_2} = \frac{a - 1}{a} \qquad (9\text{-}9)$$

Thus $P_c = P/a$ and $P_{tr} = P(a - 1)/a$, with $a > 1$ for a step-down autotransformer.

## EXAMPLE 9-1

A standard 5-kVA 2300/230-V distribution transformer is connected as an autotransformer to step down the voltage from 2530 V to 2300 V. The transformer connection is as shown in Fig. 9-3a. The 230-V winding is section $ab$, the 2300-V winding is $bc$. Compare the kVA rating of the autotransformer with that of the original two-winding transformer. Also calculate $P_c$, $P_{tr}$, and the currents.

### SOLUTION

The rated current in the 230-V winding (or in $ab$) is

$$I_1 = \frac{5000 \text{ VA}}{230} = 21.74 \text{ A}$$

The rated current in the 2300-V winding (or in $bc$) is

$$I_2 - I_1 = \frac{5000}{2300} = 2.174 \text{ A}$$

Therefore,

$$I_2 = 2.174 + I_1 = 23.914 \text{ A}$$

The secondary current $I_2$ can also be calculated from

$$I_2 = aI_1 = \frac{2530}{2300} \times 21.74 = 23.914 \text{ A}$$

Since the transformation ratio $a = 2530/2300 = 1.1$,

$$P = V_1 I_1 = V_2 I_2 = 2530 \times 21.74 = 55.00 \text{ kVA}$$

The conducted power is

$$P_c = \frac{P}{a} = \frac{50{,}000}{1.1} = 50 \text{ kVA}$$

and that transformed is

$$P_{tr} = P \frac{a - 1}{a} = 55{,}000 \frac{1.1 - 1}{1.1} = 5.0 \text{ kVA}$$

Consider now the step-up transformer of Fig. 9-3b. Following reasons similar to those above, it follows that

$$P = V_1 I_1 = V_1[I_2 + (I_1 - I_2)] \tag{9-10}$$
$$= V_1 I_2 + V_1(I_1 - I_2) \quad \text{W}$$

where we made the substitution of $I_1$ from Eq. (9-4), which really is Kirchhoff's current law applied to point $b$. To show this, note at point $b$ we have

$$I_1 + (I_2 - I_1) = I_2$$

so that

$$I_1 = I_2 - (I_2 - I_1) = I_2 + (I_1 - I_2)$$

Again, Eq. (9-10) shows us that the power supplied to the load consists of two parts,

$$P_c = V_1 I_2 \equiv \text{conducted power to load through } ab \tag{9-11}$$

and

$$P_{tr} = V_1(I_1 - I_2) \equiv \text{transformed power to load through } bc \tag{9-12}$$

In terms of total power, we have

$$\frac{P_c}{P} = \frac{V_1 I_2}{V_1 I_1} = \frac{I_2}{I_1} = a \tag{9-13}$$

and

$$\frac{P_{tr}}{P} = \frac{V_1(I_1 - I_2)}{V_1 I_1} = \frac{I_1 - I_2}{I_1} = 1 - a \tag{9-14}$$

Thus for the step-up transformer with $a < 1$, we obtain $P_c = aP$ and $P_{tr} = P(1 - a)$. As before, $P_c$ is the power directly conducted to the load and $P_{tr}$ is the portion that is transformed.

## EXAMPLE 9-2

Repeat the problem of Example 9-1 for a 2300 V-to-2530 V step-up connection as shown in Fig. 9-3b.

**SOLUTION**

As calculated in Example 9-1, the current rating of the winding ab is $I_2 = 21.74$ A, which also is the load current. The output voltage is 2530 V; thus the volt-ampere rating of the autotransformer is

$$P = V_2 I_2 = 2530 \times 21.74 = 55 \text{ kVA}$$

which is the same as in Example 9-1. The transformer ratio

$$a = \frac{2300}{2530} = 0.909$$

The conducted power is therefore

$$P_c = aP = 0.909 \times 55 \text{ kVA} = 50 \text{ kVA}$$

and the transformed power

$$P_{tr} = P(1 - a) = 55 \text{ kVA}(1 - 0.909) = 5 \text{ kVA}$$

The examples given make it clear that an autotransformer of given physical dimensions can handle much more load power than an equivalent two-winding transformer; in fact, $a/(a - 1)$ times its rating as a two-winding transformer for the step-down autotransformer or $1/(1 - a)$ for the step-up arrangement. A 5-kVA transformer is capable of taking care of 11 times its rating. These great gains are possible since an autotransformer transforms, by transformer action, only a fraction of the total power; the power that is not transformed is conducted directly to the load.

It should be noted that an autotransformer is not suitable for large percentage voltage reductions as is a distribution transformer. This is due to the required turns ratio becoming too large; hence the power-handling advantage would be minimal. Furthermore, in the unlikely but possible event that the connections to the low-voltage secondary were to fail somewhere below point b in Fig. 9-3a, the winding bc would be deleted from the circuit. This implies that the load would see the full high line voltage. Autotransformers are not used for these reasons where large voltage changes are encountered. In situations where autotransformers can be used to their full advantage, it will be found that they are cheaper than a conventional two-winding transformer of similar rating. They also have better regulation (i.e., the voltage does not drop so much for the same load), and they operate at higher efficiency. In all applications using autotransformers it should be realized that the primary and secondary circuits are not electrically isolated, since one input terminal is common with one output terminal.

## 9-4 INSTRUMENT TRANSFORMERS

It is often necessary to measure high voltages or large currents. For this it would be desirable if ordinary standard low-range instruments could be used in conjunction with specially constructed accurate-ratio transformers. These are called *in-*

*strument transformers*.  There are two types, current transformers and potential or voltage transformers.

*Current transformers*, as the name implies, are used with an ammeter to measure the current in an ac circuit.  Usually, current transformers are connected to standard 5-A ammeters.  The current transformer has a primary coil of one or more turns of heavy wire, which is always connected in series in the circuit in which the current is to be measured.  The secondary has a great many turns of comparable fine wire, which must always be connected across the ammeter terminals.  The secondary is almost always designed for a range of 5 A.  The transformer will have any of various standardized current ratios, for instance, 100 to 5 A, which implies a 20:1 current ratio.

To obtain accuracy with a reasonably sized core, current transformers should never be operated with the secondary open-circuited.  Unlike power transformers, the number of primary ampere-turns is a fixed quantity for any given primary current.  Upon open-circuiting the secondary circuit, the primary ampere-turns is not reduced.  Thus a very high flux density is produced in the core because of the absence of the "counter" ampere-turns due to the secondary current.  The resulting flux density results in a very high induced voltage in the secondary winding, with subsequent high stresses on the insulation and possible danger to operators.  Furthermore, the high magnetizing forces acting on the core may, if suddenly removed, leave behind considerable residual magnetism in the core, so that the ratio obtained after such an opening may be appreciably different from that before.  It is for this reason that even when not in use for measuring purposes, the secondary circuit is shorted when primary current is flowing.  There is, of course, no danger when the secondary winding is short-circuited, since when used in conjunction with an ammeter or wattmeter, it is practically short-circuited.  As Fig. 9-4 shows, a shorting switch is thus necessary for the removal of the ammeter.  Such switches are an integral part of these transformers.

*Potential transformers* are specially designed, extremely accurate-ratio step-down transformers.  They are used with ordinary low-range voltmeters.  In general, they differ very little from the ordinary two-winding transformer discussed,

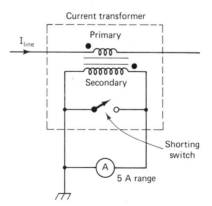

**Figure 9-4**  Current transformer connections.

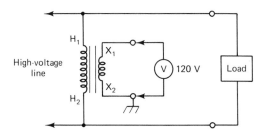

**Figure 9-5** Potential transformer connections.

except that they handle very small amounts of power. Once installed, it has a constant load, since the secondary is always connected to the same instruments. The high-voltage circuit whose voltage is to be measured may vary in potential and the transformer must be designed for the maximum possible primary voltage. Ordinarily, 150-V voltmeters are used. A potential transformer used to measure a voltage of 12,000 V will have a transformer ratio of approximately 100:1. The load on the transformer consists of a high-resistance voltmeter; thus the currents are small. Fine wire of a large number of turns is used, making the transformers smaller and lighter, since the core flux density is greatly reduced. For safety reasons the secondary side is often grounded and well insulated from the high-voltage side, as shown in Fig. 9-5.

## 9-5 PARALLEL OPERATION OF TRANSFORMERS

For satisfactory operation of transformers in parallel, the following conditions should be fulfilled:

1. Primary and secondary voltage ratings should be identical; this implies that the transformation ratios are the same.
2. The transformer connections should be made with proper regard to polarities.
3. The equivalent impedances should be inversely proportional to the respective kVA ratings.
4. The ratio of equivalent resistance to reactance $(R_e/X_e)$ should be the same.

These conditions are most easily met by paralleling transformers which are identical and of the same make and model. Figure 9-6 shows how two transformers should be connected in parallel when they have the same polarities.

    With reference to Fig. 9-6, let the transformation ratios of the transformers be $a$; their equivalent impedances $Z_1$ and $Z_2$ and the secondary currents $I_1$ and $I_2$ of transformers 1 and 2, respectively. $V_L$ and $I_L$ are the secondary load voltage and current. Applying Kirchhoff's laws yields

$$I_1 + I_2 = I_L \qquad (9-15)$$

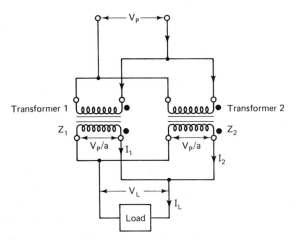

**Figure 9-6** Two transformers in parallel.

and

$$V_L = \frac{V_p}{a} - I_1 Z_1 = \frac{V_p}{a} - I_2 Z_2 \tag{9-16}$$

From Eq. (9-16) we obtain

$$I_1 Z_1 = I_2 Z_2 \tag{9-17}$$

Substituting $I_2 = I_L - I_1$ from Eq. (9-15) into Eq. (9-17) and solving for the current $I_1$ gives

$$I_1 = I_L \frac{Z_2}{Z_1 + Z_2} \tag{9-18}$$

Similarly, we can determine the current $I_2$ by substituting $I_1 = I_L - I_2$ from Eq. (9-15) into Eq. (9-17), which results in

$$I_2 = I_L \frac{Z_1}{Z_1 + Z_2} \tag{9-19}$$

It is immediately apparent from Eq. (9-18) and (9-19) that if the transformers are to share the load in proportion to their ratings, their kVA ratio should be inversely proportional to their impedance ratio, that is,

$$\frac{I_1}{I_2} = \frac{kVA_1}{kVA_2} = \frac{Z_2}{Z_1} \tag{9-20}$$

Furthermore, the transformer impedances are complex quantities, and if they are to operate at the same power factor, their impedances should have the same phase angle, or

$$\frac{X_1}{R_1} = \frac{X_2}{R_2} \tag{9-21}$$

The currents then add arithmetically and are in phase with the load current.

## EXAMPLE 9-3

The following information is given for two transformers to be connected in parallel:

| Transformer 1 | Transformer 2 |
|---|---|
| Rating = 25 kVA | Rating = 35 kVA |
| 2360/230 V | 2360/230 V |
| $Z_e = 0.3 + j3.0 \; \Omega$ | $Z_e = 0.2 + j2.0 \; \Omega$ |
| (in secondary terms) | (in secondary terms) |

Calculate the load current and kVA delivered by each transformer when supplying a load current of 180 A at a power factor of 0.8 lagging.

**SOLUTION**

$$Z_1 = 0.3 + j3.0 = 3.015 \; \angle 84.3° \; \Omega$$

$$Z_2 = 0.2 + j2.0 = 2.01 \; \angle 84.3° \; \Omega$$

$$Z_1 + Z_2 = 0.5 + j5.0 = 5.025 \; \angle 84.3° \; \Omega$$

Since

$$I_L = 180 \times (0.8 - j0.6) = 144 - j108 = 180 \; \angle -36.9° \; \text{A}$$

$$I_2 = 180 \; \angle -36.9 \times \frac{3.015 \; \angle 84.3°}{5.025 \; \angle 84.3°} = 108 \; \angle -36.9° \; \text{A}$$

and

$$I_1 = I_L - I_2 = 180 \; \angle -36.9° - 108 \; \angle -36.9° = 72 \; \angle -36.9° \; \text{A}$$

Therefore,

$$\text{kVA}_1 = 108 \times \frac{230}{1000} = 24.84$$

$$\text{kVA}_2 = 72 \times \frac{230}{1000} = 16.56$$

In Example 9-3 the turns ratios of the transformers in parallel were identical. If the turns ratios of two transformers in parallel supplying a common load are unequal, it can be shown that a circulating current will arise between the secondaries. In this book we will not consider the circulating current.

## 9-6 THREE-PHASE TRANSFORMER CONNECTIONS

All electric power generated is distributed by means of a three-phase voltage system. Three-phase power may be transformed by using suitably arranged three single-phase transformers or by using one three-phase transformer. Transformers may

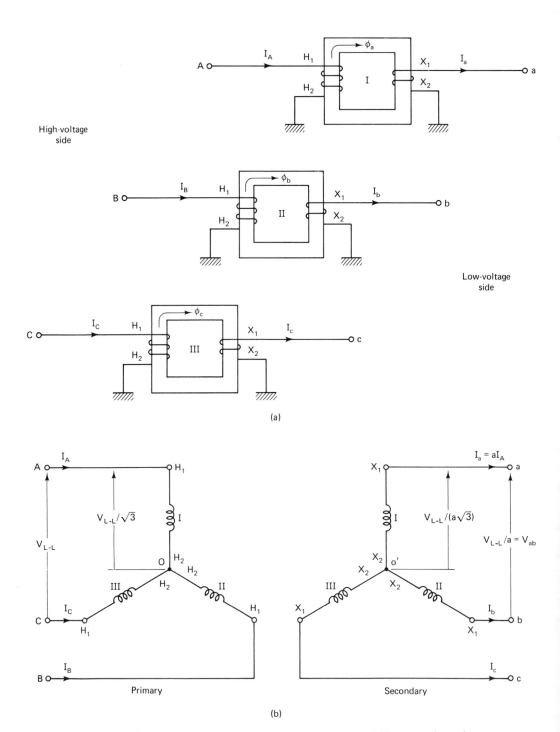

**Figure 9-7**  Three single-phase power transformers arranged in a Y-Y connected transformer bank: (a) physical setup; (b) electric circuit diagram; (c) voltage phasor diagram.

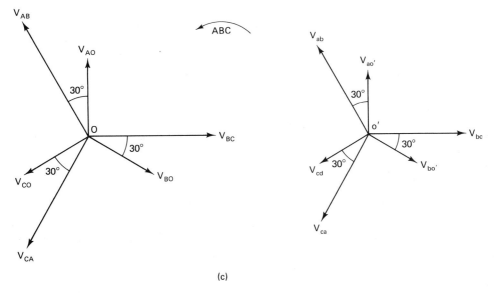

(c)

**Figure 9-7**   Continued

also be connected so as to change the number of phases in a system. Although many combinations are employed depending on the particular application, we limit our discussion to some of the commonly applied arrangements.

When we talk of a certain arrangement, it refers to the way in which the transformers have been interconnected. The primaries may be connected in either wye (Y) or delta (Δ). The secondaries can be connected in this way as well. This gives four basic connections, Y-Y, Y-Δ, Δ-Δ, and Δ-Y. In the discussion to follow it will be assumed that all transformers are identical. This is understood to mean that they have identical kVA ratings, the same transformation ratios, and the same internal impedances. As before, identical impedances also implies that their $X/R$ ratios are the same. Primary line terminals will be identified by A, B, and C, while the secondary terminals are designated as a, b, and c. The neutral points are specified by $O$ and $O'$ for the primary and secondary sides, respectively.

### Y-Y Connection

The simplest way of transforming three-phase power is shown in Fig. 9-7a. Three identical single-phase transformers are connected as shown, one in each phase. The whole arrangement is generally referred to as a *transformer bank*. The corresponding electric circuit diagram is given in Fig. 9-7b. If the primary line-to-line voltage is $V_{L-L}$, the line-to-neutral voltage on the primary side is $V_{L-L}/\sqrt{3}$. If a transformation ratio $a > 1.0$ is assumed (i.e., a step-down transformer), the line-to-neutral voltage on the secondary side will be $V_{L-L}/(a\sqrt{3})$. If the line current on the primary side is $I$, which is also the winding or phase current, the line current on the secondary side is $aI$.

The impressed set of primary voltages are 120° displaced from each other.

The induced secondary voltages are therefore also displaced by 120°. The symmetrical connections then produce a 120° phase shift in the secondary line voltages, as shown in Fig. 9-7c, hence a true three-phase system. Note that the phase sequence is unaffected.

With reference to Fig. 9-7b it can be seen that the voltage across the primary windings is only $V_{L-L}/\sqrt{3}$ volts. The secondary phase voltage is $V_{L-L}/(a\sqrt{3})$, with the secondary line-to-line being $V_{L-L}/a$. A major advantage of this connection is self-evident, namely, the transformer windings are subject to less voltage stress, only $1/\sqrt{3}$ or 57.7% of the line voltage. As we see, the line currents equal the phase currents, requiring a larger cross-sectional area. The combination of less required voltage insulation but larger cross-sectional conductor area generally results in a less expensive transformer arrangement compared to a Δ-Δ transformer bank. Practically, with an increase in conductor size the mechanical strength increases. This is beneficial for greater short-circuit protection of the transformer. We must realize that accidents do happen and transformers can be subjected to short circuits.

For instance, during adverse weather conditions lines could break and fall to the ground resulting in possible short circuits. Also, a tree falling across the wires might cause a short circuit. Although protective circuits will quickly isolate the transformer electrically, the short circuit occurs before corrective action can be taken. This interval between actual short circuit occurring and breaker operation is generally short, measured in terms of cycles of the frequency. However, the transformer should not be permanently damaged in this period. The lower phase voltages have the advantage of reducing the core size because a lower core flux is required.

Another advantage of the Y-Y connections is the ability to relate all line voltages on both sides of the transformers to the neutral point. On distribution transformers the low-voltage side can then supply single-phase loads which may have one side grounded.

The Y-Y connection has a major objection when the three-phase load is unbalanced and it is necessary to operate without a primary neutral conductor. With an unbalanced load, the electrical neutral will shift from its center to a point that will make the three voltages unequal. Without the primary neutral the third harmonic current, because of saturation of the core as we have seen in Section 8-4, cannot exist and the phase voltages on both sides of the transformer will contain third harmonic components.

Let us examine this further. In Fig. 8-8 we showed that the magnetization current in a transformer contained harmonics. If we now show the 60-Hz three-phase voltage waveforms with their third harmonic components (180 Hz) in each phase, there results the situation illustrated in Fig. 9-8. As we see, the harmonic components are all in phase.

Since the 180-Hz phase components are in phase, they add up to the neutral current of a magnitude three times that of the individual phase components. We could provide a fourth return conductor for this current back to the generator, but this is costly. Also, 180 Hz causes interference in communication networks, so this is not a good alternative.

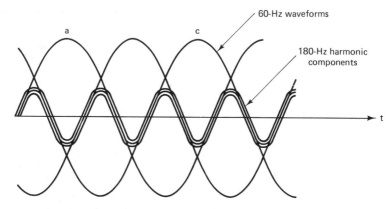

**Figure 9-8**  Voltage waveforms with 180-Hz current components.

What can be done is to isolate the neutral and prevent the formation of the harmonics. But when you do this, the third harmonic component will distort the flux waveform. If we let the magnetization current be sinusoidal, the flux and the voltage waveform cannot be sinusoidal. All this means is that since the flux is now distorted, the output voltage waveforms are distorted and hence they have a harmonic content.

The problem can be solved by using a transformation where the windings are connected in delta (Δ). The Δ-loop permits the formation of the resulting 180-Hz current components. It will circulate in the Δ without entering the network, and no distortion of the voltage waveforms takes place.

### Δ-Δ *Connection*

This arrangement is illustrated schematically in Fig. 9-9a with its corresponding circuit diagram in Fig. 9-9b. This transformer bank, or *cluster*, as it is called when they are mounted on a single pole, is generally used in systems where the voltages are not very high. It is also used where continuity of service is essential even when one of the transformers should fail. The faulty transformer may be removed from the circuit and reduced output supplied by the two remaining transformers. Operation continues on what is known as an *open-delta connection*. In this situation, part of the load will be carried by the other two transformers until the faulty one can be changed.

Contrary to unbalanced loading in a Y-Y connection, the three-phase load voltages in a Δ-Δ connected transformer bank remain substantially equal regardless of the degree of unbalance introduced. An unbalanced load causes a small unbalance in the voltages only to the extent to which the internal voltage drops in the individual transformers differ.

The line voltage ratio for the Δ-Δ transformation is equal to the transformation ratio of the individual transformers. The transformer coils are now subjected to the full line voltage and heavier insulation is needed. On the other hand, the winding currents are now $1/\sqrt{3}$ or 57.7% of the line currents, thereby requiring a smaller conductor cross section. Thus compared to the Y-Y connection, the Δ-Δ

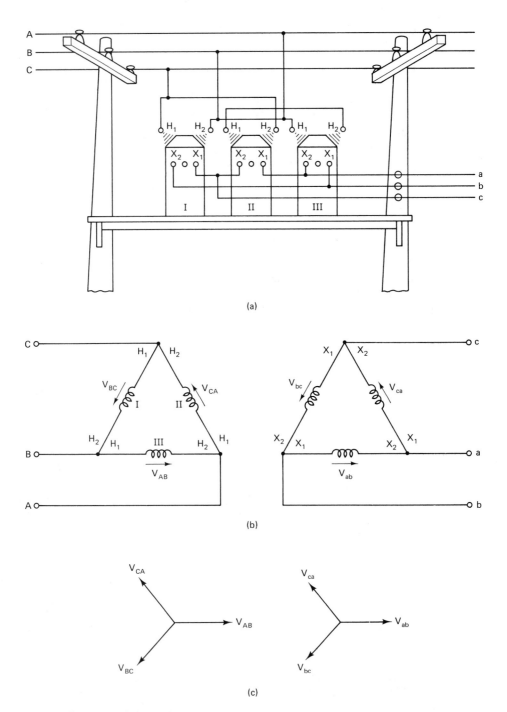

**Figure 9-9**  Δ-Δ transformer connection: (a) pole-mounted arrangement for local power distribution; (b) circuit diagram; (c) phasor diagram.

connection results in windings with greater voltage stresses. In addition to smaller wire diameter, the mechanical strength of the windings is less. Therefore, Δ-Δ transformer banks are generally used in moderate voltage applications. Another disadvantage is that only one voltage is available at the load side, since we have no neutral present. This connection does have an advantage when the third harmonic component is present. The delta connection allows it to circulate in the closed delta so that the third harmonic does not appear in the output voltage waveform.

## EXAMPLE 9-4

A 2400-V three-phase bus supplies a bank of three single-phase transformers which delivers 600 kVA to a balanced three-phase 240-V resistive load. Assuming a Y-Y transformation, determine:

(a) The voltage across each transformer winding and the current through it.

(b) The kVA rating of each transformer.

## SOLUTION

(a) For a three-phase system, the primary line current is

$$I_{L_p} = \frac{\text{kVA} \times 1000}{\sqrt{3} \times V_{L_p}} = \frac{600{,}000}{\sqrt{3} \times 2400} = 144.3 \text{ A}$$

The secondary line current is

$$I_{L_s} = \frac{\text{kVA} \times 1000}{\sqrt{3} \times V_{L_s}} = \frac{600{,}000}{\sqrt{3} \times 240} = 1443 \text{ A}$$

The primary winding or phase voltage is

$$V_{\text{ph}}(\text{prim}) = \frac{2400}{\sqrt{3}} = 1386 \text{ V}$$

and the secondary winding voltage is

$$V_{\text{ph}}(\text{sec}) = \frac{240}{\sqrt{3}} = 138.6 \text{ V}$$

Note that the transformation ratio $a = 2400/240 = 10$. Thus secondary quantities could also be obtained from

$$I_{L_s} = aI_{L_p} \quad \text{and} \quad V_{\text{ph}}(\text{sec}) = \frac{V_{\text{ph}}(\text{prim})}{a}$$

(b) The individual transformer ratings are obtained as follows:

$$\text{kVA} = \frac{V_{\text{ph}}(\text{prim}) \times I_{L_p}}{1000} = \frac{1386 \times 144.3}{1000} = 200 \text{ kVA}$$

As can be seen, each transformer carries one-third of the total load power.

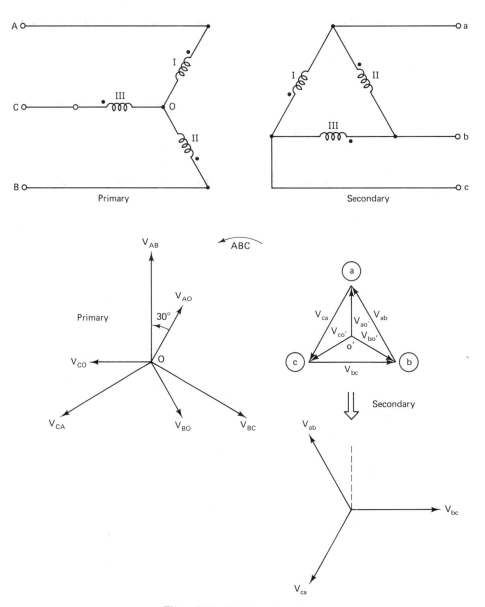

**Figure 9-10** Y-Δ transformation.

## Y-Δ and Δ-Y Connections

These connections with their associated phasor diagrams are shown in Figs. 9-10 and 9-11, respectively. They are considered the most satisfactory three-wire three-phase connections. In general, they combine the advantages of both Y-Y and Δ-Δ connections. The wye neutral can be grounded when Δ-Y connected, enabling grounding of the loads. Similarly, in the Y-Δ connection the ground may be

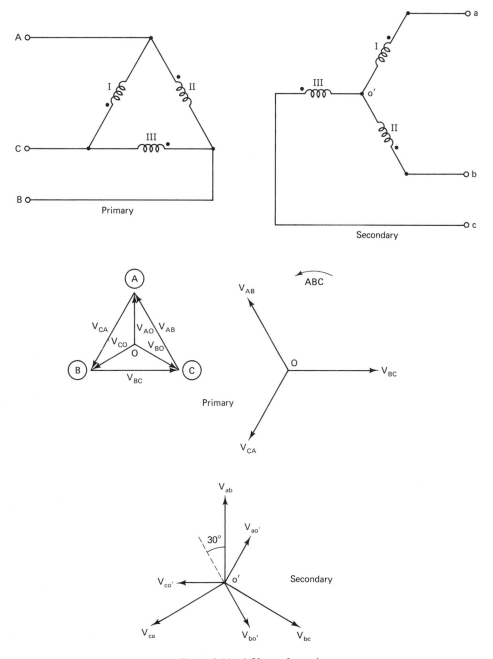

**Figure 9-11** Δ-Y transformation.

connected to the generator neutral. Furthermore, the neutral is stable and being locked by the delta.

It can be shown that the third-harmonic component present in the magnetizing current circulates in the delta, preventing a third-harmonic component from being present in the phase voltages on the Y-side; therefore, the line voltages are free of third-harmonic components.

Similarly, in the Y-Δ connection, no third-harmonic magnetizing current can be drawn from the line, hence no third-harmonic voltages will be generated because of their absence. Although they are generated, they all appear in the same direction on both the primary and secondary windings. Since the delta forms a closed loop for the third-harmonic voltage, a circulating current is created of a frequency three times that of the supply. This current will generate a flux, which in turn sets up a counter third-harmonic voltage of such magnitude as to oppose the one otherwise being present.

The Y-Δ connection is generally used where the voltage is to be stepped down, such as at the end of a transmission line. Also, it is employed in moderately low voltage distribution systems for stepping down from transmission-line voltages directly to 480 or 230 V. As this connection results in a 30° phase shift between primary and secondary voltages, they can therefore not be paralleled with either a Y-Y or Δ-Δ connected transformer bank. The Δ-Y transformer, on the other hand, is used in step-up voltage applications, where advantage is taken of the secondary windings being at $1/\sqrt{3}$ volts of that of the line voltage. This connection also results in a phase shift of 30° between primary and secondary voltages.

## EXAMPLE 9-5

Three 10-kVA 2300/460-V transformers are connected in Y-Δ to supply a balanced three-phase load of 18 kW at 460 V at a power factor of 0.8 lagging. Determine:

(a) The current in each of the primary and secondary windings of the transformers.

(b) The primary and secondary line currents.

## SOLUTION

(a) Since the load is balanced, each transformer delivers one-third of the total load,

$$\frac{18}{3} = 6 \text{ kW at a power factor of 0.8 lagging}$$

Therefore,

$$I_s(\text{transformer winding}) = \frac{6000}{460 \times 0.8} = 16.3 \text{ A}$$

and

$$I_p(\text{transformer winding}) = \frac{6000}{2300 \times 0.8} = 3.26 \text{ A}$$

(b) For the Y-connected primary side, the winding (or phase) current equals the line current; thus

$$I_{L_p} = 3.26 \text{ A} \quad \text{and} \quad V_{L\text{-}L} = \sqrt{3}\,2300 = 3984 \text{ V}$$

For the $\Delta$ connection on the secondary side, the line current is $\sqrt{3}$ times greater than the winding current; hence

$$I_{L_s} = \sqrt{3} \times 16.3 = 28.2 \text{ A}$$

As a check,

$$P_{\text{in}} = \sqrt{3}\,V_{\text{L-L}}I_L \cos\theta$$
$$= \sqrt{3} \times 3984 \times 3.26 \times 0.8 = 18 \text{ kW}$$

## 9-7 THREE-PHASE TRANSFORMERS

With increasing power levels in the transformation of three-phase power, it becomes more and more economical to use a three-phase transformer rather than, as has been discussed so far, an arrangement of three separate single-phase transformers. The unique assembly of the windings on the core makes it possible to save a great quantity of iron by interlinking the magnetic structures so that the same iron is used simultaneously by the three phases. One other main advantage is that the costly high-voltage terminals to be brought out of the transformer housing are reduced to three, rather than the six necessary in case of three separate single-phase transformers.

Figure 9-12 shows three core-type transformers placed together to form a single unit having a common path for the return flux. To simplify the illustration, only the primary windings have been shown on the outside legs. If the three transformers are identical, their fluxes will be sinusoidal and balanced. But al-

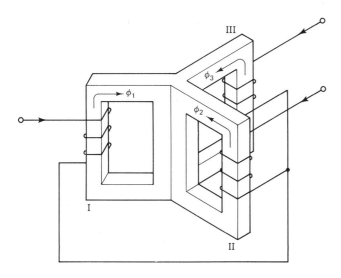

**Figure 9-12**  Core-type three-phase transformer.

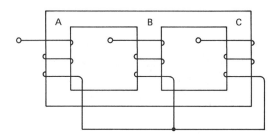

**Figure 9-13** Core-type three-phase transformer.

though they have the same magnitudes, they differ in phase by 120°. Therefore, if the fluxes merge in the common leg, the total resulting flux is

$$\boldsymbol{\phi} = \boldsymbol{\phi}_1 + \boldsymbol{\phi}_2 + \boldsymbol{\phi}_3 = \phi_1 \angle 0° + \phi_2 \angle -120° + \phi_3 \angle -240°$$

will be zero and therefore it is not necessary. Hence the common leg is omitted and the core-type three-phase transformer manufactured will look like that shown in Fig. 9-13. The result is a substantial saving in core material.

The resulting structure does not have complete symmetry since the magnetic reluctance of winding *B* differs somewhat from those of *A* and *C*. The result is a

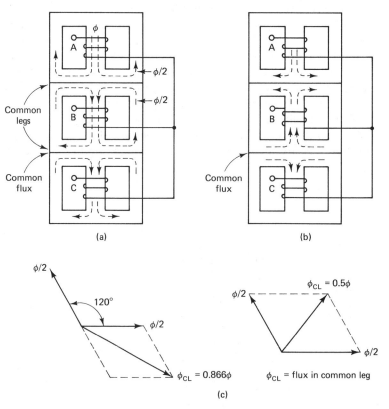

**Figure 9-14** Development of a shell-type transformer: (a) coil *B* wound in same direction; (b) coil *B* wound in opposite direction; (c) difference and sum of flux components in common legs.

slight unbalance in the excitation currents; however, this effect is negligible, especially at full load.

The shell-type three-phase transformer may be represented as in Fig. 9-14, in which three single-phase transformers have been stacked one above the other. In Fig. 9-14a, the three primary windings shown are wound in the same direction. The flux in the paths that are common between adjacent phases is thus equal to the difference of the two phase fluxes. The fluxes are 120° apart in time. Thus the mutual flux is equal to $\sqrt{3}(\phi/2)$, or 0.866 of the flux in the center leg. Maintaining the same flux density throughout the core will thus require less iron in the common legs.

In the usual shell-type construction, the center coil, phase B, has its winding direction reversed as shown in Fig. 9-14b. The flux in the common leg is now the sum of the fluxes of the two adjacent phases. The sum of the two fluxes equals the magnitude of either flux alone, which amounts to $0.5\phi$. Figure 9-14c demonstrates the common resultant flux. What it amounts to is simply this—by reversing coil B a further reduction in iron is obtained. Of course, the less iron that is necessary in the construction of a transformer, the greater the savings in costs. Thus for the same kVA rating, a three-phase transformer usually costs less than three individual transformers, weighs less, has a higher efficiency, and occupies less space. In addition, fewer connections have to be made to the external circuit.

There are, however, some disadvantages; namely, the three-phase transformer is heavier and bulkier than the single-phase transformer, but above all, if one coil happens to break down in a three-phase transformer, the whole transformer must be removed.


## REVIEW QUESTIONS

**9-1.** Under what conditions is it necessary to know the polarity markings of the transformer windings?

**9-2.** What is an autotransformer? What advantages does it have? Disadvantages?

**9-3.** Illustrate how a two-winding transformer can be connected as an autotransformer in
   (a) step-up and
   (b) step-down configuration.

**9-4.** Why can a two-winding transformer connected as an autotransformer handle so much more power?

**9-5.** What are current transformers? Potential transformers?

**9-6.** What special precautions must be taken in the use of a current transformer?

**9-7.** Give the conditions to be fulfilled to parallel two transformers.

**9-8.** Why are transformers paralleled?

**9-9.** State the four basic connections in which three single-phase transformers can be connected when employed to transform three-phase power.

**9-10.** How does the voltage ratio of a Y-Y transformer bank compare to the individual transformer turns ratios?

**9-11.** Why is grounding the neutral of a Y-Y transformer bank desirable?

**9-12.** Give a practical disadvantage of the Y-Y transformer bank connection.

**9-13.** List some advantages and disadvantages of the Δ-Δ transformation.

**9-14.** Explain why the Y-Δ transformation is particularly advantageous.

**9-15.** List the advantages and disadvantages of three-phase transformers.

**9-16.** What are the general types of three-phase transformers? Does one construction have an advantage over the other?

## PROBLEMS

**9-1.** A transformer having four windings is represented symbolically in Fig. 9-15.
(a) Place appropriate polarity markings on windings 3 and 4.
(b) How many turns has winding 4?

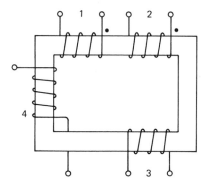

Figure 9-15

**9-2.** A 125-kVA 2400/240-V transformer has its windings connected in series to form a step-down autotransformer. What will be the voltage ratio and output when fully loaded?

**9-3.** A 10-kVA 2200/460-V transformer is connected as an autotransformer to step up the voltage from 2200 V to 2660 V. When used to transform 10 kVA, determine the kVA load output.

**9-4.** A polarity test is performed upon a 2200/440-V transformer. If the input voltage is 110 V, calculate the voltmeter reading if:
(a) The polarity is additive.
(b) The polarity is subtractive.

**9-5.** A 20:5 current transformer is connected to a 5-A ammeter that indicates 4.45 A. What is the line current?

**9-6.** A 50-hp 440-V three-phase motor with an efficiency of 0.88 and a power factor of 0.82 on full load is supplied from a 6600/440-V Δ-Y connected transformer. Calculate the currents in the high- and low-voltage transformer windings when the motor is running at full load.

**9-7.** Two 10-kVA 4600/230-V single-phase transformers $A$ and $B$ are to operate in parallel. The equivalent impedance for the transformers in secondary terms are 0.16 Ω and 0.24 Ω, respectively. They have identical $X/R$ ratios. Calculate the kVA load carried by each transformer if the total load supplied is 120 kVA.

**9-8.** A three-phase 60-Hz shell-type transformer has an iron cross section of 420 cm². If the flux density is to be limited to 1.2 T (Wb/m²), find the number of turns per phase on the high- and low-voltage windings. The voltage ratio is 12,200/660 V. The high-voltage side is star connected, and the low-voltage side delta connected.

**9-9.** Three transformers connected Δ-Y step down the voltage from 12,600 to 660 V and deliver a 55-kVA load at a power factor of 0.866 lagging. Calculate:

(a) The transformation ratio of each transformer.

(b) The kVA and kW load in each transformer.

(c) The load currents.

(d) The currents in the transformer windings.

**9-10.** A Y-Δ transformer bank supplies a balanced load of 500 kW, 1100 V, 0.85 PF lagging. Determine the primary and secondary voltages and currents. The primary line voltage is 1100 V.

**9-11.** A 13,200-V three-phase generator delivers 10,000 kVA to a three-phase 66,000-V transmission line through a step-up transformer. Determine the kVA, voltage, and current ratings of each of the single-phase transformers needed if they are connected:

(a) Δ-Δ.

(b) Y-Δ.

(c) Y-Y.

(d) Δ-Y.

**9-12.** A 30-hp 480-V three-phase motor with an efficiency of 0.9 and a power factor of 0.82 lagging on full load is supplied from a 2400/460-V Δ-Y connected transformer. Calculate the currents in the high- and low-voltage transformer windings when the motor is running at full load.

**9-13.** A Y-Δ transformer bank may be connected in parallel with a Δ-Y transformer bank. It cannot be paralleled with a Δ-Δ connection. With the aid of a phasor diagram, explain why.

# 10

# THREE-PHASE INDUCTION MOTORS

## 10-1 INTRODUCTION

Initial power distribution on a broad scale was in the form of direct current. Voltage levels were low (110 to 220 V dc), which resulted in poor regulation and efficiencies. Practical synchronous generators and transformers were known by 1885-86, but the ac motor was not.

All this changed with the invention of the induction motor in the late 1880s by Nikola Tesla (1856-1943). The whole concept of polyphase alternating current, including the induction motor, was his. The system was patented in 1888. The first large-scale application of the Tesla polyphase ac system was the Niagara Falls hydroplant, completed in 1895.

Today, most industrial electric machines are three-phase induction motors. The number of phases thus corresponds to the number of phases in commercial power systems. Induction motors are rugged, relatively inexpensive, and require very little maintenance. They range in size from a few watts to about 10,000 hp. Large induction motors (usually above 5 hp) are invariably designed for three-phase operation, the reason being that one desires a symmetrical network loading. Small fractional-horsepower motors are usually of single-phase design, which are referred to in Chapter 11.

The speed of an induction machine is nearly constant, dropping only a few percent from no-load to full-load operation. They do have certain disadvantages, however, in that:

1. The speed is not easily controlled.

2. They run at low and lagging power factors when lightly loaded.

3. The starting current is usually five to seven times full-load (rated) current.

In this chapter we study the induction motor characteristics. Starting methods and some modern speed-control concepts of induction motors are examined in Chapter 13.


## 10-2 GENERAL DESIGN FEATURES

The induction motor with a squirrel-cage rotor is the most commonly used ac motor of all rotating electrical machines. Typical design features of this motor are illustrated in Fig. 10-1. The stator is composed of laminations of high-grade steel with slotted inner surface to accommodate a three-phase winding. This winding is essentially identical to that found in the stator of synchronous generators as described in Chapter 7 (Fig. 7-4). The speed of the induction motor depends on the number of poles the winding is wound for, as discussed in Section 10-3.

The rotor winding design varies depending on the need for torque or speed control. However, the most common rotor design is shown in Fig. 10-1b, where a squirrel-cage winding is illustrated. It consists of solid copper or aluminum bars embedded in rotor slots, each bar short-circuited by end rings. In the smaller motors, the complete rotor winding, including bars, end rings, and fans, is cast in place. In the larger motors, the conductors are formed bars, generally copper, and inserted in slots in the rotor laminations. There are no slip rings or carbon brushes, making this construction virtually maintenance-free.

The rotor conductors may be nearly parallel to the machine axis, or skewed. This is to provide a more uniform torque and to reduce noise during operation. Furthermore, it prevents the rotor and stator teeth from lining up opposite each other and thus lock in place, due to what is normally referred to as the *reluctance torque*. More will be said about this in Chapter 14. This locking in is due to the magnetic circuit tending to minimize its circuit reluctance. To prevent this from happening, the rotor bars are skewed.

The second type of rotor construction is the wound-rotor induction motor. In this type of construction the rotor has a three-phase winding as well, similar to the stator and wound for the same number of poles as the stator winding. The rotor winding terminates in slip rings mounted on the rotor shaft. Brushes ride on the slip rings, and during starting they are connected externally to a three-phase resistor bank (one in each phase and wye-connected) (see Fig. 10-1c). The external resistances are shorted out simultaneously in one or more steps as the motor comes up to speed. Thus this rotor construction has the added feature of allowing variable external three-phase resistors to be added in series to the rotor winding for starting purposes and providing speed control. Details of these operations are discussed in later sections.

Regardless of the rotor construction employed, voltage is induced in the rotor winding by a rotating flux of constant magnitude set up by the stator winding. The exact mechanism of the rotating magnetic field produced by the stationary three-

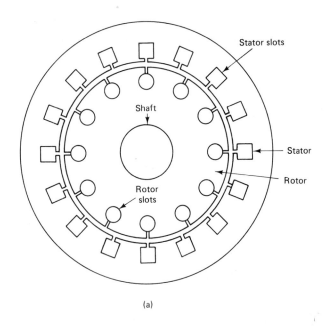

Stator slots

Shaft

Stator

Rotor

Rotor slots

(a)

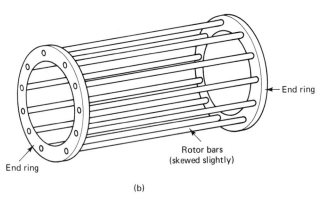

End ring

End ring

Rotor bars
(skewed slightly)

(b)

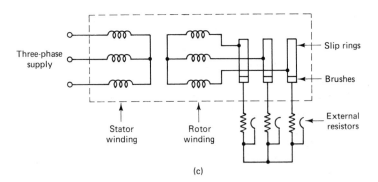

Three-phase
supply

Slip rings

Brushes

External
resistors

Stator
winding

Rotor
winding

(c)

**Figure 10-1**  Induction motor construction: (a) section of the magnetic circuit in an induction motor; (b) squirrel-cage rotor construction; (c) wound-rotor induction motor.

phase stator winding is discussed in Section 10-3. In any event, when the rotor with its closed winding is placed in the revolving magnetic field set up by the currents in the stator windings, the flux cutting the rotor bars generates EMFs in them. The resulting current flow in this rotor circuit is proportional to these induced EMFs and limited only by the rotor impedance. The induced rotor currents in turn produce an armature field having magnetic poles on the rotor surface. Magnetic forces between the rotating field and the magnetic poles on the rotor surface produce a torque, which will be in the direction of the rotating stator field. The squirrel-cage winding adjusts itself to any number of poles, and the exact number and shape of the rotor bars used is dictated by the desired motor characteristics.

At no load, the rotor runs almost as fast as the rotating field. Since the relative speed between the rotor and field is then small, the induced EMFs are small and subsequently the induced currents in the rotor bars are small. When the motor is loaded, the rotor starts to lag or slip behind the field in speed and larger rotor currents are developed to compensate for the increased load.

Current flow in the rotor circuit is by induction. Because of this it becomes important that the air gap in induction motors be made as small as possible. The magnetizing current needed to establish the pole flux is a considerable component of the current flowing in the supply lines. Usually, this component is in the order of 30 to 50% of rated current, which is significantly larger than that found in transformers. Furthermore, the leakage reactances are larger and therefore induction motors are found to operate at low power factors at light loads, improving greatly at values approaching full-load conditions.

We now turn our attention to the stator winding and "see" how a stationary three-phase winding connected to a three-phase voltage supply produces a rotating field of constant magnitude.

## 10-3 ROTATING FIELD: GRAPHICAL ANALYSIS

The three-phase stator winding is distributed around the internal surface of the core so as to create a rotating field when connected to a three-phase power supply. It will now be shown that with three currents 120° displaced in time in a three-phase winding which is distributed 120° in space, a field is created that rotates, although the physical poles were actually rotated mechanically.

When the three-phase winding is connected to a three-phase power supply, three currents tend to flow, one in each phase winding. They will be displaced from each other by 120 electrical degrees as shown in Fig. 10-2, assuming an *abc* phase sequence. Next, the following convention will be adopted. With reference to Fig. 10-3, phase winding *a* when carrying positive current will produce a resulting MMF in the positive direction of its magnetic axis, as determined by the right-hand rule. Furthermore, the positive current is assumed to enter the winding at *a*, and directed into the page as indicated by $\oplus$. When the current $i_a$ becomes negative, the current direction reverses and the MMF direction will also be opposite. Similar reasoning is applied to phase windings *b* and *c*. Referring to Fig. 10-2, four instants

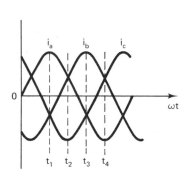

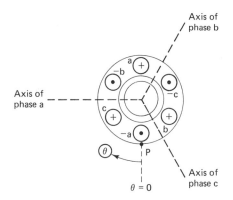

**Figure 10-2** Three-phase currents in stator windings.

**Figure 10-3** Simplified two-pole three-phase stator winding showing polarities for positive phase currents.

in time are selected and it will be shown that the resultant MMF, $\mathscr{F}_t$, is constant in magnitude and revolves the distance covered by one pole or 180 electrical degrees, in one half-cycle.

*At time $t = t_1$*: In Fig. 10-2, $i_a$ is positive maximum; $i_b$ and $i_c$ are negative and equal to one-half their respective maximums. Therefore, the MMFs due to these currents will have corresponding magnitudes, and their directions, as determined by the right-hand rule and our adopted convention, are as indicated in Fig. 10-4a. Combining the three MMFs vectorially yields a resultant MMF designated by $\mathscr{F}_t$, which is 3/2 the maximum MMF per phase and directed, according to our assumptions made, horizontally from right to left.

*At time $t = t_2$*: At this instant, $i_a$ and $i_b$ are positive and equal to one-half their maximum value; $i_c$ is maximum but negative. The MMFs will have corresponding directions and magnitudes and when combined, the resultant MMF is again $\mathscr{F}_t$, 3/2 the maximum MMF per phase and rotated 60° in a clockwise direction compared to that of $\mathscr{F}_t$ at $t = t_1$ (see Fig. 10-4b). Note that the MMF vector rotated 60° in space which follows the time increment from $t_1$ to $t_2$ equals 60 electrical degrees.

*At time $t = t_3$*: $i_b$ is positive maximum and $i_a$ and $i_c$ are negative with a magnitude equal to half their maximum value. Proceeding as before, the individual MMFs, proportional to the current magnitudes and directed as dictated by the current directions, are combined. As before, the resulting MMF is $\mathscr{F}_t$ but further rotated in the clockwise direction an additional 60°, as illustrated in Fig. 10-4c. This corresponds to the time increment from $t_2$ to $t_3$, which is also 60°.

*At time $t = t_4$*: At this instant of time, 60 electrical degrees later than $t_3$, current $i_a$ is negative maximum, while $i_b$ and $i_c$ are one-half their respective maximum and positive. Determining the corresponding MMF directions and magnitudes, they are combined as shown in Fig. 10-4d. The resulting MMF is $\mathscr{F}_t$ and rotated an additional 60° compared to its direction at $t = t_3$. The resulting MMF is thus seen to have rotated 180° in space, or one pole span. This has happened in the time that the current $i_a$ changed from positive maximum to negative maxi-

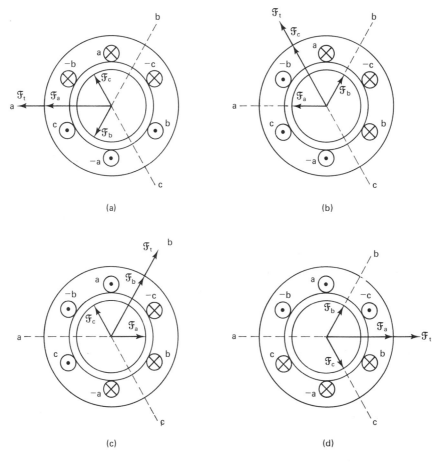

**Figure 10-4** Rotating magnetic field for the four time instants indicated in Fig. 10-2: (a) $t = t_1$; (b) $t = t_2$; (c) $t = t_3$; (d) $t = t_4$.

mum, representing one-half cycle. Therefore, the field will revolve a distance covered by two poles for each cycle of the supply frequency. Thus, continuing for a complete cycle it will be seen that the resultant MMF rotates a full cycle for the two-pole machine. In the event of a four-pole machine, the revolving field would have rotated only one-half a revolution for each cycle; and so on. This implies that the speed of the rotating field is inversely proportional to the number of pole pairs. Also, as discussed, for a two-pole stator the speed of rotation is one rotation per second or 60 rotations per minute, for a frequency of 1 Hz, two rotations for a frequency of 2 Hz, or 60 rotations for 60 Hz. The speed is thus proportional to the frequency of the supply, $f$. Expressing this by a formula gives what is known as the *synchronous speed* $n_s$, namely,

$$n_s = \frac{f}{P/2} \text{ r/s} = \frac{120f}{P} \text{ r/min} \tag{10-1}$$

By analytical analysis the concept of the rotating field can be proven. This analysis, however, will not be included here.

**EXAMPLE 10-1**

Calculate the synchronous speed of an induction motor having six poles and supplied with power from a 60-Hz supply.

**SOLUTION**

$$n_s = \frac{120 \times 60}{6} = 1200 \text{ r/min}$$

It is worth noting that reversing the phase sequence of any of the two currents in Fig. 10-2 results in an MMF rotation opposite the one discussed. This can easily be verified. Furthermore, in practice this is the procedure to reverse the direction of rotation of an induction motor, by reversing any two of the connections to the motor. The ease with which it is possible to reverse the direction of rotation constitutes one of the advantages of three-phase motors.

## 10-4 SLIP AND ROTOR SPEED

The rotor of an induction motor revolves in the same direction as that of the rotating field. It cannot do so at synchronous speed; otherwise, the rotor conductors rotate in unison with the field and no flux is cut. An examination of Fig. 10-5 will make this clear. Imagine for a moment that the rotating air-gap field is rotating clockwise and instantaneously directed as shown by $\phi_s$. As the field sweeps through a coil (indicated by $a-a'$) on the rotor it will induce an EMF in it. Because of the short-circuited rotor circuit, a current will be created in this coil which flows in the direction indicated, as determined by Lenz's law. The rotor current in turn will create its own field $\phi_r$, which for the current direction shown and as determined by the right-hand rule is directed to the left at an angle of 90 electrical degrees to the rotating field. The two fields will create an electromagnetic torque directed in such a way as to align themselves; hence the rotor will follow the direction of the rotating stator field.

We can readily confirm this by determining the direction of the developed force (hence torque) on the coil sides $a$ and $a'$. For example, applying the left-hand rule to coil side $a$ yields a force in the direction indicated by $F$. Thus, indeed, the torque is directed in such a way as to rotate the rotor in the direction of the rotating field. As shown in Fig. 10-5, this results in a tendency for the rotor to follow the main field, in order to catch up with it. As can be appreciated, there must always be relative motion between the rotating field and the rotor conductors; otherwise, there will be no rotor-induced EMFs. Consequently, no rotor-induced currents and no torque will be developed. The difference between the synchronous

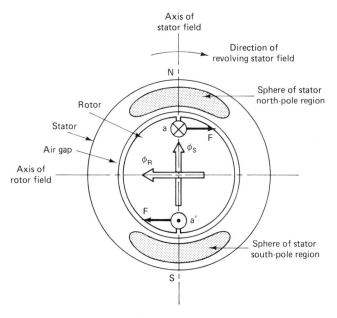

Figure 10-5 Torque development in induction motor schematically predicted for a single rotor coil $a$–$a'$.

speed $n_s$ and the rotor speed $n_r$ is called the *slip s* and is expressed as a percentage of synchronous speed:

$$\% \text{ slip} = \frac{n_s - n_r}{n_s} \times 100 \tag{10-2}$$

The rotor speed may be expressed as

$$n_r = (1 - s) \times n_s \quad \text{r/min} \tag{10-3}$$

where $s$ is expressed as a decimal.

### EXAMPLE 10-2

The induction motor of Example 10-1 has a rotor speed at full load of 1140 r/min.

(a) Determine the slip.
(b) If the load is reduced so that the slip $s = 0.02$, calculate the new rotor speed.

### SOLUTION

(a) From Example 10-1, $n_s = 1200$ r/min, so

$$\text{slip} = \frac{1200 - 1140}{1200} \times 100 = 5\%$$

(b) $n_r = (1 - 0.02) \times 1200 = 1176$ r/min.

At standstill, when the rotor is at rest, the rotating field sweeps the rotor bars at its maximum rate. Under this condition, the generated voltage in the rotor circuit will be maximum and determined by the number of stator turns, stator voltage, and the number of turns on the rotor. As the field revolves, a back EMF is generated in the stator winding, which is nearly equal to the impressed voltage. It thus follows that at standstill, the flux sweeps the stator turns at the same rate as those in the rotor. This means that the induced voltages in rotor and stator turns on a per phase basis are related by the turns ratio, as is the case in a transformer between primary and secondary. It also follows that the frequency of the rotor-induced voltage equals the line frequency when the rotor is at rest. In this situation the slip $s = 1.0$ or 100%. As the slip decreases, the rate at which the flux sweeps across the conductors decreases proportionately and the rotor EMF becomes

$$E_R = s \times E_{BR} \tag{10-4}$$

and the rotor frequency

$$f_R = s \times f \tag{10-5}$$

where
$E_R$ = rotor-induced voltage at slip $s$
$E_{BR}$ = blocked rotor-induced voltage per phase
$f_R$ = rotor frequency

**EXAMPLE 10-3**

A three-phase 60-Hz four-pole 220-V wound-rotor induction motor has a stator winding $\Delta$-connected and a rotor winding Y-connected. The rotor has 40% as many turns as the stator. For a rotor speed of 1710 r/min, calculate:

(a) The slip.
(b) The blocked rotor-induced voltage per phase $E_{BR}$.
(c) The rotor-induced voltage per phase $E_R$.
(d) The voltage between rotor terminals.
(e) The rotor frequency.

**SOLUTION**

(a) $n_s = \dfrac{120f}{P} = \dfrac{120 \times 60}{4} = 1800$ r/min

$s = \dfrac{n_s - n_r}{n_s} = \dfrac{1800 - 1710}{1800} = 0.05$

(b) $E_{BR} = 40\%$ of $\dfrac{V_{\text{stator}}}{\text{phase}} = 0.4 \times 220 = 88$ V/phase

(c) $E_R = sE_{BR} = 0.05 \times 88 = 4.4$ V
(d) $V_{\text{L-L}}(\text{rotor}) = \sqrt{3} \times 4.4 = 7.62$ V
(e) $f_R = sf = 0.05 \times 60 = 3$ H$_z$

As indicated, under normal running conditions the induced rotor voltage and frequency are quite low.

From what has been discussed so far on the three-phase induction rotor, it is apparent that it is essentially a transformer with a short-circuited secondary that is free to move continuously with respect to the primary. It is therefore anticipated that a simplified equivalent-circuit diagram resembles that of the single-phase transformer discussed. This is indeed the situation. It has been shown in Section 10-5 that the induced rotor frequency per phase is $sE_{BR}$. Since this voltage acts in the short-circuited rotor winding, it will set up currents that will be limited only by the rotor impedance. This impedance is made up of two components: (1) the rotor resistance $R_R$, and (2) the leakage reactance $sX_R$, where $X_R$ is the rotor reactance at standstill. Since the reactance is a function of frequency, the leakage reactance is proportional to the slip. As a result, the rotor current becomes

$$I_R = \frac{sE_{BR}}{\sqrt{R_R^2 + (sX_R)^2}} \tag{10-6}$$

If both numerator and denominator of Eq. (10-6) are divided by the slip $s$, we obtain

$$I_R = \frac{E_{BR}}{\sqrt{(R_R/s)^2 + X_R^2}} \tag{10-7}$$

Although this implies a simple algebraic operation, its ramifications are significant. As Eq. (10-7) indicates, the current $I_R$ can now be considered to be produced by a voltage $E_{BR}$ of line frequency, whereas the current determined by Eq. (10-6) is of slip frequency. In other words, the division by $s$ has changed the point of reference from the rotor to the stator circuit. Translating Eq. (10-7) into an equivalent electrical circuit diagram it becomes that shown in Fig. 10-6a.

Since it is convenient to deal with the actual rotor resistance $R_R$, the term $R_R/s$ is split into two components, namely

$$\frac{R_R}{s} = \frac{R_R}{s} + R_R - R_R = R_R + R_R\left(\frac{1}{s} - 1\right) = R_R + R_R\left(\frac{1-s}{s}\right) \tag{10-8}$$

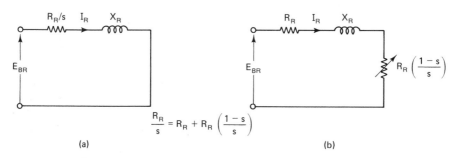

(a)

$$\frac{R_R}{s} = R_R + R_R\left(\frac{1-s}{s}\right)$$

(b)

**Figure 10-6** Rotor circuit diagram on a per phase basis. Resistive element representing rotor copper loss and rotor power developed: (a) combined; (b) separated.

and a corresponding circuit diagram is that of Fig. 10-6b. If Eq. (10-8) is now multiplied through by $I_R^2$, we obtain an equation representing power terms,

$$I_R^2 \frac{R_R}{s} = I_R^2 R_R + I_R^2 R_R \left( \frac{1 - s}{s} \right) \tag{10-9}$$

The left-hand side of the equation represents the total power input to the rotor circuit, which is made up of two components: (1) the power dissipated as copper loss in the rotor circuit $I_R^2 R_R$, and (2) the electric power that is converted into mechanical power, $I_R^2 R_R[(1 - s)/s]$. Thus on a per phase basis,

rotor power input (RPI)

$$= \text{rotor copper loss (RCL)} + \text{rotor power developed (RPD)}$$

where

$$\text{RPI} = I_R^2 \frac{R_R}{s} \tag{10-10}$$

$$\text{RCL} = I_R^2 R_R = s \text{ RPI} \tag{10-11}$$

$$\text{RPD} = I_R^2 R_R \left( \frac{1 - s}{s} \right) = \text{RPI} (1 - s) \tag{10-12}$$

It is rather interesting to note that the mechanical output power is represented in the electrical circuit by a resistance having a value $R_R(1 - s)/s$. In general, the power developed by a motor is the product of its torque and the angular velocity of the rotor. Therefore,

$$P_d = \omega_R T$$

It follows, then, that the developed torque by the motor is

$$T_d = \frac{\text{RPD}}{\omega_R} \quad \text{N} \cdot \text{m} \tag{10-13}$$

where $\omega_R = 2\pi n_R/60$ rad/s and $n_R$ = rotor speed in r/min at which the power is developed. Later it will be seen that to obtain the output torque, the losses must be accounted for.

## 10-7 COMPLETE CIRCUIT DIAGRAM

So far, the rotor circuit has been developed and it was shown that the power transformed across the air gap represents the rotor copper losses and the mechanical power developed by the motor. To complete the equivalent-circuit diagram, the stator circuit must be included. The stator phase winding, having a resistance $R_s$ and a leakage reactance $X_s$, also has a magnetizing branch. This magnetizing branch, unlike the transformer equivalent circuit, cannot be neglected here because of the presence of the air gap. However, in this book we make certain assumptions that will aid significantly in the performance calculations of the motor characteristics

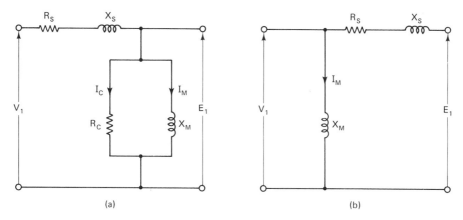

**Figure 10-7**  Stator equivalent circuit diagram of induction motor, per phase.

without sacrificing accuracy. The stator equivalent circuit can be represented by the diagram in Fig. 10-7a. The following assumptions are made:

1. The core loss is assumed constant and obtained from the no-load test to be described later. This means that the resistance $R_c$ can be deleted from the equivalent-circuit diagram. This should not be interpreted as the core loss being negligible, merely as a constant loss which in calculating the motor efficiency must be accounted for.

2. As mentioned when discussing the transformer, the magnetizing current was a negligibly small component of rated current. Here it is approximately 30 to 50% of rated current depending on the size of motor. Thus the magnetizing reactance is an essential component in the equivalent circuit. However, calculations can be simplified considerably with not much loss in accuracy, by moving the magnetizing reactance to the input terminals as shown in Fig. 10-7b.

At this point it remains to combine the rotor and stator circuit diagrams to yield an equivalent diagram on a per phase basis for the induction machine. In doing so, it must be realized that both circuits must be compatible. That is, the rotor parameters must be referred to the stator side, identical to that done for the transformer by referring secondary quantities to the primary side. In other words, Fig. 10-6a or b can be joined to Fig. 10-7b, provided that $E_{BR}$ equals $E_1$. To ensure this equality, the turns ratio between stator and rotor phase windings must be accounted for, such that $E'_{BR} = \alpha E_{BR} = E_1$, where $\alpha$ represents this ratio and $E'_{BR}$ denotes stator-referred rotor quantity. Having done so, the equivalent circuit referred to the stator side is that of Fig. 10-8.

From the equivalent diagram it can now be appreciated why induction motors at light load operate under such poor power factors. At light loads (small slip values), the resistor representing the mechanical power $R'_R(1 - s)/s$ is large. This means that $I'_R$ is relatively small compared to the magnetizing current $I_m$. Therefore, the circuit behaves largely inductive since $X_m$ is the dominating element in the equivalent circuit. Therefore, the circuit behaves almost purely inductively and the power factor is small. With increased loading, the resistance $R'_R(1 - s)/s$ decreases quickly since the slip $s$ increases. Therefore, it rapidly becomes the dominating element in its branch, with a corresponding large improvement in the overall circuit power factor with increasing loads.

The rotor current in Fig. 10-8 in stator terms is

$$I'_R = \frac{V_1}{(R_s + R'_R/s) + j(X_s + X'_R)} \tag{10-14}$$

which enables calculation of the RPI, RCL, and RPD according to Eqs. (10-10) to (10-12). The magnetizing current $I_m$ is

$$I_m = \frac{V_1}{jX_m} \tag{10-15}$$

and therefore the motor line current

$$I_1 = I_m + I'_R \tag{10-16}$$

where the boldface type indicates that the current components must be added vectorially. Since the magnetizing branch is moved to the input terminals, the stator quantities appear as part of the rotor circuit. This would result in a negligible error if $V_1 = E_1$ in Fig. 10-7b.

In practical work, this simplification in the normal operating range of the motor leads to insignificant errors and is therefore adopted as explained. However, for efficiency calculations, the copper stator loss will be calculated as follows:

$$P_{\text{Cu stator}} = 3 \times I_1^2 R_S \tag{10-17}$$

since the stator winding also carries the magnetization current, which, as was mentioned, is significant in terms of rated current. Furthermore, the output or shaft torque is

$$T = \frac{\text{RPD} - \text{mechanical losses}}{\omega_R}$$

where the mechanical losses include friction, windage, and core losses. The mechanical loss is obtained from a no-load test to be discussed yet and need not be separated further.

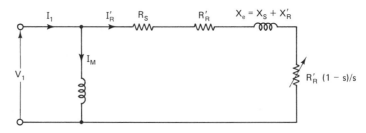

**Figure 10-8** Equivalent circuit diagram for the induction motor on a per phase basis referred to the stator.

## EXAMPLE 10-4

A three-phase 220-V 60-Hz six-pole 10-hp induction motor has the following circuit parameters on a per phase basis referred to the stator:

$$R_S = 0.344 \ \Omega \qquad R'_R = 0.147 \ \Omega$$

$$X_S = 0.498 \ \Omega \qquad X'_R = 0.224 \qquad X_m = 12.6 \ \Omega$$

The rotational losses and core loss combined amount to 262 W and may be assumed constant. For a slip of 2.8%, determine:

(a) The line current and power factor.
(b) The shaft torque and output horsepower.
(c) The efficiency.

### SOLUTION

Assuming a Y-connected stator winding, the phase voltage is $220/\sqrt{3} = 127$ V; the equivalent circuit is given in Fig. 10-9.

(a) $I'_R = \dfrac{127 \ \underline{/0°}}{0.344 + 5.25 + j0.722} = 22.52 \ \underline{/-7.4°} = 22.33 - j2.88$ A

$\quad I_m = \dfrac{127 \ \underline{/0°}}{j12.6} = -j10.08$ A

The line current is

$$I_L = 22.33 - j(2.88 + 10.08) = 25.82 \ \underline{/-30.1°} \ \text{A}$$

The power factor is

$$\text{PF} = \cos(-30.1°) = 0.865 \ \text{(lagging)}$$

(b) $n_s = \dfrac{120 \times 60}{6} = 1200$ r/min

$\quad n_r = (1 - 0.028) \times 1200 = 1166$ r/min

$\quad \omega_r = \dfrac{2\pi \times 1166}{60} = 122.1$ rad/s

$\quad \text{RPI} = \dfrac{3I'^2_R R_R}{s} = 3 \times 22.52^2 \times 5.25 = 7988$ W

$\quad \text{RPD} = \text{RPI}(1 - s) = 7988 \times (1 - 0.028) = 7764$ W

$\quad P_{\text{out}} = \text{RPD} - P_{\text{rot}} = 7764 - 262 = 7502$ W

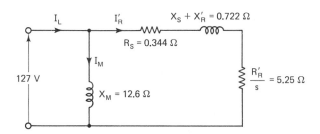

Figure 10-9  Circuit diagram for Example 10-4.

$$T = \frac{7502}{122.1} = 61.4 \text{ N} \cdot \text{m}$$

$$\text{hp} = \frac{7502}{746} = 10.1$$

(c) The losses are as follows:

| | |
|---|---:|
| Rotational + core loss | = 262 W |
| RCL: 0.028 × 7988 | = 224 W |
| Stator copper loss: 3 × 25.82² × 0.344 | = 688 W |
| Total | = 1174 W |

$$\text{efficiency } \eta = \frac{7502}{7502 + 1174} \times 100 \qquad = 86.5\%$$

Note that powers calculated are on a per phase basis; therefore, to obtain three-phase quantities they must be multiplied by 3. Furthermore, the input power can be obtained from

$$P_{\text{in}} = \sqrt{3} \, I_L V_L \cos \theta = \sqrt{3} \times 25.82 \times 220 \times 0.865 = 8510.5 \text{ W}$$

Hence calculating the efficiency from $P_{\text{out}}/P_{\text{in}}$ results in an efficiency of 88.1%. This apparent discrepancy in calculating the efficiency is due to the simplification introduced in the circuit diagram by moving the magnetizing reactance out to the input terminals and the manner in which the losses are accounted for. As stated earlier, this expedites performance calculations, which is justified in view of the small errors committed. Also, as has been indicated throughout the book, our aim is to study the various devices from a user-oriented approach. The reader interested in additional refinements at the expense of a more complex mathematical approach may wish to refer to some of the books listed in the References.

With regard to Example 10-4, the calculations may be repeated for any value of slip to obtain the complete performance characteristics as a function of speed. This will be left to the student as a programming exercise (Problem 10-7).

## 10-8 INDUCTION MOTOR CHARACTERISTICS

The variable-speed induction motor has a wound rotor so that variable external resistances may be used to vary the speed or to increase the starting torque. It is generally used where frequent starts under load are necessary, such as for hoists, cranes, and so on. When fairly constant power is required, starting is infrequent and only average starting torque is necessary, such as driving pumps, blowers, fans, and so on; the much simpler and therefore cheaper squirrel-cage motor is normally employed. Figure 10-10 shows a typical characteristic curve for a three-phase induction motor with a squirrel-cage rotor.

Common squirrel-cage motors are classified into four specific designs, A through D. Their specific characteristics are laid down by the National Electrical

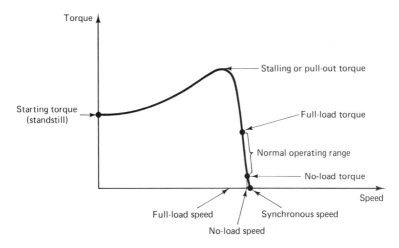

Torque

Stalling or pull-out torque

Starting torque
(standstill)

Full-load torque

Normal operating range

No-load torque

Speed

Full-load speed       Synchronous speed

No-load speed

**Figure 10-10**  Torque–speed characteristic of a squirrel-cage induction motor.

Manufacturers Association (NEMA) and result in markedly different machine characteristics.   Of the four designs, only B, C, and D are in common use and are listed in manufacturers' handbooks.   These four classes, however, cover nearly all practical applications of induction machines.

1. *Class A Motors.*   They are characterized by having a low rotor-circuit resistance and therefore operate at very small slips ($s < 0.01$) under full load. Machines in this class are suitable only in situations where very small starting torques are required.   Because of their low rotor resistance, starting currents are very high.

2. *Class B Motors.*   This is a general-purpose motor of normal starting torque and starting current.   The speed regulation at full load is low (usually under 5%) and the starting torque is in the order of 150% of rated, being lower for the lower-speed and larger motors.   It should be realized that although the starting current is normal, it generally is 600% of full-load value.

3. *Class C Motors.*   Compared to class B motors, they have higher starting torque, normal starting currents, and run at slips of less than 0.05 at full load. The starting torque is about 200% of rated value and the motors are generally designed to start at full load.   Typical applications of this class motor are for driving conveyors, reciprocating pumps, compressors, and so on.

4. *Class D Motors.*   This is a high-slip motor with high starting torque and relatively low starting current.   As a result of the high full-load slip, their efficiency is generally lower than the other classes of motors.   The peak of the torque characteristic is moved to the zero point of the torque–speed curve, resulting in a starting torque of about 300%, which is identical to the stalling torque.

As discussed, it is evident that the characteristics of an induction motor can be influenced considerably.   The question arises: How is this accomplished?   The answer is simple: by shaping the rotor conductor bars.   Figure 10-11 illustrates

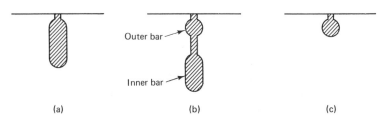

Outer bar

Inner bar

(a)                  (b)                 (c)

**Figure 10-11** Some rotor-bar constructions: (a) low inpedance; (b) double-cage construction; (c) high resistance rotor.

what is meant; it shows various rotor slot configurations to achieve the desired torque–speed motor characteristics. The rotor construction as depicted in Fig. 10-11b has, in effect, two squirrel-cage windings. The low-resistance but high-reactance winding is the one that is embedded deeply into the rotor. The top one has a higher resistance (smaller conductor cross section) but a lower reactance (leakage reactance is lower). At startup and lower speeds, the frequency of the current in the rotor bars is high and the current tends to flow in the upper, low-reactance but high-resistance winding, therefore producing high starting torque. As the motor speed increases and subsequently the rotor frequency decreases, the rotor current will essentially flow in the deeper low-resistance winding. Figure 10-12 shows the influence of the various parameters discussed on the induction motor characteristic.

### Starting Torque

Referring to Fig. 10-10, certain specific points, such as the starting torque and stalling torque, are identified. The starting torque occurs with the rotor at standstill (i.e., the slip $s = 1.0$). If this value of slip is substituted when calculating the

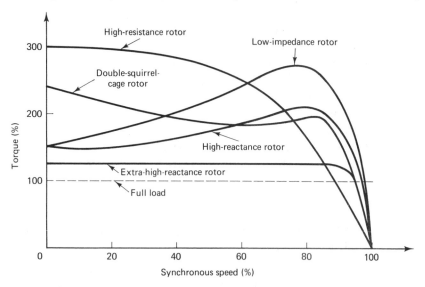

**Figure 10-12** Torque–speed characteristics of squirrel cage induction motors with various rotor constructions.

                                     Three-Phase Induction Motors    Chap. 10

rotor current at standstill, we obtain

$$I'_{R_{st}} = \frac{V_1}{\sqrt{(R_S + R'_R)^2 + (X_S + X'_R)^2}} \quad (10\text{-}18)$$

then $\text{RPI}_{st} = 3I'^2_{R_{st}}R'_R$ can be determined, which is the rotor power input at the instant of starting. From this the starting torque can be obtained.

**EXAMPLE 10-5**

Calculate the starting torque of the motor in Example 10-4.

**SOLUTION**

$$I'_{R_{st}} = \frac{127}{\sqrt{(0.344 + 0.147)^2 + (0.498 + 0.224)^2}} = 145.45 \text{ A}$$

$$\text{RPI}_{st} = 3 \times (145.45)^2 \times 0.147 = 9330 \text{ W}$$

$$T_{st} = \frac{\text{RPI}_{st}}{\omega_s} = \frac{9330}{2\pi \times 1200/60} = 74.2 \text{ N} \cdot \text{m}$$

## *Maximum Torque*

As the load on the motor is increased, the slip increases just enough to meet this new load requirement. This increased loading can generally be carried somewhat beyond the value the motor is designed for, but at the expense of overheating. Assuming that the load on the machine is increased further, there will be a point at which the motor cannot produce any additional torque, and the motor will stall. It can be shown by the familiar impedance matching principle in circuit theory that this will happen when the impedance of $R'_R/s$ matches the impedance looking back into the source. Accordingly, the slip at maximum torque is,

$$s_{mt} = \frac{R'_R}{\sqrt{R_S^2 + (X_S + X'_R)^2}}$$

Since the stator resistance $R_S$ is normally small compared to the equivalent reactance $X_e = X_S + X'_R$, we can say

$$s_{mt} = \frac{R'_R}{X_e} \quad (10\text{-}19)$$

where $s_{mt}$ is the slip at which maximum torque occurs. Using this value of slip, the rotor current can be found from Eq. (10-14), which then enables calculation of the rotor power input at the point of maximum torque, namely,

$$I'_{R_{mt}} = \frac{V}{\sqrt{2}\, X_e} \quad (10\text{-}20)$$

and

$$\text{RPI}_{mt} = 3I'^2_{R_{mt}} \frac{R'_R}{s_{mt}} = \frac{3V^2}{2X_e} \quad (10\text{-}21)$$

which can then be used to determine the rotor-developed power as RPD = $(1 - s)$ RPI, from which $P_{out}$ = RPD − $P_{rot}$, representing the output at the shaft. The stalling or maximum torque is obtained from $T_{mt} = P_{out}/\omega_R$.

It is interesting to see that the maximum output torque on the shaft of the motor, as derived from Eq. (10-21), is independent of the rotor winding resistance $R'_R$. Increasing this resistance increases the slip at which maximum torque occurs as determined by Eq. (10-19), but its magnitude is unchanged. This is illustrated in Example 10-7.

**EXAMPLE 10-6**

Calculate the maximum torque that the motor of Example 10-4 can develop and the speed at which this occurs.

**SOLUTION**

$$s_{mt} = \frac{R'_R}{X_e} = \frac{0.147}{0.498 + 0.224} = 0.20$$

$$n_{R_{mt}} = 1200(1 - 0.20) = 960 \text{ r/min}$$

$$\text{RPI}_{mt} = \frac{3V^2}{2X_e} = \frac{3 \times 127^2}{2(0.722)} = 33,509 \text{ W}$$

$$\text{RPD}_{mt} = \text{RPI}_{mt}(1 - s_{mt}) = 33,509 \times (1 - 0.20) = 26,807 \text{ W}$$

$$T = \frac{\text{RPD} - P_{rot}}{\omega_{R_{mt}}} = \frac{26,807 - 262}{2\pi \times 960/60} = 264 \text{ N} \cdot \text{m}$$

The maximum torque is 264 N · m at a speed of 960 r/min. Any increase in motor load will stall the motor.

## 10-9 WOUND-ROTOR MOTOR

As discussed earlier, the basic difference between the wound-rotor induction motor and the squirrel-cage induction motor is basically in the rotor construction. Both motors use similar stators. The squirrel-cage motor has the limitation of constant rotor resistance. Under normal running conditions a high efficiency is desired which calls for a low rotor resistance, but this in turn implies high starting currents and low starting torque. Thus the rotor design must be a compromise.

The wound-rotor motor, on the other hand, provides an effective way to circumvent this compromise. This is made possible by connecting the rotor winding to slip rings which are in contact with brushes. Thus provisions are made to connect external resistances in series with the rotor winding. At startup and with added resistance, the starting current is reduced, the starting torque increased, and the power factor is improved. As is apparent from Eq. (10-19), the slip at which maximum torque occurs is directly related to the rotor resistance. Thus by adding the appropriate amount of resistance into the rotor circuit, the startup torque can be made equal to the maximum torque if a high starting torque is desired. As the motor speeds up, the external resistances are decreased in such a manner as to provide maximum torque throughout the accelerating range. Once the motor reaches operating speed the external resistances are shorted out at the brushes.

Since the maximum torque can be maintained throughout the accelerating range, the wound-rotor motor is desirable when starting high-inertia loads. Furthermore, this motor also has the advantage that during startup the heating effect in the motor is less, since the $I^2R$ losses are dissipated mainly in the external resistances rather than in the rotor itself. Although this may help the motor, the energy wasted is considerable in large motors. Therefore, schemes are available, called *slip-recovery systems*, to recover the energy that would otherwise be lost in the external resistances.

To show the effects of varying rotor resistance on the torque–speed characteristics, the following example is included. To simplify the procedure, calculations will be limited to the rotor circuit only, according to Fig. 10-6. This, of course, is justified, since rotor parameters can be determined by measurement. The transformation ratio can also be obtained; therefore, by referring the rotor quantities to stator terms and including the magnetizing reactance, the line current, power factor, and input power can be calculated, if desired.

### EXAMPLE 10-7

The wound-rotor induction motor of Example 10-3 has a rotor reactance $X_R$ = 0.352 $\Omega$ and a rotor resistance $R_R$ = 0.088 $\Omega$.

(a) Determine: the torque–speed characteristic for this machine.
(b) Repeat with various additional rotor resistances $R$ inserted of value:
    (i) 0.1 $\Omega$.
    (ii) 0.264 $\Omega(R + R_R = X_{BR})$.
    (iii) 0.8 $\Omega$.

### SOLUTION

From Example 10-3 the following data apply.

$$E_{BR} = 88 \text{ V} \qquad n_s = 1800 \text{ r/min}$$

Because of the repeated calculations, this exercise was referred to the accompanying computer program.

### Program for Example 10-7

```
2    REM THIS IS "INDUCTION MOTOR T-N CURVE"
5    PRINT "☒"
10   PRINT "SYNC SPEED="; TAB (25) : INPUT SN
20   PI = 3.14
30   PRINT "BLOCKED ROTOR VOLTAGE ="; TAB (25) : INPUT EBR
32   PRINT "ROTOR RESISTANCE ="; TAB (25) : INPUT RR
33   PRINT "ROTOR REACTANCE ="; TAB (25) : INPUT XR
34   PRINT
```

```
35    PRINT "   S   IROT   NROT   TORQ"
36    FOR I=1 TO 20
38    S=I/20
40    RI=EBR/SQR ((RR/S)*(RR/S) + XBR*XBR)
42    RPI=3*RI*RI*RR/S
44    RPD=RPI*(1 - S)
46    RN=SN*(1 - S)
48    W=2*PI*RN/60
49    IF RN=0 THEN 51
50    T=RPD/W
51    GO TO 53
52    T=RPI/(2*PI*SN/60)
53    PRINT S;RI, RN, T
54    NEXT I
```

The results calculated are represented graphically in Fig. 10-13. It is apparent that for increased rotor resistances the slip at which maximum torque occurs moves toward the origin. With the total rotor circuit resistance equal to the rotor reactance, $R + R_R = X_R$, the startup torque equals the maximum torque. Further increasing $R$ gives curve 4 in Fig. 10-13. As can be seen, this characteristic has a negative slope throughout the operating range. This negative slope represents positive damping in control system applications, which is desirable from a stability point of view. Two-phase induction motors having such very high rotor resistance designs are usually referred to as servomotors, to be discussed later.

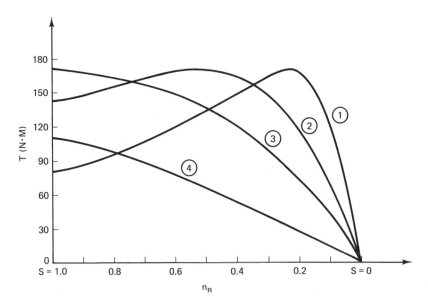

**Figure 10-13**  Representation of calculated data of Example 10-7 using a computer program for (1) $R_r = 0.088\Omega$; (2) $R_r + R = 0.188\Omega$; (3) $R_r + R = X_{BR}$; (4) $R_r + R = 0.888\Omega$.

To use the equivalent-circuit diagram developed for the induction motor in the calculation of performance characteristics, values for the circuit parameters must be assigned. For this, measurements are required from tests, of which the no-load test and the blocked-rotor test are the main evaluations to be carried out. These tests are comparable to the open-circuit and short-circuit tests on a transformer, respectively.

### No-Load Test

As implied, the motor is run with the load uncoupled and rated voltage supplied to the stator. In case of a wound-rotor induction motor, the rotor terminals are shorted out. Measure

$$V_{NL} = \text{line-to-line stator voltage}$$

$$I_{NL} = \text{line current}$$

$$P_{NL} = \text{three-phase power input}$$

Since the motor at no load runs at a very low value of slip, the no-load rotor copper loss is negligible. Therefore, the input power consists of core loss $P_c$, friction and windage loss $P_{rot}$, and stator copper loss, namely,

$$P_{NL} = P_c + P_{rot} + 3I_{NL}^2 R_s \qquad (10\text{-}22)$$

This permits the sum of the rotational losses and core loss to be evaluated:

$$P_{rot+c} = P_{NL} - 3I_{NL}^2 R_s \qquad (10\text{-}23)$$

where the stator resistance $R_s$ is obtained from a resistance measurement at the stator terminals. If the stator winding is assumed to be Y-connected, the value of $R_s$ per phase is half of the resistance value measured between terminals. The power factor under no-load conditions is low, so that the circuit behavior is essentially reactive. The input current is at least 30% of rated value, depending on the size motor. These facts suggest that the magnitude of $R_s$ is small compared to $X_m$. Also, the resistance element $R_{rot+c}$, representing the rotational losses plus core loss as indicated in Fig. 10-14, must then be large compared to $X_m$. Fur-

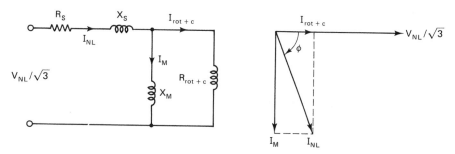

**Figure 10-14**   Induction motor no-load test.

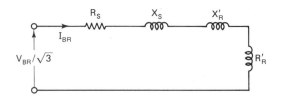

Figure 10-15 Induction motor blocked-rotor test.

thermore, for conventional induction machines, $X_m \gg X_s$. This implies that under the stated assumptions the input impedance at no load is approximately

$$X_m = \frac{V_{NL}}{\sqrt{3}\, I_{NL}}$$  (10-24)

which is the value of the magnetizing reactance.

### Blocked-Rotor Test

As the name implies, the blocked-rotor test is performed with the rotor blocked such that the rotor is prevented from turning. Since it cannot turn, $n_R = 0$ and the slip $s = 1.0$ or $100\%$. This corresponds to the condition at startup and we would expect currents that are five to six times their rated value. It is for this reason, as with transformers during the short-circuit test, that the applied stator voltage is reduced to such a voltage permitting rated stator current to flow. Furthermore, at this greatly reduced input voltage of about 10 to 20% of rated value the air-gap flux is relatively small, implying that $X_m$ is much larger than normal. Therefore, the magnetizing reactance is neglected and the equivalent circuit with $R_R'/s = R_R'$ reduces to that of Fig. 10-15. During the test, measure

$$V_{BR} = \text{line-to-line stator voltage}$$

$$I_{BR} = \text{line current}$$

$$P_{BR} = \text{three-phase power input}$$

Assuming a Y-connection,

$$Z_e = \frac{V_{BR}}{\sqrt{3}\, I_{BR}} = \sqrt{(R_s + R_R')^2 + (X_s + X_R')^2}$$  (10-25)

$$R_e = \frac{P_{BR}}{3I_{BR}^2} = R_s + R_R'$$  (10-26)

$$X_e = \sqrt{Z_e^2 - R_e^2} = X_s + X_R'$$  (10-27)

where $Z_e$, $R_e$, and $X_e$ are equivalent motor impedance, resistance, and reactance values per phase, respectively, in stator terms.

Since $R_s$ was measured separately, the rotor resistance in stator terms is

$$R_R' = R_e - R_s$$  (10-28)

The division between $X_s$ and $X_R'$ is relatively unimportant, since this cannot be determined on a squirrel-cage motor. For wound-rotor induction machines it is generally accepted to assume that

$$X_s = X'_R = 0.5X_e \qquad (10\text{-}29)$$

When using a wound-rotor induction motor, it is intended to add external resistance to the rotor circuit for the purpose of starting and speed control. For this application it is necessary to determine the effective turns ratio between rotor and stator, in order to arrive at the actual rotor reactance. The rotor resistance $R_R$, of course, is measured on the slip rings in a fashion similar to how $R_s$ is measured on the stator side.

The turns ratio is readily determined by measuring the voltage at the slip rings after having opened the rotor circuit. The rotor does not turn during this test since no rotor currents can flow. The motor then behaves as a transformer with the secondary open-circuited. If the measured voltage between slip rings is $V_{sr}$ with rated voltage $V_{L\text{-}L}$ applied to the stator, the transformation ratio is

$$\alpha = \frac{V_{L\text{-}L}}{V_{sr}} = \frac{N_s}{N_r} \qquad (10\text{-}30)$$

where    $N_s$ = stator turns per phase
         $N_r$ = rotor turns per phase

The actual rotor reactance is then

$$X_R = \frac{X'_R}{\alpha^2} \qquad (10\text{-}31)$$

Similarly, the actual rotor resistance when obtained from Eq. (9-38), namely,

$$R_R = \frac{R'_R}{\alpha^2} \qquad (10\text{-}32)$$

should agree with that measured from the resistance measurement.

## EXAMPLE 10-8

A 5-hp 220-V 60-Hz four-pole three-phase induction motor was tested and the following data were obtained:

No-load test: $V_{NL} = 220$ V    $P_{NL} = 340$ W    $I_{NL} = 6.2$ A

Blocked-rotor test: $V_{BR} = 54$ V    $P_{BR} = 430$ W    $I_{BR} = 15.2$ A

The resistance of the stator winding gives a 4-V drop between terminals with rated dc current flowing. Calculate the efficiency of the motor when operating at a slip of 0.04.

## SOLUTION

From the dc resistance test,

$$R_{dc} = \frac{4}{2 \times 15.2} = 0.132 \ \Omega/\text{phase}$$

assuming a Y-connection.

The effective ac resistance is taken to be 1.4 times the dc value, consistent with earlier work. Thus

$$R_s = 1.4 \times 0.132 = 0.18 \ \Omega$$

From the no-load test:

$$X_m = \frac{220}{\sqrt{3} \times 6.2} = 20.5 \ \Omega$$

$$P_{\text{rot}+c} = 340 - 3 \times 6.2^2 \times 0.18 = 340 - 21 = 319 \ \text{W}$$

From the blocked-rotor test:

$$R_e = \frac{430}{3 \times 15.2^2} = 0.62 \ \Omega$$

$$R_R' = 0.62 - 0.18 = 0.44 \ \Omega$$

$$Z_e = \frac{54}{\sqrt{3} \times 15.2} = 2.05 \ \Omega$$

$$X_e = \sqrt{2.05^2 - 0.62^2} = 1.96 \ \Omega = X_S + X_R'$$

Thus $X_S = 0.98 \ \Omega$ and $X_R' = 0.98 \ \Omega$. The resulting equivalent circuit in stator terms is as shown in Fig. 10-16.

The performance for $s = 0.04$ can now be calculated following the procedure as indicated in Example 10-4. Only the essential parameters are calculated here to arrive at the efficiency $\eta$.

$$\frac{R_R'}{s} = \frac{0.44}{0.04} = 11.0 \ \Omega$$

$$I_R' = \frac{127 \ \angle 0°}{0.18 + 11.0 + j1.96} = 11.2 \ \angle -9.9° = 11 - j1.93 \ \text{A}$$

$$I_m = \frac{127 \ \angle 0°}{j20.5} = -j6.2 \ \text{A}$$

The line current is

$$I_L = 11.0 - j(1.93 + 6.2) = 13.7 \ \angle -36.5° \ \text{A}$$

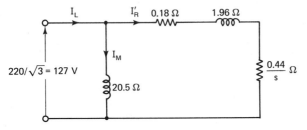

**Figure 10-16** Equivalent circuit for Example 10-8.

Three-Phase Induction Motors    Chap. 10

Further,

$$RPI = 3 \times 11.2^2 \times 11.0 = 4140 \text{ W}$$

$$RPD = 4140 \times (1 - 0.04) = 3974 \text{ W}$$

$$RCL = 0.04 \times 4140 = 166 \text{ W}$$

$$P_{out} = 3974 - 319 = 3655 \text{ W}$$

$$\text{Total losses: } 319 + (3 \times 13.7^2 \times 0.18) + 166 = 579 \text{ W}$$

$$\text{Efficiency: } \eta = \frac{3655}{3655 + 579} \times 100 = 86.3\%$$

As the example indicates, the performance data for the machine can be determined from performing relatively simple tests, in order to establish the equivalent-circuit parameters. With regard to the efficiency, as explained before, it can be determined by measuring the input and output directly. In the event of large machines, it may become impossible to simulate an actual load, and therefore it is more expedient to resort to measuring the losses as indicated, to arrive at the motor efficiency. Although straightforward, it is not completely accurate. From the no-load test, the friction, windage, and core losses are determined. They are assumed to be substantially constant in the normal operating range of the motor, which is correct since the speed regulation is good. In the blocked-rotor test, however, it may be necessary to impress a relatively large voltage compared to rated voltage of the motor. Contrasting this with less than 10% for a transformer, we see that the core loss during this test may not be as negligible as first thought. In addition, the rotor is blocked, causing rotor frequencies equal to the applied line frequency. This has the effect of modifying the rotor resistance compared to that at the full-load slip value. Subsequently, the rotor current will not be the same. Despite the discrepancy in rotor currents during a test and actual operating conditions, the two values are remarkably close. Thus efficiencies calculated from the no-load and blocked-rotor tests are not far apart from those determined by more accurate methods. In case of wound-rotor induction motors, the rotor current and resistance can be measured directly, thereby eliminating possible errors.

## 10-11 SELECTING AN INDUCTION MOTOR

As discussed previously, the NEMA class B induction motor is industry's general-purpose motor. It has good starting properties and is suitable for ventilating equipment such as fans and blowers, for pumps of the centrifugal type, and so on. If the torque–speed curve of a centrifugal pump is superimposed on the curve for the induction motor, as shown in Fig. 10-17, it can be seen that the load is easily started and there is ample accelerating torque. Thus full speed is quickly reached.

If this motor were to drive a load that requires a high starting torque, such as a loaded conveyor belt, the motor may not be able to start the belt, as Fig. 10-18 shows. This implies selection of a motor with higher starting torque capabilities,

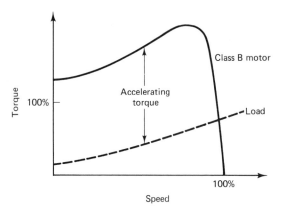

Figure 10-17 Class B induction motor with load torque–speed characteristic.

such as the class C induction motor. While the class D motor would also provide the required higher starting torque, it would not be a proper selection since it runs at a higher slip and therefore at lower efficiency. Even though the class C motor is the clear choice here, it still must be checked to make sure that the motor meets the startup torque requirements if voltage drops are prevalent in the supply system. This is important since the net torque developed by an induction motor is proportional to the square of the applied voltage. The rationale for this is as follows. The torque developed is proportional to the flux and the current, but the flux itself is proportional to the voltage. Therefore, the developed torque $T \propto V^2$. To illustrate this dependency, the starting torque of the motor in Example 10-8 will be determined for full-voltage starting, considering the following assumption.

At the instant of starting, the rotor is not turning and the friction and windage loss is therefore zero. The iron loss at starting, however, is greater than normal because the rotor frequency equals the stator frequency. It therefore seems reasonable to assume that the increased iron loss caused by the rotor compensates for the rotational loss, under which condition the no-load loss may be considered constant.

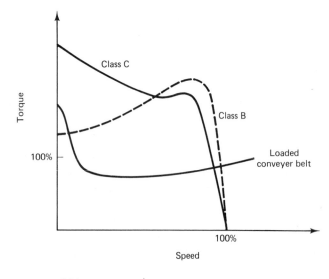

Figure 10-18 Showing high starting torque of class C motor.

Three-Phase Induction Motors    Chap. 10

**EXAMPLE 10-9**

Calculate the starting torque of the motor in Example 10-8 when started at full voltage.

**SOLUTION**

From Example 10-8,

iron loss = 319 W

$$\text{line current at 220 V, rotor blocked} = \frac{220}{54} \times 15.2 = 61.9 \text{ A}$$

$$\text{power input at 220 V, rotor blocked} = \left(\frac{220}{54}\right)^2 \times 430 = 7137 \text{ W}$$

stator copper loss = $3 \times (61.9)^2 \times 0.18 = 2071$ W

RPI = $7137 - 2071 - 319 = 4747$ W

$$T_{\text{st}} = \frac{4747}{2\pi \times 1800/60} = 25.2 \text{ N} \cdot \text{m}$$

## 10-12 MULTISPEED MOTORS

The squirrel-cage induction motor is essentially a constant-speed motor. Although ideal for most applications, it is often desirable to have a motor that can run at other fixed speeds. Considering that the synchronous speed of an induction motor

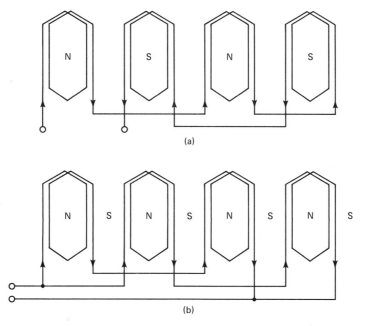

(a)

(b)

**Figure 10-19**  Single-winding two-speed motor: (a) standard connection for four poles; (b) consequent-pole connection for eight poles.

is inversely proportional to the number of poles on the machine (Eq. 10-1), it is apparent that the speed may be changed by changing the number of poles. A multispeed motor is one in which the speed is changed by rearranging the stator winding connection. Restricting ourselves to a two-speed motor, the stator winding of the usual four-pole machine is connected such that they produce alternate poles, as indicated in Fig. 10-19a. Using the same coils, the end connections of each group can be changed so that successive coils have the same polarity; this is shown in Fig. 10-19b, showing four successive north poles. As a consequence, four additional poles of the opposite polarity will be produced between them, at the same time. That is, by connecting all the coil groups for the same polarity, the stator acts as though it has twice as many poles. The additional poles so created are called *consequent poles* and the winding is known as a *consequent-pole winding*. It should be realized that this type of multispeed motor winding always provides one speed that is half as much as the other.

## 10-13 BRAKING METHODS

An induction motor can be subjected to a braking condition by reversing the direction of the rotating magnetic field. This braking method, commonly referred to as *plugging*, results in the field and rotor rotating in opposite directions, giving rise to a slip greater than unity. Once the motor reaches near zero speed, it is disconnected by means of a plugging switch mounted on the motor to sense speed. Figure 10-20 illustrates the torque–speed curve for a typical motor when operating as a motor, when plugging, and when generating. The latter operation is not discussed in this textbook.

Other forms of electrically braking an induction motor are known as dynamic braking, regenerative braking, and braking by using an eddy current clutch. The eddy current brake or clutch is discussed in Chapter 14.

In dynamic braking the motor is disconnected from the main supply and a dc voltage is applied as shown in Fig. 10-21a. At the moment the dc is applied the motor brakes and slows down, giving a braking torque characteristic as indicated in Fig. 10-21b. The braking curve can be controlled by adjusting the dc voltage. Other advantages are that the heating effects are less than with plugging and the

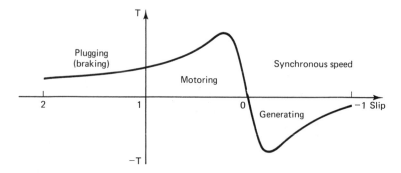

**Figure 10-20** Complete torque–speed characteristic of an induction motor.

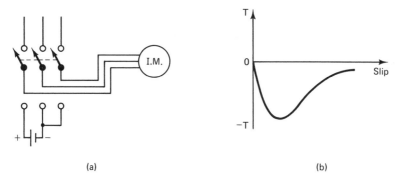

**Figure 10-21** Dynamic braking: (a) circuit; (b) braking characteristic.

motor cannot reverse. This method is, however, at the expense of additional equipment needed.

Regenerative braking occurs when the load overhauls the motor; the load drives the rotor in the direction of the field. As such, the machine in effect operates as an induction generator, the slip becomes negative, and no extra switching is necessary. As the motor slows down but not to a complete stop, it will reach a point on the torque–speed characteristic where the motoring torque is compensated by the load. In the multispeed motor discussed earlier, when switching from the high-speed to the low-speed winding takes place, a braking torque is exerted until the motor reaches its lower speed. This braking action is uncontrolled, however. In addition to electrical braking, there are, of course, mechanical brakes, where energy is dissipated in the mechanical parts in the form of heat. Some of these include shoe brakes or disk brakes, being an integral part of the motor.

## REVIEW QUESTIONS

**10-1.** What types of rotor construction are used in induction motors? What is the advantage of each construction?

**10-2.** How can the direction of the revolving stator field be reversed?

**10-3.** Explain why the rotor of a squirrel-cage induction motor rotates in the direction of the magnetic field.

**10-4.** At what speed does the revolving field rotate? How can this be determined? On what factors does the speed depend?

**10-5.** Why is it impossible for the rotor speed to equal that of the revolving-field speed?

**10-6.** What is slip? How is it calculated? Why does it increase as the load increases?

**10-7.** How is the rotor frequency related to that of the stator current frequency?

**10-8.** How is it possible to represent electrically the mechanical load of an induction motor? What advantage is gained in doing so?

**10-9.** Define operating torque, starting torque, and maximum torque. Which is the largest?

**10-10.** Why do induction motors run at low power factors when lightly loaded?

**10-11.** Show clearly how a three-phase induction motor develops a torque. What is the effect on the torque of changing the value of the rotor resistance? Does this affect the maximum value of the torque? What is the condition to obtain maximum torque in an induction motor at startup?

**10-12.** By inserting external rotor resistances in the rotor circuit of a wound-rotor induction motor, the starting torque may be increased.
(a) What resistance value is needed to make the starting torque equal to the maximum torque?
(b) How is the starting current affected?

**10-13.** How are induction motors generally classified? State some particular characteristics of the motors in each group.

**10-14.** How can the shape of the torque–speed characteristic of a squirrel-cage induction motor be "shaped" in the design stage? Compare two basic rotor conductor designs.

**10-15.** Describe a method to make an induction motor a two-speed motor. What two speeds are generally obtained?


## PROBLEMS


**10-1.** At what speed will a 12-pole 60-Hz induction motor operate if the slip is 0.06?

**10-2.** A 60-Hz induction motor runs at 860 r/min at full load. Determine:
(a) The synchronous speed.
(b) The frequency of the rotor currents.
(c) The rotor speed relative to the revolving field.

**10-3.** A three-phase induction motor runs at almost 1200 r/min at no-load and 1140 r/min at full load when supplied with power from a 60-Hz three-phase line.
(a) How many poles has the motor?
(b) What is the percent slip at full load?
(c) What is the corresponding frequency of the rotor voltages?
(d) Determine the corresponding speed of:
   (i) The rotor field with respect to the rotor.
   (ii) The rotor with respect to the stator.
   (iii) The rotor field with respect to the stator field.
(e) What is the rotor frequency at the slip of 10%?

**10-4.** If the phase sequences of the currents in Fig. 10-2 are reversed, for instance by reversing $i_b$ and $i_c$ (*acb* sequence), demonstrate graphically, using the convention in the text as depicted in Fig. 10-3, that the resultant flux rotates counterclockwise.

**10-5.** A three-phase 60-Hz six-pole 220-V wound-rotor induction motor has its stator connected in delta and its rotor in star. There are 80% as many rotor conductors as stator conductors. For a slip of 0.04, calculate the voltage between slip rings at the rotor.

**10-6.** A three-phase 60-Hz six-pole 220-V wound-rotor induction motor has its stator connected in Δ and its rotor in Y. The rotor has half as many turns as the stator. Calculate the voltage and frequency between slip rings if normal voltage is applied to the stator and:
(a) The rotor is at rest.

**(b)** The rotor slip is 0.04.

**(c)** The rotor is driven by another machine at 800 r/min in the direction opposite to that of the revolving field.

**10-7.** For slip values of $s = 0.05$ to 1.0 in suitable steps, calculate the performance characteristics for the motor in Example 10-4. Plot efficiency, line current, and power factor as a function of developed horsepower. Write a computer program to do these calculations.

**10-8.** A three-phase 15-hp 220-V 60-Hz six-pole Y-connected squirrel-cage induction motor has the following parameters per phase in stator terms: $R_s = 0.126$ $\Omega$, $R'_R = 0.094$ $\Omega$, $X'_e = 0.46$ $\Omega$, and $X_m = 9.8$ $\Omega$. The rotational and core losses combined are 560 W. For a slip of 3% find:

**(a)** The line current and power factor.

**(b)** The horsepower output and shaft torque.

**(c)** The efficiency.

**10-9.** A three-phase 125-hp 440-V 60-Hz eight-pole Y-connected induction motor has the following electric circuit parameters on a per phase basis referred to the stator:

$$R_s = 0.068 \ \Omega \qquad X_s = X'_R = 0.224 \ \Omega$$

$$R'_R = 0.052 \ \Omega \qquad X_m = 7.68 \ \Omega$$

The rotational losses are 2400 W. Determine for a slip of 3%:

**(a)** The line current and power factor.

**(b)** The output horsepower and torque.

**(c)** The efficiency.

**10-10.** For the machine of Problem 10-9, calculate:

**(a)** The slip at which maximum torque occurs.

**(b)** The speed at which this occurs.

**(c)** The line current under this condition.

**10-11.** For the machine of Problem 10-9, determine the starting torque and the value of the starting current.

**10-12.** The nameplate of a squirrel-cage induction motor has the following information: 25 hp, 220 V, three-phase, 60 Hz, 830 r/min, 64 A per line. If the motor takes 20.8 kW when operating at full load, calculate:

**(a)** The slip.

**(b)** The power factor.

**(c)** The torque.

**10-13.** A 7.5-hp 220-V four-pole three-phase induction motor was tested and the following data were recorded:

No-load test: $\quad V_{nL} = 220$ V $\quad P_t = 320$ W $\quad I = 6.4$ A

Blocked-rotor test: $\quad V_{BR} = 46$ V $\quad P_t = 605$ W $\quad I = 18$ A

The effective ac resistance between stator terminals is 0.64 $\Omega$. The slip at full load is 4%. Determine:

**(a)** The equivalent electrical circuit on a per phase basis for this motor.

**(b)** The input current and power factor when delivering full load.

**(c)** The efficiency and motor speed in part (b).

**10-14.** For the motor in Problem 10-13, calculate the starting torque when it is started at 127 V (line voltage).

**10-15.** A 25-hp 440-V squirrel-cage induction motor has a starting torque of 112 N · m and a full-load torque of 83 N · m. The starting current of the motor is 128 A when rated voltage is applied. Determine:

(a) The starting torque when the line voltage is reduced to 300 V.

(b) The voltage that must be applied in order to develop a starting torque equal to the full-load torque.

(c) The starting current when the voltage is reduced to 300 V.

(d) The voltage that must be applied at startup in order not to exceed the rated line current of 32 A.

# 11

# SINGLE-PHASE MOTORS

## 11-1 INTRODUCTION

Everyone is well aware of the numerous and diverse applications of single-phase motors, both in the home and in industry. It is probably safe to say that single-phase motor applications far outweigh the three-phase motor applications. In the home, only single-phase is provided, since power was originally generated and distributed to provide lighting. For this reason, early motor-driven appliances in the home depended on the development of single-phase motors. Nowadays, single-phase motors are equally common in industry and the home. In the home we have various types of single-phase motors, performing tasks in washing machines, furnace oil burner pumps, record players, clocks, hand power tools, fans, and so on. As the applications indicate, small- or fractional-horsepower motors are employed. Generally, the term "small motor" refers to one having a horsepower rating of less than 1 hp (i.e., a fractional-horsepower motor). Most single-phase motors fall into this category, although single-phase motors are also manufactured in standard horsepower sizes of 1.5, 2, 3, 5, 7.5, and 10 hp for both 120 and 220 V.

According to the American National Standards Institute (ANSI) and the National Electrical Manufacturers Association (NEMA), a small motor is defined as having a frame size smaller than that having a continuous rating of 1 hp, open type, at 1800 r/min. Thus a small motor, generally considered a fractional-horsepower motor, is really classified as such by its frame size. A comparison should illustrate this definition. Assume that we have a $\frac{3}{4}$-hp 900-r/min motor. It may appear that this is a fractional-horsepower motor since its rating is less than 1 hp. However, it is not considered a fractional-horsepower motor because if its

frame is used for an 1800-r/min motor, it would have a rating of 0.75 hp ×
1800/900 = 1.5 hp, hence more than 1 hp. Similarly, a 1.5-hp 3600-r/min motor
is a fractional-horsepower motor because of its frame size, since if used for an
1800-r/min motor, it would yield a rating of 1.5 hp × 1800/3600 = 0.75 hp.

Although various constructional differences do exist among the various types,
they can be divided into four basic groups: (1) single-phase induction motors, (2)
shaded-pole motors, (3) universal motors, and (4) single-phase synchronous motors.
In this chapter we deal with some of the more common types and their operating
principles.

## 11-2 PRODUCTION OF TORQUE

As we may recall from the study of the three-phase induction motor, the three-
phase distributed stator winding sets up a rotating magnetic field which is fairly
constant in magnitude and rotates at synchronous speed. In the single-phase
induction motor we have only a simple field winding excited with alternating current; therefore, it is not inherently self-starting, since it does not have a true
revolving field. Various methods have been devised to initiate rotation of the
squirrel-cage rotor, and the particular method employed to start the motor will
designate the specific type.

At first let us examine the behavior of the magnetic field as set up by an ac
current in the single-phase field winding. With reference to Fig. 11-1, we have
current flowing in the field winding. If this current is sinusoidal, then neglecting
saturation effects of the magnetic iron circuit, the flux through the armature will
vary sinusoidally with time. The magnetic field created is as shown at the particular
instant in time, it will reverse during the next half-cycle of the ac supply voltage
cycle. Since the flux is pulsating, it will induce currents in the rotor bars, which

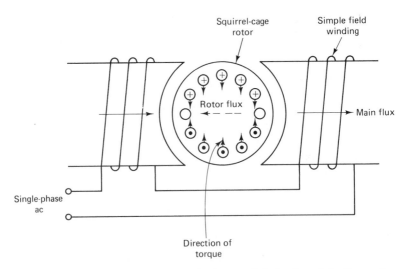

**Figure 11-1** Torque produced in squirrel cage of single-phase induction motor
having a simple field winding.

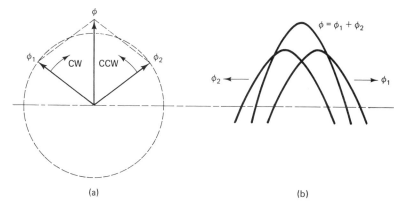

**Figure 11-2** Pulsating field resolved into two oppositely rotating fields.

in turn will create a rotor flux which by Lenz's law opposes that of the main field. From this the current direction in the rotor bars can be determined, as shown in Fig. 11-1, as well as the torque created between the field and rotor currents. It is apparent that the clockwise torque produced is counteracted by the counter-clockwise torque; hence no motion results. Since the field is pulsating, the torque is pulsating, although no net torque is produced over a full cycle of the ac supply frequency.

However, any pulsating field can be resolved into two components, equal in magnitude but oppositely rotating vectors, as shown in Fig. 11-2a. The maximum value of the component fields equals one-half of $\phi_{max}$. As can be observed, the resultant fields of $\phi_1$ and $\phi_2$ in Fig. 11-2a, as they rotate at an angular velocity dictated by the supply frequency, must always lie on the vertical axis. The resultant value of these two vectors at any instantaneous time equals the value of the magnetic field as it actually exists. A physical interpretation of the two oppositely rotating field components is as predicted in Fig. 11-2b. Each component field glides around the air gap in opposite directions and equal velocities; their instantaneous sum represents the instantaneous resultant magnetic field, which changes from $\phi_{max}$ to $\phi_{min}$. This method of field analysis is commonly known as the double revolving field theory. Each field component acts independently on the rotor and in a similar fashion as does the rotating field in a three-phase induction motor. Except that here there are two fields, one tending to rotate the rotor clockwise, the other tending to rotate counterclockwise.

Considering the clockwise flux component by itself, it would produce a torque–speed characteristic labeled $T_{cw}$ in Fig. 11-3, while the counterclockwise flux component produces the torque $T_{ccw}$. Observe that at standstill ($s\cdot= 1.0$) the two torque components produced are equal but directed oppositely. Although the net torque produced at standstill is zero, as can be seen, if the rotor is advanced in either direction, a net torque will result and the motor will continue to rotate in the direction in which it has been started. For instance, if we assume that the rotor is in some way started in the clockwise direction, the torque $T_{cw}$ will exceed $T_{ccw}$ immediately and the rotor will accelerate in the direction of $T_{cw}$. Steady-state speed will be reached near synchronous speed at a slip dictated by the load.

Sec. 11-2    Production of Torque

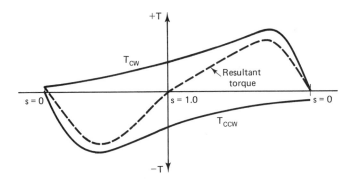

**Figure 11-3** Torque–speed characteristic of a single-phase squirrel-cage induction motor.

It is interesting to note that at this slip speed, $T_{cw}$ predominates over $T_{ccw}$, which is fairly small but exists nevertheless. Also, the rotor operates at a small slip value as far as $T_{cw}$ is concerned, but the slip is nearly 2, as regards to $T_{cw}$. This implies induced rotor currents due to $T_{ccw}$ which are at double the line frequency. These rotor currents do not produce any significant countertorque because of their high frequency, since the rotor reactance is many times its value at slip frequency.

In the three-phase induction machine the rotating field strength does not vary appreciably as it rotates. This means that the locus traced by the field vector arrow head is a circle, as shown in Fig. 11-4a, and the rotating field produced is said to be circular. The field set up by a single-phase motor is usually elliptical in shape. To see why this occurs, we have to examine the rotor field more closely.

When the rotor is rotating, voltages are induced in the rotor conductors, which are in phase with the stator field. Since these voltages are speed dependent, they are referred to as "speed EMFs," as opposed to "transformer EMFs," which are produced by transformer action. Both, of course, are produced by a changing flux, the speed EMF as a result of relative motion between the field and conductor, the transformer EMF as a result of a pulsating field. Therefore, since rotor-induced currents flow in the rotor bars which represent almost entirely a reactive impedance,

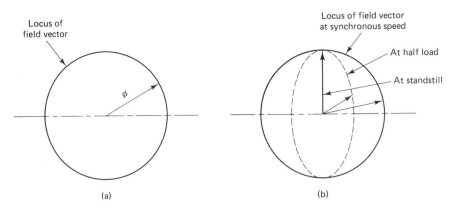

**Figure 11-4** Comparison of air-gap fields in three-phase and single-phase induction motors: (a) three-phase motor field (circular); (b) single-phase motor field.

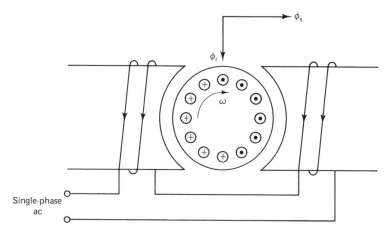

**Figure 11-5** Cross field $\phi_r$ created by rotor rotation.

these rotor currents will lag the rotor-induced voltage by nearly 90°. The field in turn created by the rotor currents is likewise displaced by 90° and is known as a *cross field*, as indicated in Fig. 11-5.

Thus the stator sets up a pulsating field while the rotating rotor sets up a second field which pulsates 90° behind the stator field in space and time. When the rotor rotates at nearly synchronous speed, these two fields will virtually be equal and will combine to produce a nearly circular field. Upon loading the single-phase induction motor, its speed will drop, thereby reducing the speed EMF in the rotor. This in turn reduces the cross field and the resulting field becomes elliptical, as shown in Fig. 11-4b. Further decreasing the speed until finally standstill is reached, the resulting field will only pulsate in value along the stator axis and thus not rotate.

Thus, once started, the single-phase motor having a simple winding will, as explained, continue to run in the direction in which it is started. The non-self-starting is not a desirable feature in practice, and modifications are introduced to obtain the torque required to start. To accomplish this, a quadrature flux component in time and space with the stator flux must be provided at standstill. Auxiliary windings normally placed on the stator have proved effective in developing this starting torque. The methods employed to accomplish this will now be dealt with.

## 11-3 SPLIT-PHASE INDUCTION MOTORS

One of the most widely used types of single-phase motors is the split-phase induction motor. Its service includes a wide variety of applications, and is used in refrigerators, washing machines, portable hoists, many small machine tools, blowers, fans, and centrifugal pumps, among others.

The essential parts of the split-phase motor are shown in Fig. 11-6a. It shows the auxiliary winding, also called the starting winding, in space quadrature (i.e., 90 electrical degrees displacement) with the main stator winding. The rotor is

Sec. 11-3    Split-Phase Induction Motors    **301**

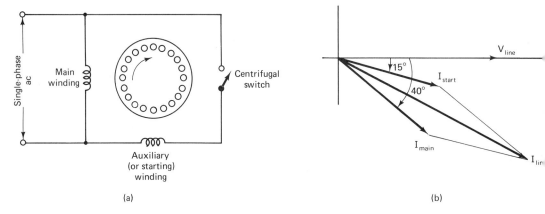

**Figure 11-6** Split-phase motor: (a) schematic representation; (b) phasor diagram at instant of starting.

normally the normal squirrel-cage type. The two stator windings are connected in parallel to the ac supply voltage. A phase displacement between the winding currents is obtained by adjusting the winding impedances, either by inserting a resistor in series with the starting winding or as is generally the practice, by using a smaller-gauge wire for the starting winding. A phase displacement between the currents of 30° can be achieved at the instant of starting. A typical phasor diagram for this motor at startup is illustrated in Fig. 11-6b.

When the motor has come up to about 70 to 75% of synchronous speed, the starting winding may be opened by a centrifugal switch and the motor will continue to operate as a single-phase motor. At the point where the starting winding is disconnected, the motor develops nearly as much torque with the main winding alone as with both windings connected, as can be observed from the typical torque–speed characteristic for this type of motor (see Fig. 11-7). The starting winding is designed to take the minimum starting current for the required torque. The locked rotor starting current is typically in the range 5 to 7 times rated current, while the starting torque is about 1.5 to 2 times rated torque. The high starting current, as such, is not objectionable, since once started it drops off almost instantly. The major disadvantages are the relatively low starting torque and the high slip it

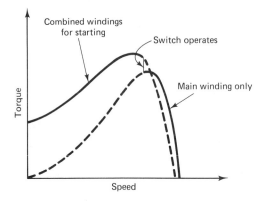

**Figure 11-7** Typical torque–speed characteristic of a general-purpose split-phase motor.

operates at when heavily loaded. As can be appreciated from the earlier discussion, when the speed drops significantly (of course, not to the extent that the centrifugal switch operates), the speed EMF is similarly reduced. This results in an elliptical or pulsating torque, which makes this motor somewhat noisy. It is precisely for this reason that the split-phase motor is operated and employed where the drive loads themselves are noisy.

Unlike the three-phase induction motor, which may start in either direction, the split-phase motor is factory connected and as such its direction of rotation is fixed (counterclockwise when viewed from the opposite end of the shaft extension). To reverse the direction of rotation it is necessary to reverse the connection to the starting winding. Again in contrast to its three-phase counterpart, this reversal (plugging) cannot be done under running conditions, since the split-phase motor torque will be much less than the torque developed by the single main winding, and rotation will not reverse.

In the event that the centrifugal switch contacts fuse, the starting winding will be permanently connected in the circuit during normal operation. Although in principle this does not affect motor operation significantly, it must be realized that this winding is designed for intermittent operation (for startup only). Therefore, when permanently connected in the circuit due to switch failure, it will quickly heat up and raise the motor temperature excessively and eventually burns out the winding.

As Fig. 11-6b shows, the starting winding current $I_s$ lags the supply voltage by about 15°, and the running or main winding current $I_{main}$ lags the voltage by about 40°. Although the currents are not equal, their quadrature or in-phase components with the voltage are nearly the same. This can be illustrated in the following example.

### EXAMPLE 11-1

A $\frac{1}{4}$-hp 120-V split-phase loaded motor draws at the instant of starting a current of 4 A in its starting winding while the main winding current takes 5.8 A, lagging the supply voltage by 15° and 45°, respectively. At startup, determine:

(a) The line current and the power factor.
(b) The in-phase components of the currents with the supply voltage.

### SOLUTION

(a) $\mathbf{I}_{start} = 4 \angle -15° = 3.86 - j1.04$ A
$\mathbf{I}_{main} = 5.8 \angle -45° = 4.10 + j4.10$ A
$\mathbf{I}_{line} = \mathbf{I}_{start} + \mathbf{I}_{main} = 7.96 - j5.14 = 9.48 \angle -33°$
power factor $= \cos(-33°) = 0.84$ lagging

(b) From the calculated results in part (a) we see that the in-phase components of currents with the line voltage, being the real parts of the respective currents, are 3.86 A and 4.10 A for the starting winding and main winding current, respectively. As shown, these components are practically equal at the instant of starting.

## 11-4 CAPACITOR-START MOTORS

In the split-phase motor the phase shift between stator currents was accomplished by adjusting the impedances of the windings by making the starting winding a relatively higher resistance. This resulted in a phase shift of nearly 30°. Since the developed torque of any split-phase motor is proportional to the pole flux produced and the rotor current, it is also dependent on the angle between the winding currents. This implies that if a capacitor is connected in series with the starting winding, the starting torque will increase. This is indeed the case. By proper selection of the capacitor, the current in the starting winding will lead the voltage across it and a greater displacement between winding currents is obtained. This results in a significantly greater starting torque than that obtained in split-phase motors, as Example 11-2 will illustrate. Typical starting torques may be in the range of four times rated torque.

Figure 11-8 shows the capacitor-start motor and its corresponding phasor diagram, indicating a typical displacement between winding currents of about 80°. The value of capacitor needed to accomplish this is typically 135 μF for a ¼-hp

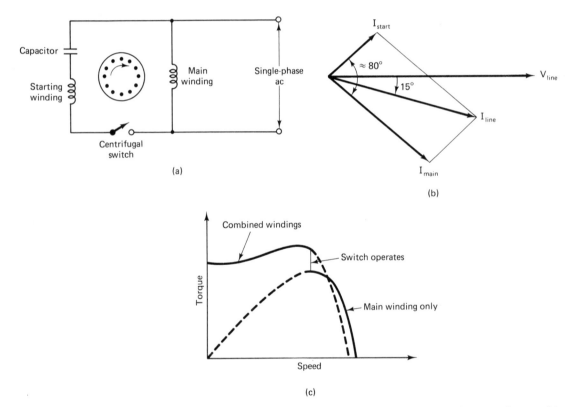

**Figure 11-8** Capacitor-start induction motor: (a) circuit diagram; (b) phasor diagram; (c) torque–speed characteristic.

motor and 175 μF for a $\frac{1}{3}$-hp motor. Since they are rated for ac line voltages, their size is about $1\frac{1}{2}$ in. in diameter and $3\frac{1}{2}$ in. long. Contrary to the split-phase motor discussed, the capacitor-start motor under running conditions is reversible. If temporarily disconnected from the supply line, its speed will drop, allowing the centrifugal switch to close. The lead connections to the starting winding are reversed during this interval and the motor reconnected to the supply once the centrifugal switch closes. The resulting rotating field will now rotate opposite to the direction in which the motor rotates. Since the current displacement between windings is much larger in this motor compared to the split-phase motor, the torque being proportional to this will also be much larger and exceed the torque produced by the rotor. Therefore, the motor will slow down, stop, and then reverse its direction of rotation. Once up to about 75 to 80% of synchronous speed, the centrifugal switch opens and the motor will reach a speed as dictated by the load.

Because of their higher starting torques, capacitor-start motors are used in applications where not only higher starting torques are required, but also where reversible motors are needed. Applications of capacitor motors are in washing machines, belted fans and blowers, dryers, pumps, and compressors.

**EXAMPLE 11-2**

A capacitor is added to the starting winding of the motor in Example 11-1, with the result that its current now leads the voltage by 40°. The main winding remains as is.

(a) With this added capacitor, determine at the instant of starting: the line current and the power factor.

(b) Compare the results with those calculated in Example 11-1.

**SOLUTION**

(a) $I_{\text{start}} = 4\angle 40° = 3.06 + j2.57$ A
$I_{\text{main}} = 5.8\angle -45° = 4.10 - j4.10$ A
$I_{\text{line}} = I_{\text{start}} + I_{\text{main}} = 7.16 - j1.53 = 7.32\angle -12°$ A
power factor $= \cos(-12°) = 0.98$ lagging

(b) The line current has been reduced from 9.48 A to 7.32 A and the power factor improved. The motor starting torque, being proportional to the sine of the angle between the winding currents, has also been increased and becomes maximum with minimum starting current. It can be shown that the starting current of the motor with the added capacitor compared to that without increases by a factor of

$$\frac{T_c}{T} = \frac{\sin[40° - (-45°)]}{\sin[45° - (15°)]} = 1.99$$

where the subscript $c$ indicates the developed torque with the added capacitor, and $T$ that without.

## 11-5 CAPACITOR MOTORS

The capacitor-start motor just discussed still has a relatively low starting torque, although as we have seen, it is considerably better than the split-phase motor. For many applications this does not present a serious limitation.

In cases where high starting torques are required, best results will be obtained if a large value of capacitance is used at startup, which is then gradually decreased as the speed increases. In practice, two capacitors are used for starting and one is cut out of the circuit by a centrifugal switch once a certain speed is reached, usually at about 75% of full speed. The starting or intermittent capacitor is of fairly high capacity (usually on the order of 10 to 15 times the value of the running capacitor, which remains in the circuit). Figure 11-9 illustrates the connection diagrams for the capacitor motor, showing two methods generally encountered.

The first method, shown in Fig. 11-9a, uses an electrolytic capacitor in the starting circuit, which is not built to be left in the circuit continuously, since its leakage is too high. The second capacitor, being oil filled, remains in the circuit; it has little leakage and is therefore suitable for continuous operation.

The second circuit (see Fig. 11-9b) uses an autotransformer but only one oil-filled high-voltage capacitor. This method utilizes the transformer principle of reflected impedance from secondary to the primary. This, as we have discussed, is proportional to the square of the secondary-to-primary turns ratio. For instance, an autotransformer with 180 turns tapped at the 30-turn point would reflect an 8-$\mu$F running capacitor to the primary as

$$\left(\frac{180}{30}\right)^2 \times 8 \ \mu\text{F} = 288 \ \mu\text{F}$$

representing an increase of about 36 times. Thus a running oil-filled capacitor may be used for starting purposes as well, thereby eliminating one capacitor in lieu of the autotransformer, which is of comparable cost. Care must be exercised to ensure that the capacitor can withstand the stepped-up voltage, which is $180/30 = 6$ times the rated voltage at startup. For instance, a 120-V motor would have a capacitor voltage at the instant of starting of 720 V. Typically, a 1000-V ac rating capacitor is required.

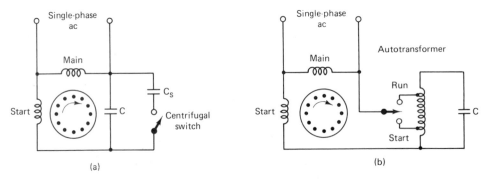

(a)  (b)

**Figure 11-9**  Capacitor motors: (a) two-value capacitor motor; (b) capacitor motor with autotransformer.

As is the case with the capacitor-start motor, the capacitor motor may be damaged for identical reasons if the centrifugal switch fails to operate properly. The primary advantage, then, of a two-value capacitor motor is its high starting torque, good running torque, and quit operation. Reversing the line leads to one of the windings in the usual manner and causes motor operation in the opposite direction. It is therefore classified as a reversible-type motor. In operations requiring frequent reversals it is preferred to use a single-value capacitor-type motor using no centrifugal switch.

## 11-6 SHADED-POLE MOTORS

Like any other induction motor, the shaded-pole motor is caused to run by the action of the magnetic field set up by the stator windings. There is, however, one extremely important difference between the polyphase induction motor and the single-phase induction motors discussed so far. As discussed, these motors have a truly rotating magnetic field, either circular, as in the three-phase machine, or of elliptical shape, as encountered in most of the single-phase motors. In the shaded-pole motor the field merely shifts from one side of the pole to the other. In other words, it does not have a rotating field but one that sweeps across the pole faces.

An elementary understanding of how the magnetic field is created may be gained from the simple circuit in Fig. 11-10, illustrating the shaded-pole motor. As can be seen, the poles are divided into two parts, one of which is "shaded"; that is, around the smaller of the two areas formed by a slot cut across the laminations, a heavy copper short-circuited ring, called the *shading coil*, is placed. That part of the iron around which the shading coil is placed is called the *shaded part* of the pole. When the excitation winding is connected to an ac source, the magnetic field will sweep across the pole face from the unshaded to the shaded portion. This, in effect, is equivalent to an actual physical motion of the pole; the result is that the squirrel-cage rotor will rotate in the same direction.

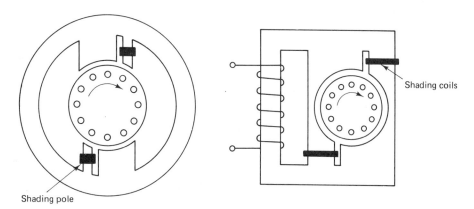

**Figure 11-10** Shaded-pole motors.

To understand how this sweeping action of the field across the pole face occurs, let us consider the instant of time when the current flowing in the excitation winding is starting to increase positively from zero, as illustrated in Fig. 11-11a. In the unshaded part of the pole the flux $\phi$ will start to build up in phase with the current. Similarly, the flux $\phi_s$ in the shaded portion of the pole will build up, but this flux change induces a voltage in the shading coil which causes current to flow. By Lenz's law, this current flows in such a direction as to oppose the flux change that induces it. Thus the building up of flux $\phi_s$ in the shaded portion is delayed. It has the overall effect of shifting the axis of the resultant magnetic flux into the unshaded portion of the pole. When the current in the excitation coil is at or near its maximum value as indicated in Fig. 11-11b, the flux does not change appreciably. With an almost constant flux, no voltage is induced in the shading coil, and therefore it, in turn, does not influence the total flux. The result is that the resultant magnetic flux shifts to the center of the pole. Figure 11-11c shows the current in the excitation coil decreasing. The flux in the unshaded portion of the pole decreased immediately. However, currents induced in the shading coil tend to oppose this decrease in flux; consequently they try to maintain the flux. The result of this action translates into a movement of the magnetic flux axis toward the center of the shaded portion of the pole. Hence the flux $\phi_s$ continues to lag behind the flux $\phi$ during this part of the cycle.

It can similarly be reasoned that at any instant of the current cycle, the flux $\phi_s$ lags behind $\phi$ in time. The net effect of this time and space displacement is to produce a gliding flux across the pole face and consequently in the air gap, which is always directed toward the shaded part of the pole. Therefore, the direction of rotation of a shaded-pole motor is always from the unshaded toward the shaded part of the pole.

Simple motors of this type cannot be reversed but must be assembled so that the rotor shaft extends from the correct end in order to drive the load in the proper direction. There are specially designed shaded-pole motors which are reversible. One form of design is to use two main windings and a shading coil. For one direction of rotation one main winding is used and for the opposite rotation the other. Such an arrangement is adaptable only to distributed windings; hence this necessitates a slotted stator.

Another method employed is to use two sets of open-circuited shading coils, one set placed on each side of a pole. A switch is provided to short-circuit either shading coil, depending on the rotational direction desired. Offsetting the simple construction and low cost of this motor are the low starting torque, little overload capacity, and low efficiencies (5 to 35%).

These motors are built in sizes ranging from $\frac{1}{250}$ hp up to about $\frac{1}{20}$ hp. Typical applications of shaded-pole motors are where efficiencies are of minor concern, such as in toys and fans. Since the applied voltage to the motor greatly affects its speed under load, as the slip increases with reduced voltage, advantage is taken of this fact, particularly when driving fans. Practically, this is generally done by providing line voltage taps to the excitation winding. With fewer turns on this winding, the volts per turn, as well as the current, are larger than with a full

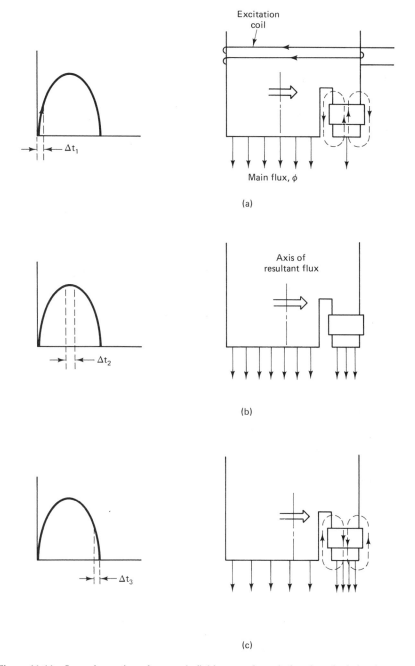

**Figure 11-11** Sweeping action of magnetic field across the pole face in a shaded-pole motor, during positive half of excitation current cycle: (a) excitation current increasing, eddy current flux, $\phi_s$, opposing main flux; (b) excitation current approximately constant, negligible eddy current flux; (c) excitation current decreasing, eddy current flux aiding main flux.

winding. Since the volts per turn is proportional to the flux, it in turn increases and the motor runs at a greater speed, since a larger torque is developed.

## 11-7 HYSTERESIS MOTORS

In a hysteresis motor, torque is produced by the effect of hysteresis losses induced in the rotor. The motor itself is basically a synchronous motor having a specially constructed rotor that carries no winding. Figure 11-12 illustrates its basic rotor construction. As indicated, it consists of a nonmagnetic core which carries a layer of special magnetic material. This outer layer may be composed of a number of thin rings assembled to give a built-up laminated rotor. In the smaller motors a solid single ring is used. Thus there are no salient poles and it is the material of which the rings are made that possesses the special magnetic properties essential for the required developed torque. The stator construction can be almost of any other conventional motor, the prime prerequisite being that it produces a rotating magnetic field. As has been discussed in previous chapters, polyphase stators set up a more uniform rotating field than those designed for single phase, such as the shaded-pole or capacitor-type stators. The stator construction ultimately chosen will designate the type of hysteresis motor.

When started, the stator current will be relatively large, as is the case with induction motors, and the rotor ring material is cycled around its hysteresis loop. The induced rotor flux lags in phase behind the rotating stator flux and the angle $\delta$ between these magnetic axes is responsible for the creation of the developed torque. The angle $\delta$ (see Fig. 11-12) depends only on the shape of the hysteresis loop and not on the frequency at which it is cycled. It is therefore essential to select a material that possesses a wide hysteresis loop. In other words, the coercive force $H_c$ should be relatively large and the residual flux density $B_r$ should be predominant. An ideal material would have a rectangular hysteresis loop, as indicated by loop 1 in Fig. 11-13. Materials of the cobalt–vanadium type, for instance, have hysteresis loops according to curve 2 in Fig. 11-13 which approximate the desired characteristics. Ordinary electric steels have loops that are more like curve 3 and are not suitable for this type of motor.

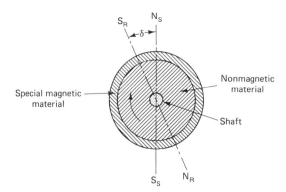

**Figure 11-12** Rotor construction for hysteresis motor.

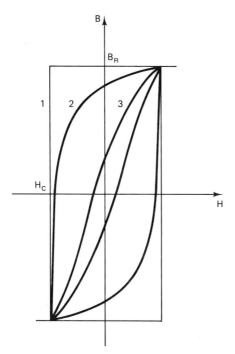

**Figure 11-13** Various hysteresis loops for different materials.

When synchronous speed is reached, the stator current falls off and the motor will run as a permanent-magnet type of machine. An idealized torque–speed curve for the hysteresis motor is shown by curve 1 in Fig. 11-14. As can be seen, it is quite distinct from that of the induction motor. The torque developed by an induction motor decreases to zero at synchronous speeds, whereas in the "ideal" hysteresis motor it is constant at all speeds, including synchronous speed. Thus from the curve it is apparent that locked-rotor, starting, and pull-out torques are all equal. This is their valuable property, in that they can pull into synchronism a high inertia load.

In practical hysteresis motors, however, the shape of the torque–speed characteristic will not be identical to the ideal, but is somewhat modified. This is mainly due to the presence of harmonics in the rotating field and other irregularities. For instance, in capacitor-type hysteresis motors the rotating field is not uniform

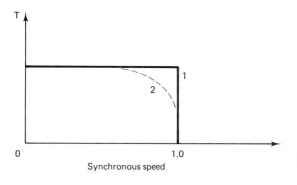

**Figure 11-14** Torque–speed characteristic of (1) an ideal and (2) a practical hysteresis motor.

(i.e., cylindrical) but rather is elliptical in shape. Furthermore, the shape of the ellipse varies with loading conditions. These effects translate into a departure from the ideal curve as illustrated by curve 2 in Fig. 11-14.

Of all the various types of hysteresis motors available, the shaded-pole hysteresis motor is probably most widespread. It is used for all kinds of timer applications and in drives requiring smooth starts and free from torque oscillations (jerks). They are usually applied in conjunction with high-ratio-reduction gearboxes. Motor efficiencies are generally low and power outputs are generally limited to a few watts.

## 11-8 RELUCTANCE MOTORS

A reluctance motor is similar in construction to a synchronous motor, except that it operates without dc excitation and has salient poles on its rotor so that the reluctance varies over the air gap. Torque is created by the tendency of the rotor to align itself with the revolving magnetic field in order to minimize the reluctance path. The small reluctance motor is usually built from induction motor parts except that some of the teeth are cut out of the squirrel-cage rotor, as shown in Fig. 11-15. Even the large synchronous dc excited motors start as induction motors by means of the damper winding. This damper winding is also necessary in reluctance motors, which start as induction motors but run synchronously once up to speed.

The operating principles of a reluctance motor can be understood by examining Fig. 11-15, which illustrates the essential characteristics. The rotating stator field is replaced by a permanent magnet here to facilitate the discussion and the moving part, the rotor, is free to rotate between the poles. With reference to Fig. 11-16a, the rotor is positioned such that the angular deflection β is 0°. The magnetic flux will tend to close upon itself, taking the path of lowest reluctance, and it is not deformed. The reluctance torque $T_r = 0$. The rotor is in a stable position. Giving the rotor a slight angular displacement β degrees, say in a clockwise direction as indicated in Fig. 11-16b, the flux again will take the path of lowest reluctance, but now the flux lines are deformed. As a result, the resilient properties of the magnetic field lines trying to "straighten" themselves out produce a torque tending to return the rotor to its initial position. Naturally, the rotor will stay at an angle β if the external torque is balanced by this reluctance torque. Removing the

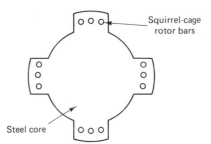

**Figure 11-15** Rotor arrangement for synchronous reluctance motor.

external torque results in the rotor returning to the β = 0 position. Turning the rotor through an angular displacement of β = 90° also results in a position where the flux lines are rectilinear and $T_r$ = 0. However, as is evident from Fig. 11-16c, the reluctance of the magnetic circuit is now much greater. Another important difference in the position β = 90° as compared to that of β = 0 is the fact that any minor deviation from this position will tend to return the rotor to its original position. In other words, the β = 0 position is stable while the β = 90° position is unstable. Thus the reluctance torque is always directed such as to return the rotor to a position of minimum reluctance, which is at β = 0 and β = 180°. The torque produced is thus due to the variation of reluctance with rotor position—hence the name *reluctance motor*. The variation of reluctance ℜ as a function of rotor position may be represented graphically as done in Fig. 11-17.

Replacing the permanent magnet in Fig. 11-16 now with a stator winding that will provide a rotating field, the motor will start as an induction motor. As the rotor approaches synchronous speed, the salient poles of the rotor slip by the poles

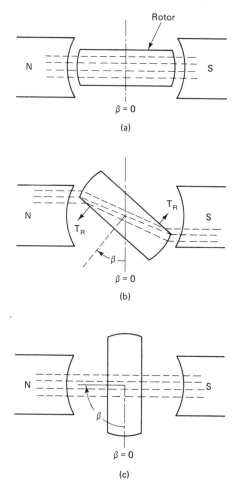

Figure 11-16  Reluctance motor principle.

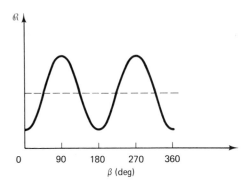

Figure 11-17 Relationship between circuit reluctance and angular rotor position.

of the rotating field at a constantly slower rate. As the rotor passes and lags behind the revolving poles, there will be a torque exerted tending to pull the rotor in step with the revolving field. This torque momentarily accelerates the rotor above its induction motor speed, because it is pulled in the direction of rotation. Once pulled into step it will rotate at synchronous speed. Reluctance motors like the hysteresis motor discussed may use any type of stator construction, such as polyphase, split-phase, or capacitor-type stator windings. The absence of the dc field excitation reduces the maximum output of these motors considerably as compared to synchronous motors. Also, they have low power factors and efficiencies. The low power factor is due to the relatively large magnetizing current as a result of the high reluctance. This relatively large stator current causes high power losses in the stator winding subsequently giving rise to low efficiencies. These drawbacks, however, are easily offset in many applications by the simplicity of its construction, since no slip rings, brushes, or dc field windings are required.

## 11-9 RELUCTANCE-START MOTORS

Another form of single-phase induction motor with only one winding is the reluctance-start motor. Here the pole tips are modified to provide some form of field rotation, similar to that of the shaded-pole motor. Figure 11-18 shows the constructional details of the pole tips for a four-pole motor. As shown, the air-gap length under the pole faces is not uniform but is wider under one side as compared to the other. As is known from our study of magnetic circuits, the reluctance in an iron part is greatly increased by the introduction of an air gap. Similarly, if the air gap is increased, the reluctance increases.

The direction of rotation of this motor is always from the wider air-gap region to the narrow air-gap region of the pole face. When the field coils are excited from an ac source, the flux will build up in the core and causes two out-of-phase flux components. The high-reluctance path has a flux component essentially in phase with the current; the low-reluctance portion will have a flux component which lags the voltage by nearly 90°. As a result, the motor acts as though there

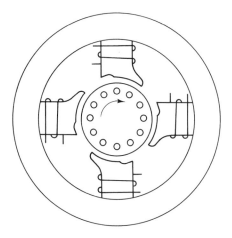

**Figure 11-18** Typical reluctance-start motor.

are two windings displaced in space that carry currents which are out of phase in time. This, as we know, is an essential criterion for the production of a rotating magnetic field and hence motor action results.

Reluctance-start motors are not used as much as shaded-pole motors, since their efficiencies are even worse. They have starting torques of about 50% of maximum torque, which in turn is slightly higher than full-load torque. They operate at large slip speeds and can therefore be speed controlled by varying the voltage across the motor using a tapped auto-transformer. Reversing the direction of rotation is not possible with this motor, for obvious reasons.

## 11-10 UNIVERSAL (SERIES) MOTORS

Motors that can be used on ac as well as dc sources of voltage supply are called universal motors. All these motors are of the dc series motor type. The direction of the developed torque is determined by both field polarity and the direction of the current through the armature. Since the same current flows through the field and the armature, it follows that ac reversals from positive to negative, or vice versa, will simultaneously affect both the field direction and the current direction through the armature. This means that the direction of the developed torque will remain the same and rotation will continue in the same direction.

Although the torque is in the same direction, it will pulsate in magnitude at twice the line frequency. The torque–speed characteristic at ac and dc operation differ somewhat, as illustrated in Fig. 11-19. This is attributable to the relatively large voltage drop across the series field due to its high reactance at ac operation, which reduces the output power. The reduction in output power can be compensated for by providing more armature winding turns. However, this in turn increases commutation problems. Thus a universal motor has to be designed carefully to minimize these drawbacks. Factors such as increased brush area, brush material, and pressure are fairly critical.

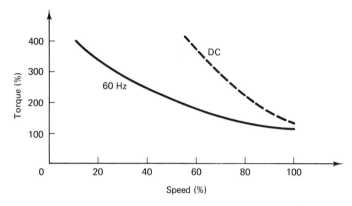

**Figure 11-19**  Typical torque–speed characteristic of a universal motor.

Another difference between the universal motor and ordinary dc series motor is in the parts of the field structure. On ac operation the field flux induces eddy currents in the solid parts of the field structure, such as the yoke and cores, generating excessive heating and thereby lowering the efficiency. This is overcome by laminating these parts and operating the machine generally at a lower flux density and using very short air gaps. Because of the relatively small number of field ampere-turns and the low flux density, a short pole results having a large cross section.

Rotor speeds of universal motors are generally in the range 5000 to 20,000 r/min, the higher speeds occurring at no load. These high no-load speeds are not destructive since the attached gearbox usually prevents the motor from overspeeding and thus reaching these speeds, or because the load is directly attached. Universal motors for small power applications such as portable hand drills and food mixers operate at high speeds but are geared down to their loads so that it operates at conveniently low values. The motors in these applications operate at these high speeds because, other things being equal, at higher speeds better cooling is obtained. When properly designed the gear losses are kept to a minimum. In other applications, the actual motor speed is the load speed; an example of this is the vacuum cleaner.

Since all commutator-type motors, especially those operating at high speeds, interfere with radio reception, universal motors are usually provided with a small capacitor connected directly across the line terminals inside the motor. In addition to the applications mentioned above, there are numerous applications where universal motors are used, such as portable drills, hair dryers, hedge trimmers, polishers, routers, sewing machines, and grinders, just to name a few. Applications must take into consideration its large speed regulation as well as the large horsepower developed for its size.

Speed control of universal motors is best obtained by *solid-state devices*. These electronic devices consist of thyristors or triacs in wave-chopping circuits to provide voltage variation to the series motor. As a consequence, speeds from full load to almost zero can be achieved with the motor developing full-load torque.

At this point it is worthwhile examining why line voltage speed control lends itself to single-phase induction motors but not to three-phase induction motors. This can be explained in terms in which the single-phase induction rotor develops its torque as compared to its three-phase counterpart. The latter machine, as we have discussed, tends to maintain its rotor excitation constant with decreasing voltage. Consequently, it requires a considerable reduction in line voltage before the slip increases appreciably. But as we have seen, since the developed torque is proportional to the voltage squared, it thereby greatly decreases the horsepower rating of the machine.

The torque developed in the single-phase induction motor, on the other hand, is more sensitive to changes in excitation since it is produced by two oppositely rotating magnetic fields. This can be illustrated as was done with the three-phase motor, by examining the effect of the stator field excitation on the developed torque–speed characteristic for a split-phase induction motor, for example, as in Fig. 11-20. For two values of applied voltages, the curve shows that for a given load a reduction in voltage produces a corresponding reduction in speed. The torque, changing as the square of this applied voltage, reduces to about 50% of its rated value if the applied voltage is reduced to 70% of rated.

As an example of how the voltage reduction is practically achieved to obtain speed control, we refer to Fig. 11-21, showing the tapped winding method. Here, highest speed is obtained when full voltage is applied to the smaller section of the winding and lowest speed when applied to the full winding. This is apparent from the transformer equation $V = 4.44 \, \phi \, Nf$. At constant frequency,

$$\frac{V}{N} = 4.44 \, f \, \Phi \propto \Phi \qquad (11\text{-}1)$$

Hence the excitation flux $\phi$ is proportional to $V/N$, the effective stator volts per turn. Thus, by increasing the number of turns to which the voltage is held constant, the excitation flux decreases. Since for any motor based on the induction principle

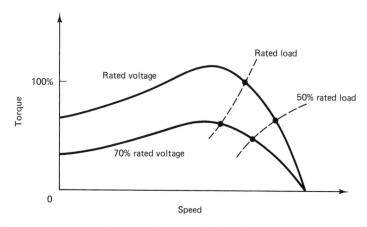

**Figure 11-20** Adjustable speed characteristic of a split-phase induction motor.

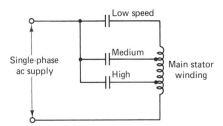

Figure 11-21 Tapped winding speed control method for single-phase induction-type motors.

the rotor current and torque are a function of the magnetic flux. It follows that for a given load, less torque ($T = k\phi I$) at a lower speed is produced with a decrease in the volts/turn ratio. The tapped stator winding, even for the smallest section, is uniformly distributed around the stator and designed to carry the full applied voltage without overheating.

Another scheme to control the applied voltage is that of using a tapped reactor coil which is externally connected in series with the single-phase motor, as shown in Fig. 11-22. The greatest voltage drop for any given load occurs with the entire reactor in series with the motor. Thus when normally-open contact Low is closed, the greatest reduction in motor voltage is obtained and the motor speed will be lowest. The advantage of this method is its adaptability to any single-phase motor, including those discussed, with the exception of single-phase synchronous motors.

The disadvantage, however, is its poor speed regulation because at low- or medium-speed operation a change in load will affect the line current. Hence the voltage drop across the reactor, subsequently the speed, is affected. To overcome this disadvantage in part, a tapped autotransformer is used, which improves the speed regulation somewhat.

The methods discussed so far may also be applied to universal motors, as stated. From dc motor theory we know that the speed equation

$$n = \frac{V_A - I_A(R_A + R_{\text{series}})}{k\phi} \tag{11-2}$$

This implies that both the armature voltage and armature current are a function of the supply voltage. The series field voltage drop is generally small compared to the overall armature circuit voltage drop. For instance, Fig. 11-23 illustrates how the tapped autotransformer, as mentioned above, may be used to achieve speed control of a universal motor operating from a fixed ac voltage supply.

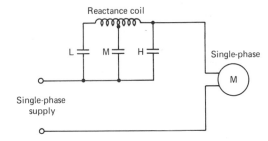

Figure 11-22 Single-phase motor with tapped series coil speed control scheme.

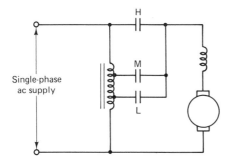

Single-phase
ac supply

**Figure 11-23** Speed control method for universal motor using a tapped auto-transformer.

So far we have discussed some methods to control the speed of single-phase motors by varying the amplitude of the applied voltage. Naturally, single-phase induction motors can also be speed controlled by varying the frequency, as is the case with polyphase induction motors. This usually involves, as we will see, electronic techniques using variable-voltage fixed-frequency or variable-voltage variable-frequency techniques. (Refer to Chapter 13.)

A number of simple methods employing electronic control to vary the voltage are illustrated in Fig. 11-24. They are suitable for use with the single-phase motors discussed in this chapter. Figure 11-24a shows a *thyristor* in series with a single-phase motor. As shown, this type of control, although simple, allows conduction only during the positive half-cycles of the supply voltage. It therefore has a dc component as well as an ac component of fundamental frequency in addition to higher-order harmonics. The amplitude of the dc and ac components is a function of the firing angle of the thyristor. This method of control is more suitable for universal motors since the dc component does not contribute to any useful torque in the other types of motors.

To improve on this scheme, two thyristors may be used as indicated in Fig. 11-24b, which produce full-wave control and therefore have no dc component present. Figure 11-24c shows a single semiconductor device, called a *triac*, which performs the same function as the two back-to-back thyristors shown in Fig. 11-24b. The triac has a single gate and is capable of conduction in both directions with either a positive or a negative gate signal, depending on the polarity of the supply voltage.

Because of its widespread use of the universal motor around the home, such as in portable tools, mixers, blenders, and electric drills, requiring speed control, a relatively inexpensive and extensively used speed-control scheme will be discussed. With reference to Fig. 11-25, the circuit uses a triac and a *diac*. Like the triac, which functions as a back-to-back pair of thyristors, the diac performs identically to a pair of back-to-back diodes. Full-wave voltage control is provided with maximum motor speed when the triac conducts during both half-cycles of the ac supply voltage. The variable resistor R controls the time constant RC of the charging circuit consisting of the resistor and capacitor C. This, in turn, controls the firing angle of the triac, since it will fire when the voltage across C exceeds the breakdown voltage of the diac. With the triac positively biased it is turned on by a positive gate signal, and when negatively biased the gate signal will be negative for conduction to occur. The triac turns off every time the input supply drops to zero.

The combination of solid-state devices and universal motor provides an economical controllable motor. The firing angle can be adjusted manually as in a trigger-controlled electric drill or can be controlled by a speed-control circuit, in which case the circuit usually employs some sort of feedback scheme to reduce the drooping torque–speed characteristic inherent in universal (series) motors.

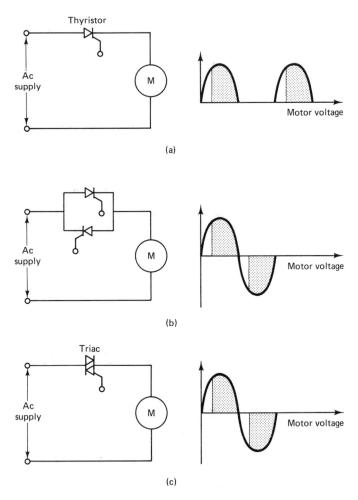

**Figure 11-24** Voltage control schemes for single-phase ac motors: (a) half-wave voltage control; (b) full-wave voltage control using thyristors; (c) full-wave voltage control using triac.

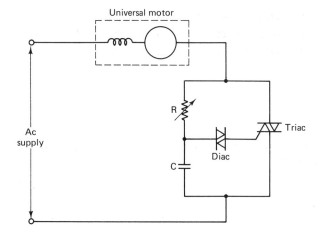

**Figure 11-25** Speed-control circuit for universal motor.

# REVIEW QUESTIONS

**11-1. (a)** Explain why a single stator winding in a single-phase motor produces no rotation.

**(b)** If the rotor in part (a) is accelerated in a certain direction, it will continue to operate as a motor. Explain.

**11-2. (a)** What is a circular field?

**(b)** What is an elliptical field, and under what load conditions does it occur?

**11-3. (a)** What design features are incorporated in a split-phase motor to make it self-starting?

**(b)** Why is the starting torque relatively small in this type of motor?

**(c)** With the starting winding disconnected once full speed is reached, is it possible to control the speed? Explain how, assuming that the frequency remains fixed.

**11-4.** The starting line current of a capacitor start motor is less than a corresponding split-phase motor. To what is this attributed?

**11-5.** When viewing split-phase, capacitor-start, and capacitor motors, how can you tell one from the other?

**11-6.** What is the advantage of using a capacitor-start motor over a split-phase motor?

**11-7.** Why is a shaded-pole motor recommended over a reluctance-start motor for the same applications?

**11-8. (a)** State three single-phase motors which have a centrifugal switch and three without.

**(b)** Explain why a shorted centrifugal switch may burn out a motor.

**11-9.** What property of the universal motor makes this motor so popular?

**11-10.** What type of motor would you use in the following applications: domestic oil burner, desk fan, sewing machine, food mixer, dishwasher, washing machine, portable electric drill. State your reasons.

**11-11.** Why can single-phase induction motors be voltage controlled whereas three-phase induction motors do not lend themselves to this kind of control?

**11-12.** Why are high speeds often desirable in the operation of universal motors? What limits the speed?

**11-13.** **(a)** How can the speed of a shaded-pole motor be controlled?

**(b)** Explain why the field produced in this motor is not revolving in the usual sense.

**11-14.** Describe a voltage speed-control method for a universal motor.


## PROBLEMS

**11-1.** With reference to the definition given for integral- and fractional-horsepower motors, classify the following motors:
**(a)** 1 1/4 hp, 1200 r/min.
**(b)** 3/4 hp, 900 r/min.
**(c)** 1-3/4 hp, 3600 r/min.
**(d)** 2/3 hp, 900 r/min.

**11-2.** A small motor is rated 10 W at 3050 r/min. What would be its developed torque in newton-meters, pound-feet, and ounce-inches?

**11-3.** A single-phase 115-V, 1/3-hp $n = 1720$ r/min motor draws a curtrent of 5.2 A when fully loaded. Calculate the efficiency at the full load.

**11-4.** A single-phase induction motor has a main winding impedance of 6 $\angle 45°$ Ω and an auxiliary winding impedance of 10 $\angle 15°$ Ω. What capacitor value is needed in series with the starting winding to produce a 90° phase shift between winding currents at startup? With this added capacitor, what are the winding currents and line current at starting? The line voltage is 120 V.

**11-5.** What capacitor value must be added in series with the starting winding of Example 11-1 to achieve a 90° phase shift between winding currents?

**11-6.** A four-pole shaded-pole motor has 165 turns per coil wound for 115 V. All four coils are connected in series. The motor is to be operated from a 230-V source, necessitating reconnection of the windings. What changes would be made if the volts per turn are to remain the same?

**11-7.** An autotransformer and a 4-μF capacitor combination are used in a two-value capacitor motor. If the transformer ratios on start and run are 6 and 1.5, respectively, determine the effective capacitor values when viewed from the primary side of the transformer, and the minimum ac voltage rating needed. The line voltage is 120 V.

**11-8.** A 115-V universal motor is loaded and takes 6.8 A at a power factor of 0.886 lagging. The output torque is measured to be 0.5 N · m at a speed of 4000 r/min. A resistor used to control the speed is connected in series with the motor, resulting in a line current of 8.5 A at a power factor of 0.80 lagging. The produced torque is maintained but at a lower speed. The motor impedance is 0.6 Ω. Determine:
**(a)** The speed at which the motor is operating with the series resistor added.
**(b)** The efficiency at which the motor is operating with and without this added resistor.

**11-9.** A universal motor has a resistance of 15 Ω and an inductance of 0.21 H. When connected to a 120-V dc supply and loaded, it takes 0.8 A and runs at 1800 r/min. When connected to a 120-V 60-Hz supply and loaded, it takes the same current but at a lagging power factor of 0.866. Determine:
**(a)** The speed at which the motor will run.
**(b)** The shaft torque in newton-meters if the rotational losses amount to 10 W.

# 12

# THREE-PHASE SYNCHRONOUS MOTORS

## 12-1 INTRODUCTION

In our discussions on dc machines we have seen that generators and motors are electrically and magnetically identical. As a matter of fact, a motor can be operated as a generator, and vice versa. Only minor changes in nameplate ratings are involved in doing so.

In the same way, we expect that a synchronous generator can be operated as a synchronous motor. If the prime mover is uncoupled from the synchronous generator, and a three-phase supply of constant frequency connected to its armature winding while the field is provided with dc excitation, torque will be developed at synchronous speed. Putting a mechanical load on the shaft will not cause the motor speed to change. The speed regulation is thus zero, and we have a constant-speed machine. Industrial applications that require constant speed are very common and include compressors, pumps, and mill machinery.

Another important advantage of the synchronous motor is its inherently high and adjustable power factor. In fact, it is often this property for which they are preferred. It can provide power factor correction in low-power-factor systems produced by other types of motors.

Induction motors and other electrical apparatus always operate at a lagging power factor. The power factor is fixed and determined by the mechanical load. For a synchronous motor we can vary its power factor merely by changing the dc field excitation current. When the motor power factor is unity, the dc excitation is said to be normal. Overexcitation causes the motor to operate at a leading power factor and underexcitation causes it to operate at a lagging power factor.

Operating a synchronous motor at no load and a greatly overexcited field causes it to take a current that leads the voltage by nearly 90°. This is like capacitor behavior. When we operate the machine this way, the synchronous motor is often referred to as a *synchronous condenser*. Such applications are common in large industrial plants and factories, which are penalized by the electrical utility for having low power factors (e.g., <0.85, depending on the utility). Thus power factor correction becomes essential.

The starting torque of a synchronous motor is negligible. We must provide a means to bring the machine up to or nearly up to (within a few percent of) synchronous speed. Then the motor can be synchronized before a mechanical load is applied. There are various ways to start the motor, the most practical being to use a damper winding, as we will see.

The main advantages offered by synchronous motors are:

1. Power factor variation and control (lagging and leading)
2. Absolutely constant speed

These are discussed later in this chapter. Starting methods and control schemes for synchronous motors are postponed until Chapter 13.


## 12-2 CONSTRUCTION

The constructional details of a three-phase synchronous motor are essentially the same as those of a synchronous generator, as illustrated in Fig. 12-1. The three-phase armature winding is on the stator and is wound for the same number of poles as the rotor. The required dc excitation source for the rotor field in a synchronous motor can be provided by means similar to those discussed for syn-

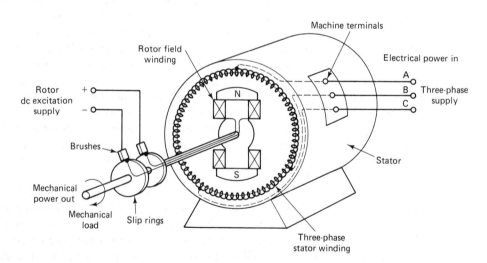

**Figure 12-1** Simplified diagram of a two-pole synchronous motor.

chronous generators (see Chapter 6). Usually, synchronous motors have slip rings and brushes on the rotor. The sliding contacts are not always wanted or permitted: for example, in chemical plants or in "hot zones" in nuclear power stations. They can be built with brushless excitation systems and started remotely by means of automatic starters. The rotor in synchronous motors is normally of the salient-pole type. There is a difference, however. In addition to the excitation winding, the rotor has an added winding in the pole faces which resembles a squirrel-cage winding. This winding is normally referred to as a *damper winding* (also known as an *armortisseur winding*). Figure 12-2 shows the details. Without the addition of this winding, the synchronous motor would not be self-starting, as we will see in Chapter 13.

The damper winding (Fig. 12-2) is placed in slots in the pole faces parallel to the shaft. The ends of the copper bars are short-circuited in the same manner as discussed for a squirrel-cage induction motor (Chapter 10).

As we mentioned, the armature winding is identical to that of the synchronous generator, and of course, identical to the stator winding of an induction motor as well. Therefore, when a three-phase voltage source is applied to the armature winding, a rotating field of constant magnitude is produced in the air gap. As you will recall from Chapter 10, [Eq. (10-1)], the speed of this rotating field depends

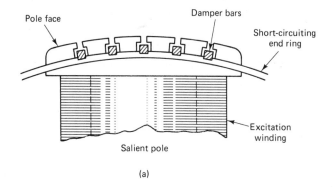

(a)

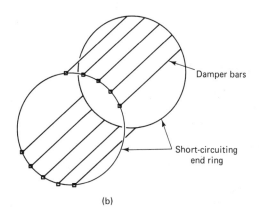

(b)

**Figure 12-2** Synchronous-motor rotor pole with five damper bars, short circuited by end rings: (a) location in pole; (b) circuit of winding for a two-pole rotor.

Sec. 12-2 Construction

on the line frequency $f$ and the number of poles $p$ the armature is wound for. It is

$$n_s = \frac{120f}{p} \quad \text{r/min} \tag{12-1}$$

where $n_s$ is the synchronous speed. Unlike the induction motor, a synchronous motor receives its excitation from two different voltage sources. One is the three-phase ac source to its stator winding and the other is a dc source for the rotor. It is the interaction of these two fields in the air gap that produces the torque, provided that those fields revolve at the same speed.

## 12-3 PRINCIPLE OF OPERATION

The important feature of synchronous motors that distinguishes them from induction motors is that they need two sources: a three-phase ac source to the armature winding and another dc source to supply the rotor. When this is done, torque can be developed but only at one speed, the synchronous speed. At any other speed the average developed torque is zero. To illustrate this point, let us consider the two-pole synchronous motor shown in Fig. 12-3. A three-phase ac voltage is applied to the stator winding and a dc source supplies the rotor field. The rotor is assumed stationary. The stator field rotates at synchronous speed, which according to Eq. (12-1) is 3600 r/min, for a two-pole machine supplied with a conventional 60-Hz three-phase voltage system. The dc field is stationary, because we assumed that the rotor is not turning.

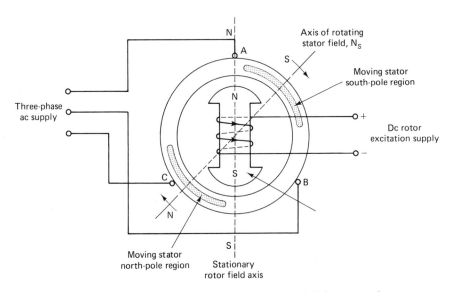

**Figure 12-3** Instantaneous position of rotor and stator field in a two-pole synchronous motor. Rotor assumed stationary.

Three-Phase Synchronous Motors    Chap. 12

It can be seen that to develop a continuous torque, the two fields must be stationary with respect to each other. This, as we can now appreciate, occurs only when the rotor is also turning at synchronous speed. It is only then when the stator and rotor fields lock in—hence the name *synchronous motor*. For the situation shown in Fig. 12-4, we see that the south pole of the rotor will lock in with the stator north pole, and vice versa. There may be momentary fluctuations in speed, but on the average the speed is constant. If the average speed of the rotor differs from the synchronous value, even by a small amount, the poles lose their "grip." No torque will then be developed and the machine will come to a standstill. The bond between stator and rotor poles is lost, an essential criterion for the development of torque.

Before we proceed further, let us give the situation in Fig. 12-3 a little thought. We have a pair of revolving armature poles that pass by the stationary rotor poles at great speed. In the position shown, there will momentarily be a positive torque developed between the pairs of poles of opposite magnetic polarity, positive being defined as the tendency for the rotor north pole to follow the stator south pole. Thus the rotor tends to rotate clockwise. However, because of the rotor inertia, even without a mechanical load on its shaft, the stator field "slides" by so fast that the rotor cannot follow it. Half a cycle later, which amounts to reversing the magnetic polarities of the stator field in Fig. 12-3, the tendency of the rotor is to move in the opposite direction. This is due to like magnetic poles repelling each other. As we can visualize, the rotor does not move. We then say that the average torque is zero. From a practical point of view this amounts to a zero starting torque.

How do we start a synchronous motor, then? This is where the squirrel-cage winding comes in (however, we investigate this in Chapter 13). Let us assume that the rotor is turning at synchronous speed as well. Figure 12-4 predicts this condition, which is repeated in Fig. 12-5 to compare with the no-load condition.

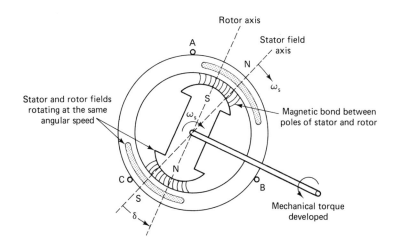

**Figure 12-4** Torque development in synchronous motor.

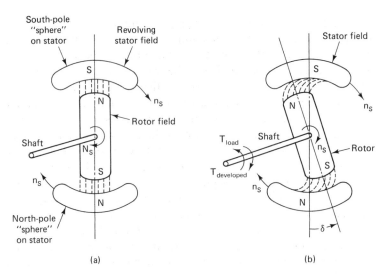

**Figure 12-5** Schematic of revolving rotor position with respect to revolving stator field position with increased shaft load: (a) no-load condition; (b) full-load condition (note power or torque angle δ, indicating amount rotor is slipped back).

For a two-pole machine, the revolving magnetic poles on the stator are schematically indicated by the north- and south-pole regions (see Fig. 12-5a). With no load on the motor shaft, the rotor poles lock in with the opposite stator poles, as shown. The entire arrangement rotates at synchronous speed and the torque angle δ is zero. When a mechanical load is now applied to the shaft, it tends to slow the rotor down somewhat. The mechanism involved will be discussed shortly, but let us assume that at full load the position is as shown in Fig. 12-5b.

Again, this entire arrangement rotates at synchronous speed. The magnetic bond between rotor and stator fields is still maintained, except for the rotor displacement by the torque angle δ. The developed torque, $T_d$, depends on the angle δ, and of course it must be sufficient to overcome the shaft torque applied, $T_{load}$. We discuss in Section 12-4 how the synchronous motor adjusts itself. But the essence of synchronous motor action is the necessary establishment of a magnetic bond between rotor and stator fields. Loss of this bond leads to a loss of synchronism and motor action ceases.

## 12-4 LOADING A SYNCHRONOUS MOTOR

Like all motors, the synchronous motor acts as a generator as well as a motor. This is due to a counter or back EMF which is created when the motor is in operation. As we have seen in our discussion on dc machines, when the shaft load on a dc shunt motor, for example, is increased, the speed of the motor decreases with a consequent reduction in counter EMF, $E_C$. Since the supply voltage V remains constant (assuming an infinite bus), the voltage necessary to overcome the armature voltage drop $I_A R_A$ increases and the current becomes

$$I_A = \frac{V - E_C}{R_A} \quad A \tag{12-2}$$

Thus, when the counter EMF is reduced, more current flows through the armature to compensate for the increased torque and power by the load. It is, in fact, the counter EMF that limits the armature current.

Contrary to the dc machine, the synchronous motor runs at absolute constant average speed, regardless of the load. The question then becomes: How does the synchronous motor adjust itself to an increased shaft load? Before answering this, the following points should be noted illustrating the fundamental differences in behavior. First, the counter EMF $E_C$ in a synchronous motor does not have to be smaller than the applied voltage $V$ as it is in a dc motor. In fact, it can be equal to or even larger than the applied voltage. This should not come as a surprise, as we have already encountered a similar situation. As you will recall, in the synchronous generator the generated voltage can be larger or smaller than the terminal voltage when the machine operates on an infinite bus system. Second, the phase relation between $E_C$ and $V$ in a synchronous motor is not fixed at 180° (i.e., they are not directly opposite). Third, the armature current $I_A$ in a synchronous motor lags behind the resultant voltage by nearly 90°, whereas in a dc motor it is always in phase. These points will be clarified shortly.

Going back to our original question of how the synchronous motor adjusts itself to load changes, let us assume that the shaft load increases. Since the average speed must remain constant, it cannot draw increased line current in the same manner as the dc motor does, (i.e., operate at reduced speed). To see what adjustments do take place in a synchronous motor, we refer to Fig. 12-6. Two salient rotor poles are shown at an instant when they are opposite two conductors in the stator. The conductors are assumed to be a pole pitch apart (see Fig. 12-6a). The counter EMF induced in the conductors is as shown, being maximum when that conductor is opposite the pole center and zero when midway between. For any other conductor position with respect to the pole faces, the induced voltage is as shown by curve $E_C$.

Upon applying the increased load to the motor shaft, a momentary slowing down of the rotor results, since it takes time for a motor to take increased power from the line. In other words, the rotor, although still rotating at synchronous speed, will have slipped back in space as a result of increased loading. This action may be visualized as follows. Visualize the magnetic bond between stator and rotor fields as rubber bands. When increasing the shaft load the rubber bands are stretched by an amount dictated by the load torque. This causes the rotor to slip back in space by an angle δ behind the stator field, but it continues to rotate at the same speed provided that the rubber bands are intact. The stretched rubber bands will eventually break if the load on the machine continues to be increased. When this happens, we speak of the machine "losing synchronism." In other words, we expected too much of our motor. The point at which this occurs is generally referred to as the *pull-out torque*. The induced EMF corresponding to

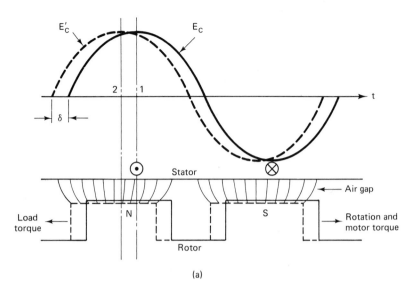

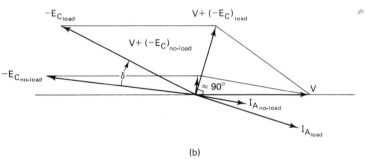

**Figure 12-6** Effect of increased shaft load on (a) rotor position and (b) armature current in a synchronous motor.

this new rotor position with respect to the stator field is shown in Fig. 12-6a as $E'_C$. Also, the pole center is shown to be moved from position 1 to position 2, by the same angle $\delta$. This can be illustrated further by the use of a phasor diagram. At light loads, $E_C$ and $V$ are almost directly opposite, while with increasing load the rotor poles slip back by an angle $\delta$. As indicated in Fig. 12-6b, with increasing load the phasor $E_C$ slips back from its no-load position by the angle $\delta$. At full load this angle may be as much as 60 electrical degrees for a two-pole machine.

The diagram tells us further that the net voltage $V + (-E_C)$ depends on the position of $E_C$. The value of $I_A$ is directly proportional to this vector sum. Also since the stator winding is predominantly inductive, the current will lag behind the net voltage by an angle of approximately 90°. In analogy to the dc shunt motor, we can state that the armature current $I_A$ in the synchronous motor is

$$I_A = \frac{V + (-E_C)}{Z_S} \qquad (12\text{-}3)$$

where $Z_S$ is the impedance per phase of the stator winding; the bold-faced quantities are to remind us that we are considering vector quantities, (i.e., their proper phase relationship must be accounted for). Thus by loading the machine, the rotor assumes an angular position back from its no-load position. This causes the motor to take increased power from the line to compensate for the increased shaft load, without changing its average speed. The total power supplied to the motor per phase is

$$P = VI_A \cos \theta \qquad \text{W/phase} \qquad (12\text{-}4)$$

where $V$ is the phase voltage, in volts and $\theta$ is the angle between $V$ and $I_A$. The total mechanical power developed is

$$P_d = E_C I_A \cos \alpha \qquad \text{W/phase} \qquad (12\text{-}5)$$

where $\alpha$ is the phase angle between the vectors $E_C$ and $I_A$. The net shaft power is less than $P_d$ by an amount equaling the frictional plus rotational and core losses. The difference between $P$ and $P_d$ is the armature copper loss, $I_A^2 R_A$.

The reader more familiar with mechanical power transmissions may appreciate the torque principle in a synchronous motor from Fig. 12-7. The small-diameter shaft is analogous to the magnetic coupling between the stator and rotor fields.

The driven pulley $B$ (the rotor) lags behind the driving pulley $A$ (the stator field) by the twist in the shaft (the $\delta$ angle). The rotating speed of both pulleys remains the same. An excessive load (pull-out torque) on pulley $B$ will cause the shaft to break (lose synchronism) and pulley $B$ will stop.

Before we go any further, it should be noted that, strictly speaking, the theory developed for the synchronous motor in this chapter is applicable to the cylindrical rotor machine. A similar approximation was applied in the theory of synchronous generators. For salient-pole machines, the theory is more complicated, to take into account the nonuniformity of the air gap. For practical purposes the added

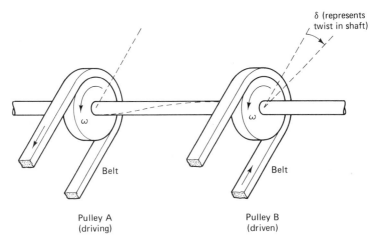

**Figure 12-7**  Mechanical power transmission analogous to synchronous motor torque.

complexity is not warranted. The nonsalient or cylindrical rotor theory is applied to salient-pole machines in practically all cases in practice, since it simplifies calculations considerably without appreciable loss in accuracy. Therefore, we will adopt this approach under the assumption that pole saliency is not important for all practical purposes. Readers who wish to pursue this matter further may refer to some of the advanced electrical machines texts listed in the References.

## 12-5 ADJUSTING THE FIELD EXCITATION

When the dc excitation current is increased, since the speed of the synchronous motor remains constant, its induced EMF $E_C$ must increase because of the strengthened rotor field. It may appear that the motor would stall, or start to act as a generator when the induced EMF becomes equal to or greater than the line voltage, respectively. The synchronous motor, however, continues to operate as a motor, even though the counter EMF exceeds the line voltage. In this condition, the motor is said to operate overexcited, with the result that it takes power from the line with a leading power factor. Figure 12-8 shows us how this is accomplished. Assume that the motor is adjusted such that it is operating at unity power factor (i.e., the armature current $I_A$) is in phase with the line voltage, as indicated in Fig. 12-8a. As usual, all calculations are made on a per phase basis. From Eq. (12-3) we see that the armature current is proportional to the vector sum of the line voltage $V$ and the counter EMF $E_C$, limited by the impedance $Z_S$. Under our assumption of increased field excitation, $E_C$ increases, but since the load on the machine has not increased, the angle δ remains unchanged. Since the vector sum of $V$ and $E_C$ increases and the impedance remains constant, the armature current must increase. However, the load power remains constant, which implies that the phase angle must change in such a manner that the in-phase component of the armature current with the terminal voltage remains unchanged; that is, $I_A \cos \theta$ is constant. This can be accomplished only if the armature current assumes a leading position, as indicated in Fig. 12-8b. Considering next a decrease in the field excitation, the counter EMF $E_C$ will become smaller. As before, the in-phase component of the armature current with the line voltage $V$ must remain the same since the power delivered is constant. This means that the vector sum of $V$ and $E_C$ again becomes larger compared to the unity-power-factor case. The armature current will increase depending on this voltage increase and can do so only by changing its power factor to a lagging angle, as illustrated in Fig. 12-8c.

The reader should note the construction difference in phasor diagram Fig. 12-8 compared to Fig. 12-6. In Fig. 12-8, $E_C$ is represented, whereas in Fig. 12-6, $-E_C$ was represented. Of course, this does not represent anything different; the minus sign associated with the phasor $E_C$ simply indicates a 180° phase shift. In both instances, as Eq. (12-3) testifies,

$$I_A Z_S = V + (-E_C) = V - E_C \quad \text{V} \qquad (12\text{-}6)$$

where $I_A Z_S$ is the impedance drop in the armature circuit. The synchronous impedance $Z_S$ of a synchronous motor is, of course, obtained in fashion similar to that

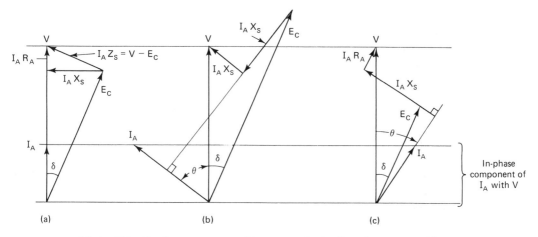

**Figure 12-8** Synchronous motor with constant load adjusted to operate (a) at unity, (b) leading, and (c) lagging power factor, by adjusting the field excitation current.

for the synchronous generator. As expected, the electric circuit representation is also identical, but the power flow is reversed, as we will see shortly. Thus an increase in dc excitation results in a leading power factor (overexcited), whereas a decrease in field excitation results in a lagging power factor (underexcited) with no appreciable increase in input power to the machine.

## 12-6 PHASOR DIAGRAMS

Expanding on the phasor diagram of Fig. 12-6b, Fig. 12-9a depicts a load condition with the excitation of the field such that a unity power factor results. The diagram shows the armature current and voltage drop due to the impedance of the motor. In vector notation,

$$V = E_C + I_A Z_S \qquad (12-7)$$

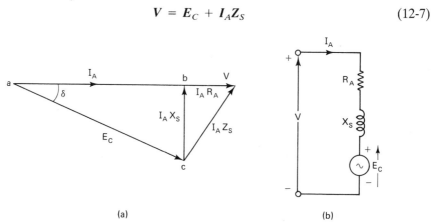

**Figure 12-9** (a) Phasor diagram of a synchronous motor (per phase) operation at unity power factor; (b) equivalent-circuit diagram.

Sec. 12-6    Phasor Diagrams    **333**

It should be noted that balanced conditions are assumed; therefore, only one phase of the three-phase motor need be considered.

The $I_A R_A$ voltage drop is due to the armature resistance, and the reactive voltage drop $IX_S$ consists of armature reaction and armature reactance, similar to the synchronous generator. As expected and as shown in Fig. 12-9b, the equivalent-circuit diagram of the synchronous motor is similar to that of the synchronous generator, except that the power flow is reversed since the motor takes power from the line. As we will see, this implies that $V$ leads $E_C$ in the phasor diagram. The counter EMF, $E_C$, is a function of the dc excitation current.

If the field excitation is now increased, the counter EMF increases, causing the current to assume a leading power factor. The phasor diagram corresponding to this condition (assuming no change in load) is represented in Fig. 12-10a. Fig. 12-10b indicates the situation where the synchronous motor is underexcited and the line current lags the applied voltage by an angle $\theta$.

Let us compare the phasor diagrams for the three kinds of excitation: unity power factor, as depicted in Fig. 12-9a, and overexcited and underexcited, as in Fig. 12-10. It is apparent that with constant torque developed by the motor as expressed by keeping the torque angle $\delta$ the same, the phase angle can be adjusted from leading to lagging, or vice versa, merely by changing the dc field excitation current. This, in turn, modifies the counter EMF as explained before, which rotates the voltage-drop triangle formed by the $I_A R_A$, $I_A X_S$, and $I_A Z_S$ vectors. When the phase angle departs from unity power factor the size of this triangle increases, reflecting an increase in line current. Because the torque delivered is constant, the in-phase component of the current and voltage must remain constant.

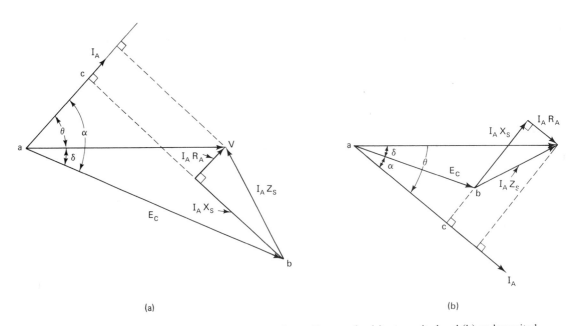

(a)                                                                          (b)

**Figure 12-10** Synchronous motor phasor diagrams for (a) overexcited and (b) underexcited conditions.

In reality, the power angle is not fixed even for constant loads, but does depend to some extent on the field excitation. For an increased field current the rotor pole field becomes stronger; therefore, less slippage occurs between the two fields. Using our mechanical analogy, this corresponds to an increased shaft diameter in Fig. 12-7, resulting in less shaft twist. Reducing the excitation has the opposite effect. However, we will assume in our study that $\delta$ is affected only by the shaft torque applied.

To solve the diagrams trigonometrically, we proceed as was done before in the transformer and synchronous generator calculations. We project the voltages on the current vector to form a right triangle; then $E_C$ is the hypotenuse, to be found as shown in the figures by the crosshatched right triangles *abc*. These calculations are clarified in the following example.

### EXAMPLE 12-1

A 75-hp 440-V 900-r/min three-phase Y-connected synchronous motor has an effective armature resistance of 0.15 $\Omega$ and a synchronous reactance of 2.0 $\Omega$/phase. At rated load and 0.8 power factor leading current, determine:

(a) The induced counter EMF $E_C$.
(b) The power angle $\delta$.
(c) The developed mechanical power $P_d$.

Assume that the motor has an efficiency (excluding the field loss) of 90%.

### SOLUTION

(a) The motor input power is

$$P_{in} = \frac{\text{hp} \times 746}{\eta} = \frac{75 \times 746}{0.90} = 62{,}167 \text{ W}$$

The line current is

$$I_A = \frac{P_{in}}{\sqrt{3} \, V_L \cos \theta} = \frac{62{,}167}{\sqrt{3} \times 440 \times 0.8} = 102 \text{ A}$$

and

$$\text{voltage per phase} = \frac{440}{\sqrt{3}} = 254 \text{ V}$$

$$\theta = \cos^{-1}(0.8) = 36.9° \text{ leading}$$

From Fig. 12-10a,

$$E_C = [(V \cos \theta - I_A R_A)^2 + (V \sin \theta + I_A X_S)^2]^{1/2}$$

$$= \{[(254 \times 0.8) - (102 \times 0.15)]^2$$

$$+ [(254 \times 0.6) + (102 \times 2.0)]^2\}^{1/2} = 403 \text{ V}$$

As a check, $E_C$ may also be determined by using complex notation, namely

$$E_C = V - I_A Z_S$$

$$= V - I_A(\cos \theta + j \sin \theta)(R_A + jX_S)$$

$$= 254 - 102(0.8 + j0.6)(0.15 + j2.0)$$

$$= 364 - j172 = 403 \ \angle -25.3°$$

The negative sign indicates that $E_C$ lags $V$, which was taken as the reference vector, by 25.3°, which also represents the power angle $\delta$.

(b) Another way to obtain $\delta$ is as follows. By inspection of Fig. 12-10a we see that

$$\tan \alpha \equiv \tan (\theta + \delta) = \frac{V \sin \theta + I_A X_S}{V \cos \theta - I_A R_A} = \frac{356.3}{187.9} = 1.90$$

from which $(\theta + \delta) = \tan^{-1}(1.90) = 62.2°$ and $\delta = 62.2° - \theta = 25.3°$, in agreement with the calculation in (a).

(c) The mechanical power developed follows from Eq. (12-5), namely,

$$P_d = 3E_C I_A \cos \alpha$$

$$= 3 \times 403 \times 102 \times \cos 62.2° = 57,484 \ W$$

which is also equal to the power input minus the armature resistance loss, or

$$P_d = P_{in} - 3I_A^2 R_A$$

$$= 62,167 - 3 \times 102^2 \times 0.15 = 57,484 \ W$$

Note that, as before, the factor of 3 arises from the fact that there are three phases. The power developed at the motor shaft is less than $P_d$ by an amount equal to the rotational losses, friction, windage, and core losses. A simple calculation shows that this amounts to $P_d - P_{out} = 1534 \ W$ for this motor.

From the phasor diagrams it is seen that $E_G$ is lagging the terminal voltage, consistent with the calculated results. This is opposite to synchronous generators, for which $E_G$ leads $V$. Let us examine this in more detail.

## 12-7 POWER FLOW

As can be seen from the phasor diagrams for the synchronous motor (e.g., Fig. 12-10), the terminal voltage $V$ of the machine is leading the counter EMF, $E_C$. It is independent of the power factor at which the motor is operating. On the other hand, the reader will remember from our study of synchronous generators that the terminal voltage is constantly lagging the generated EMF. Again this is independent of the power factor; only the amount it is leading or lagging depends on the power factor. We know that the power factor can be controlled in both machines by adjustment of the dc field excitation current.

We can now appreciate what is taking place. In a synchronous generator when the prime-mover torque is increased, it tends to speed up the generator away from the system bus. We have seen that it is locked onto the bus, and rather than pulling away, the rotor advances a certain angle with respect to the resultant air-gap field. The generated EMF waveform will follow this rotor advancement. This is shown in Fig. 12-11. The angle $\delta$, the power angle, is a measure of the real power delivered by the machine. It is positive when $E_G$ leads $V$ (i.e., power flows out of the machine).

If we apply a negative torque, by lowering our prime-mover torque from that value as required by the load, the network would then pull the machine. Another way of putting this is to say that it acts as a motor—the machine will pull some of the load. In that case $E_C$ would lag $V$ and the resultant air-gap flux would now start to lag the rotor position. This implies power flowing into the machine. Figure 12-11 predicts this action where $\delta < 0$.

As we change the field current in either operating condition, no change can take place in the real power or torque delivered by the prime mover. An indirect change will take place in an actual machine amounting to a slight adjustment of the power angle, as we have discussed. However, to simplify matters, we will ignore this change. Figure 12-12 summarizes the important differences between generator and motor action.

Consider now the situation where the torque on the motor is being increased. Under this assumption the developed power will reach a maximum when $\delta$ equals $-90°$ for a motor or $+90°$ for a generator, as discussed in Section 6-7. Figure 12-12c shows the developed power $P$ as a function of $\delta$. As we have seen, this curve follows from Eq. (6-10), where it was assumed that the armature resistance is negligible. Increasing the torque beyond this point will result in a loss of synchronism. The motor will stall. As you recall from our mechanical analogy of Fig. 12-7, this amounts to a shearing of the shaft between the two pulleys. From Eq. (6-10) we have

$$P = \frac{E_C V}{X_S} \sin \delta \qquad \text{W/phase} \qquad (12\text{-}8)$$

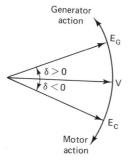

**Figure 12-11** Power flow in synchronous machine ($\delta > 0$, generator action: $\delta < 0$, motor action).

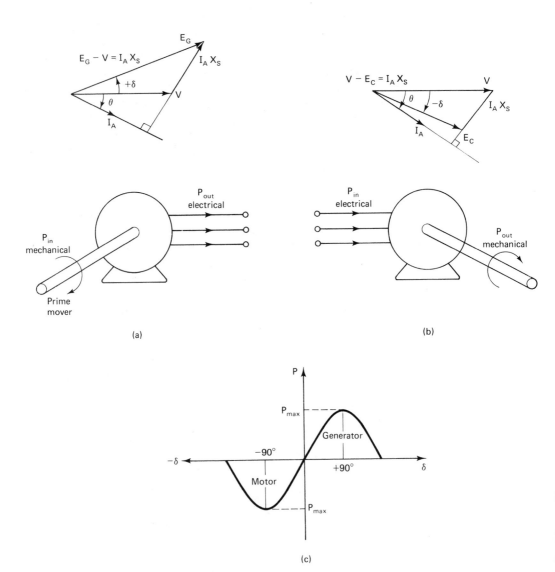

**Figure 12-12** Motor and generator action of a synchronous machine ($R_A$ is assumed negligible): (a) generator action; (b) motor action; (c) developed power versus torque (or power) angle δ.

The maximum power or pull-out power $P_{max}$ occurs when $\sin \delta = 1.0$ ($\delta = 90°$), or

$$P_{max} = \frac{E_C V}{X_S} \qquad \text{W/phase} \qquad (12\text{-}9)$$

under the assumption that $R_A$ is negligible.

Losing synchronism implies that the magnetic bond between the stator and rotor fields is lost. The rotor will "skip" poles and tries to run as an induction motor if the starting winding can support this load. It will continue to do so until the excessive load is removed. In the meantime, excessively large currents will flow in the stator winding at the moment that stator and rotor poles align. This goes hand in hand with dangerous heating effects, which must be avoided.

### EXAMPLE 12-2

A 2200-V three-phase Y-connected synchronous motor has a synchronous reactance $X_S$ = 2.6 Ω/phase. The armature resistance is assumed to be negligible. The input power is 820 kW, and the field excitation is such that the counter EMF is 2800 V. Calculate:

    (a) The torque angle.
    (b) The line current.
    (c) The power factor.

### SOLUTION

First we construct a phasor diagram (Fig. 12-13) to see what is happening. From Fig. 12-12b we see that the synchronous motor $V$ leads $E_C$; furthermore, here $E_C > V$, resulting in a leading power factor. Since the motor is Y-connected,

$$V = \frac{2200}{\sqrt{3}} = 1270 \text{ V/phase}$$

and

$$E_C = \frac{2800}{\sqrt{3}} = 1617 \text{ V/phase}$$

    (a) From Eq. (12-8),

$$P = \frac{E_C V}{X_S} \sin \delta \qquad \text{W/phase}$$

Therefore,

$$\frac{820,000}{3} = \frac{1617 \times 1270}{2.6} \times \sin \delta$$

or

$$\sin \delta = 0.346 \qquad \text{and} \qquad \delta = \sin^{-1}(0.346) = 20.2 \text{ electrical degrees}$$

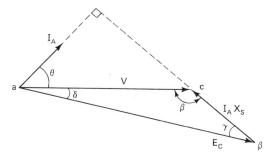

**Figure 12-13** Phasor diagram for solution of Example 12-2.

(b) To obtain the line current (= phase current) $I_A$, we see by inspection that triangle $abc$ in the phasor diagram is completely defined, since two sides ($V$ and $E_C$) and an angle ($\delta$) are known. This calls for application of the cosine rule from trigonometry to solve for $I_A X_S$. Namely,

$$(I_A X_S)^2 = V^2 + E_C^2 - 2VE_C \cos \delta$$

so that

$$(I_A X_S) = (1270^2 + 1617^2 - 2 \times 1270 \times 1617 \times \cos 20.2°)^{1/2}$$

$$= 610.8 \text{ V}$$

from which

$$I_A = \frac{610.8}{2.6} = 234.9 \text{ A}$$

(c) For the power factor calculation we need the angle $\theta$. Realizing that $\beta = 90° + \theta$, we can proceed in two ways. We can follow a procedure similar to that used in part (a), or apply the sine rule to triangle $abc$. Applying the latter, there results

$$\frac{E_C}{\sin \beta} = \frac{I_A X_S}{\sin \delta} = \frac{V}{\sin \gamma}$$

Thus

$$\frac{1617}{\sin \beta} = \frac{610.8}{\sin 20.2°}$$

from which

$$\sin \beta = \frac{1617}{610.8} \times \sin 20.2° = 0.914$$

and

$$\beta = \sin^{-1}(0.914) = 114°$$

Hence

$$\theta = 114° - 90° = 24°$$

The power factor is

$$PF = \cos 24° = 0.914 \text{ (leading)}$$

*Note:* Since $\sin(90° + \theta) = \cos \theta$, it follows that $\sin \beta \equiv PF$.


## 12-8 EFFICIENCY

To complete our picture of the power flow, we must consider the losses occurring in a synchronous motor. As expected, this is similar to the synchronous generator. Figure 12-14 indicates pictorially the losses that occur in a synchronous motor.

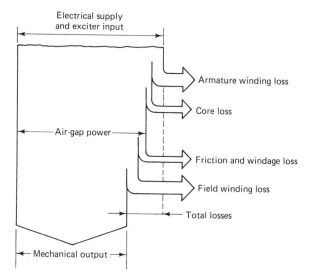

Electrical supply
and exciter input

Armature winding loss

Core loss

Air-gap power

Friction and windage loss

Field winding loss

Total losses

Mechanical output

**Figure 12-14** Power flow for a synchronous motor.

The motor losses subtracted from the electrical input give efficiencies generally between 80% and the high 90's. Losses appear as heat, which must be removed. Efficiency figures can vary widely depending on who is doing the figuring; some people do not include auxiliary needs such as exciter losses.

To calculate the efficiency, we proceed as before:

$$\eta = \frac{P_{\text{out}}}{P_{\text{out}} + \text{losses}} \times 100\% \qquad (12\text{-}10)$$

where $P_{\text{out}}$ is the output shaft power, which equals the shaft torque times the radial speed at which it is developed,

$$P_{\text{out}} = \omega T_{\text{shaft}} \qquad \text{W} \qquad (12\text{-}11)$$

The input power is simply

$$P_{\text{in}} = \sqrt{3} \, V_{\text{L-L}} I_L \cos \theta$$

so that Eq. (12-10) can also be written as

$$\eta = \frac{\omega T_{\text{shaft}}}{\omega T_{\text{shaft}} + 3I_A^2 R_A + I_F V_F + P_{\text{core}} + P_{\text{fr}+\text{w}}} \times 100\% \qquad (12\text{-}12)$$

with the various symbols representing the quantities discussed earlier.

Following the method discussed for synchronous generators, we will normally exclude the auxiliary losses. In that case, $P_{\text{out}}$ becomes

$$P_{\text{out}} = P_d - P_{\text{fr}+\text{w}+\text{core}} \qquad \text{W} \qquad (12\text{-}13)$$

from which the shaft torque (in N · m) can be established from Eq. (12-11).

**EXAMPLE 12-3**

A 415-V eight-pole 60-Hz Δ-connected three-phase synchronous motor has its field excitation adjusted, resulting in an induced EMF of 520 V and a torque angle of 12 electrical degrees. The armature impedance $Z_S = 0.5 + j4.0$ Ω. The friction + windage + core losses amount to 2000 W. Determine:

(a) The line current.
(b) The power factor.
(c) The output power.
(d) The efficiency.
(e) The shaft horsepower and torque.

**SOLUTION**

The phasor diagram for the given values is shown in Fig. 12-15. From Eq. (12-3) (or by inspection of the phasor diagram), we have

$$I_A Z_S = V \angle 0° - E_C \angle -\delta° = 415 - 520 \angle -12°$$

$$= 415 - (508.6 - j108.1) = 143.0 \angle 130.9° \text{ V}$$

since

$$Z_S = 0.5 + j4.0 = 4.031 \angle 82.9° \text{ Ω}$$

there results

$$I_A = \frac{I_A Z_S}{Z_S} = \frac{143 \angle 130.9°}{4.031 \angle 82.9°} = 35.5 \angle 48° \text{ A}$$

Therefore,

(a) $I_L = I_A \sqrt{3} = 35\sqrt{3} = 61.5$ A (armature Δ-connected)

and

(b) $\theta = 48°$     PF $= \cos 48° = 0.669$ leading
(c) The input power can now be calculated:

$$P_{in} = \sqrt{3} V_{L-L} I_L \cos \theta = 3 V I_A \cos \theta$$

$$= 3 \times 415 \times 35.5 \times 0.669 = 29,573 \text{ W}$$

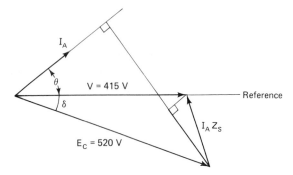

**Figure 12-15**  Phasor diagram for the solution of Example 12-3 (on a per phase basis).

The armature copper loss is

$$P_{Cu} = 3I_A^2 R_A = 3 \times 35.5^2 \times 0.5 = 1890 \text{ W}$$

The developed air-gap power is then

$$P_\alpha = P_{in} - P_{Cu} = 3 E_C I_A \cos(\theta + \delta)$$

$$= 29.573 - 1890 = 27{,}683 \text{ W}$$

The output power is

$$P_{out} = P_d - P_{fr+w+core} = 27{,}683 - 2000 = 25{,}683 \text{ W}$$

(d) The efficiency is

$$\eta = \frac{P_{out}}{P_{in}} \times 100 = \frac{25{,}683}{29{,}573} \times 100 = 86.8\%$$

(e) The shaft horsepower is

$$hp = \frac{25{,}683}{746} = 34.4$$

and the load torque is

$$T_{shaft} = \frac{25{,}683}{2\pi \times 900/60} = 273 \text{ N} \cdot \text{m or } 201 \text{ lb f} \cdot \text{ft}$$

## 12-9 V-CURVES

As has been demonstrated, the power factor of a synchronous motor can be controlled by variation of the field current. We have seen from the examples that not only does the power factor change but the line current is affected as well. Under the assumption of a constant load, the power input to the motor does not change appreciably and can be assumed to remain the same. Of course when the current increases the copper losses increase somewhat, as does the core loss, due to a slight change in flux. However, these increased losses do not change the input power too much.

To show how the field current controls the power factor, consider a large reduction in field excitation, which will reduce the induced EMF $E_C$ and cause the line current to lag. Also, we assume the motor to operate at no load. Increasing the field current from this small value, we see that the line current decreases until a minimum line current occurs, indicating that the motor is operating at unity power factor. Up to this point the motor was operating at a lagging power factor. Continuing to increase the field current, the line current increases again and the motor starts to operate at a leading power factor. Plotting the relationship of armature current versus field excitation current, the lowest curve in Fig. 12-16 is obtained. Repeating this procedure at various increased motor loads results in a family of curves, as shown.

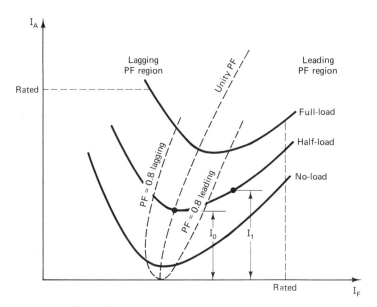

**Figure 12-16** V-curves of a synchronous motor.

Because of their resulting shape, these curves are commonly referred to as V-curves. The point at which unity power factor occurs is at the point where the armature current is minimum. This is indicated by $I_0$, for example, for the curve taken when the synchronous motor delivers 50% of its rated load (see Fig. 12-16). Connecting the lowest points of all V-curves results in the dashed curve indicated by "unity PF" in the figure. This curve is commonly called a *compounding curve*. Others can be drawn for different power factors, as shown.

Taking the necessary data during these tests, the power factor may also be plotted for each point. Plotting the power factor versus field current results in a family of curves shown in Fig. 12-17, which are inverted V-curves. The highest point on each of these curves indicates unity power factor. It shows that if the

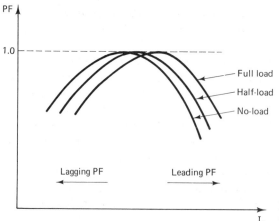

**Figure 12-17** Power factor versus field current at different loads.

excitation is adjusted at a certain load for unity power factor, increasing that load results in a lagging power factor.

These PF curves can also be constructed from the V-curves. Referring again to the 50% rated load curve in Fig. 12-16, let $I_1$ be a value of armature current at some power factor $\theta$. The power per phase is $P = VI_1 \cos \theta \equiv 1/2 -$ load, but $I_1 \cos \theta = I_0$ for all values of $\theta$, since the power delivered by the machine for this curve is constant. The excitation corresponding to the value of armature current $I_0$ is called the *normal excitation*. Reducing this excitation, the motor takes a lagging current and is said to be *underexcited*, as we have discussed. Increasing the excitation results in a leading current and the motor is said to be *overexcited*. Thus the power factor $\cos \theta = I_0 / I_1$ can be found for any value of armature current $I_1$ and given power. The corresponding field current $I_f$ at the selected values of armature current can be read from the curve, providing the data necessary to plot the curves in Fig. 12-17.

## 12-10 SYNCHRONOUS CONDENSERS

When a synchronous motor is operated without load, it still takes some power from the line to overcome its internal losses, such as rotational, friction, and windage and core losses. As we have seen, these losses are minimal and therefore the in-phase component of the armature current is rather small, although the current may be large, depending on the dc excitation. If the motor is operated overexcited, it will take a very low leading power factor current and behave like a capacitor. This can be verified by referring to Fig. 12-18, which shows the armature current to lead the supply voltage by nearly 90°.

In plants where primarily induction motors are used, advantage is taken of this property of the synchronous motor by paralleling it with the induction motors, so that part or all of the lagging component of the load is counteracted. The synchronous machine is usually placed in parallel at the incoming power lines to the plant. Hence the overall power factor of the incoming line to the plant is improved. Such a connection is indicated schematically in Fig. 12-19. When a synchronous motor is employed this way it is commonly referred to as a *synchronous condenser* or *synchronous capacitor*. It will operate at no load but with a strong dc field excitation.

### Power Factor Correction in Industrial Plants

When an industrial plant buys power from the electric utility, the total amount of electrical energy is measured as well as the maximum rate at which this energy is consumed. These measurements (or readings) are normally taken on a monthly

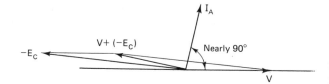

**Figure 12-18** Overexcited synchronous motor at no load.

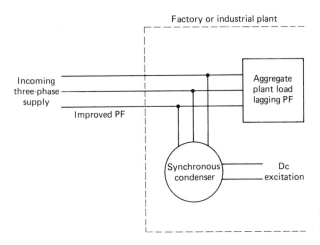

**Figure 12-19** Power factor improvement by using a synchronous condenser.

basis, and a portion of the bill reflects the rate of consumption or *demand charge*, as it is called.

The demand charge stems from the way the plant uses this energy; that is, it relates directly to the resistive component and the reactive component. Only the resistive component provides useful work, but the reactive component is a necessary ingredient in most machines since it makes things happen. The relationship between the two components is, of course, the power factor.

The reactive power needed, as we have seen in our study of the various machines, is a function of the type of motor and transformer installed. For example, when lightly loaded, induction motors (as well as transformers) have poor power factors. This implies that the reactive components are relatively large. There is not much we can do about this, but it is possible within the plant to supply a negative reactive component which will cancel out some that we otherwise have to buy. That is where capacitive action, in particular the synchronous motor, comes in. When the plant takes such corrective action, we usually refer to it as *power factor correction*. Figure 12-19 demonstrated this in principle and we will shortly see how we can determine what is needed.

Before going into these details, let us consider what benefits are attained. Naturally, it saves energy, which translates into a lower monthly bill; otherwise, why bother? In addition, it can reduce the peak demand and thus add capacity to our supply transformer or even to the plant feeders. An additional benefit may be a somewhat increased voltage. The power factor is, in principle, best corrected at the source; practically, this implies at the plant site, at the billing meters. There are various schemes to measure consumption (kWh) and peak demand (kW or kVA), such as graphic or digital meters. They all do the same thing, which is to record the maximum average demand, typically over a 15-minute period. To get a feeling for this, let us consider an indicating-type meter. This type of meter has a red and a black pointer. The red pointer is the active one and rises and falls with the average plant load. As it rises, it pushes the black pointer ahead, which remains at the highest position reached. When the meter is read at the end of the month, the meter is reset after the readings are taken. The black pointer reading is multiplied by the meter constant or multiplier to convert it to a true kVA or kW reading. This multiplier is either shown on the outside of the meter or is available from the utility office.

For the real rate of work, the plant is measured electrically in kW and is billed as such. But machines draw power with a reactive component, which really is "workless" current; that is, it does not relate to actual work done. Figure 12-20 illustrates this graphically.

The reader should note that the quantities kVA (apparent power), kVAR (reactive kVA), and kW (real power) are scalar quantities. They are related by a right-angle triangle. Usually, the analysis is shown by phasors because of the two types of kVARs. In such instances the inductive kVAR is taken as positive, the capacitive kVAR as negative.

The right-triangle theorem tells us that

$$(kVA)^2 = (kW)^2 + (kVAR)^2 \qquad (12\text{-}14)$$

from which we see that if we know kW and kVA, we can solve for kVAR. This is the quantity we want to know. From trigonometry we know that

$$PF = \cos \theta = \frac{kW}{kVA}$$

where $\cos \theta$ is the power factor. It is thus evident that two meters' readings, the kVA and kW in a plant, provide all the information we need. Also, the amount of corrective kVAR needed is readily seen from Fig. 12-20.

As we realize at this point, kVAR can flow in both directions. In induction motors we call it *lagging*; in capacitors we call it *leading*. If it is lagging we say that it is consumed; when leading it is supplied or generated. There is nothing to indicate whether a power factor is leading or lagging. The kVA meter does not indicate this and the kW meter (which is unaffected by kVAR) simply keeps step with the increasing load.

The utility therefore needs the two meters and bills you for the highest. However, in all fairness the utilities realize that plants must operate and use kVAR; they permit consumers a certain quantity of kVAR free. A matter of fact, in most areas (depending on the utility) they allow up to 48% of the kW load "free." A simple calculation shows that this amounts to a power factor of 90%. The moral of the story is this: Pay for the kW you use or 90% of the kVA demand, whichever is higher.

Let us examine in a practical way the effect of a power factor on the power bill. Then we can proceed to see what can be done to reduce it.

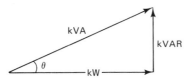

**Figure 12-20** The power triangle.

## EXAMPLE 12-4

The peak reading (black hand) on a plant's kW meter is 821, and on the kVA meter it is 1080. What is the plant billed for if an 85% power factor is allowed? Assume a multiplier of 1200.

### SOLUTION

$$PF = \frac{reading\ kW\ meter}{reading\ kVA\ meter} = \frac{820}{1080} = 0.76$$

or

$$0.76 \times 100\% = 76\%$$

The bill will be based on a 76% power factor. The peak demands are

$$\frac{820 \times 1200}{1000} = 984\ kW$$

$$\frac{1080 \times 1200}{1000} = 1296\ kVA$$

The plant is billed for true kW or 85% of kVA, whichever is larger. 85% of 1296 = 1102 kVA is larger than the kW consumption. This means that the billed kW becomes 1102. The peak kW demand was only 984; thus we pay for the extra (1102 − 984) 120 kW because of the poor power factor. This extra charge is due to what is called a *power factor penalty*.

To correct a plant's overall power factor, capacitor banks can be installed or a synchronous motor which will be operated overexcited. We will concern ourselves with the synchronous motor, although the same thing can be extended to power factor correction using capacitors. It should be made clear that the synchronous motor must remain on-line continuously. Leaving it off for only 15 minutes during the load peak period means that you may as well leave it off for the remainder of the month.

A convenient way of showing power factor correction applications is by means of solving problems, illustrating the use of vector diagrams. We demonstrate this in the examples that follow. In dealing with power factor correction the reader may notice that we use kVA and kW rather than volts and amperes. Furthermore, we will deal with total quantities now rather than on a per phase basis. This is following standard practical calculations as applied in the "field"; it makes things easier. In the end we want to specify what rating our machine should have to do the job. This should not be confused with the previous calculations we have performed on a per phase basis. When dealing with equivalent circuits and associated phasor diagrams, using one-phase power only simplifies our work considerably. Knowing the three-phase power, the power factor, and the voltage level we are working at provides us with knowledge of the line currents.

## EXAMPLE 12-5

An industrial plant represents a load of 1600 kVA at a lagging power factor of 0.60. Determine:

(a) The kVA rating of the synchronous condenser to be installed in parallel with this load to raise the line power factor to unity.
(b) The total kW load.

### SOLUTION

The overall circuit diagram is as depicted in Fig. 12-19, and the corresponding vector diagram is represented in Fig. 12-21.

(a) For an overall plant power factor of unity, the synchronous condenser will have to counteract the vertical component (reactive power) of the plant load, which is

$$1600 \times \sin 53° = 1280 \text{ kVAR}$$

(b) Total load $= 1600 \times \cos 53° = 960 \text{ kW}.$

We see that the supply rating is reduced from 1600 kVA to 900 kVA (PF = 1.0) at the expense of a 1280-kVA synchronous condenser. In practice it is seldom necessary to raise the power factor above approximately 85%, since very little is gained beyond this in reducing the line current. The expense of the additional synchronous condenser rating normally does not warrant it. This is indicated by the following example.

## EXAMPLE 12-6

Referring to the plant data in Example 12-5, determine:

(a) The kVA rating of the synchronous condenser required to raise the power factor to 0.85 lagging.
(b) The total kVA load.

### SOLUTION

(a) $\theta_t = \cos^{-1}(0.85) = 31.8°$

$$\sin \theta_t = 0.527 \quad \text{and} \quad \tan \theta_t = 0.620$$

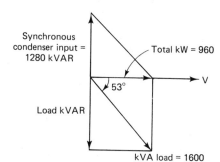

Synchronous condenser input = 1280 kVAR

Total kW = 960

53°

V

Load kVAR

kVA load = 1600

**Figure 12-21** Vector diagram for solution of Example 12-5.

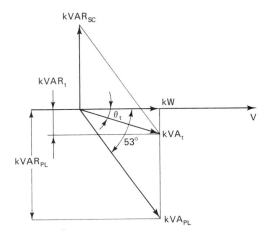

**Figure 12-22** Vector diagram for Example 12-6.

The vector diagram is shown in Fig. 12-22. From the diagram we see that the new or total $kVAR_t$ is the difference between the kVAR of the plant and that of the synchronous condenser. Therefore,

$$kVAR_{sc} = kVAR_{pl} - kVAR_t$$

Now $kVAR_{pl} = 1600 \times \sin 53° = 1280$ kVA, and since the total power does not change,

$$kVAR_t = kW \times \tan \theta_t = 960 \times 0.620 = 595$$

and

$$kVAR_{sc} = 1280 - 595 = 685$$

(b) $kVA_t = \dfrac{kW_t}{PF} = \dfrac{960}{0.85} = 1129$

Comparing Examples 12-5 and 12-6, it is evident that the size of the synchronous condenser is considerably less, namely 685 kVA as compared to 1280 kVA. This represents a reduction in capacity of approximately 87%, while at the same time the line current is only 1/0.85, or about 18% as large as that compared to unity power factor.

**EXAMPLE 12-7**

As a further example we can illustrate the power factor improvement calculation if in addition to providing power factor improvement the synchronous condenser is supplying a load. To do so, let us assume that we have the system illustrated in Fig. 12-23, showing a line diagram of a transformer supplying an existing load of 1600 kVA, 0.6 power factor lagging. To this plant we wish to add a 750-hp synchronous motor with 0.8 PF leading and an efficiency of 90%. The questions arise: (1) Will this added load overload

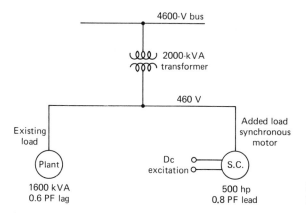

4600-V bus

2000-kVA transformer

460 V

Existing load

Added load synchronous motor

Plant

Dc excitation

S.C.

1600 kVA
0.6 PF lag

500 hp
0.8 PF lead

**Figure 12-23** Line diagram of supply line to plant.

the transformer? (2) What will be the power factor after the motor has been added?

**SOLUTION**

With reference to Fig. 12-24, which shows the vector diagram, triangles $OAB$, $BCD$, and $ODE$ are right-angle triangles, so that simple trigonometric relations apply. Triangle $OAB$, the existing plant load, has the sides

$$OA = 1600 \times \cos 53° = 960 \text{ kW}$$

$$EC = AB = 1600 \times \sin 53° = 1280 \text{ kVA}$$

The new motor that is added has a kW input (vector $BC = AE$) of

$$AE = \frac{750 \times 0.746}{0.90} = 622 \text{ kW}$$

The leading reactive component (vector $DC$) when the motor operates at full load and excitation is

$$DC = BC \tan 37° = 466.5 \text{ kVAR}$$

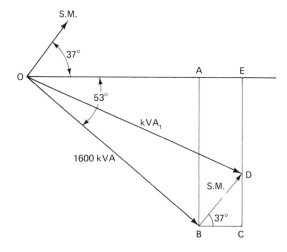

S.M.

37°

O

53°

A    E

kVA$_t$

1600 kVA

D

S.M.

37°

B    C

**Figure 12-24** Vector diagram for Example 12-7.

Sec. 12-10    Synchronous Condensers

**351**

This enables calculation of the new total kVA as

$$\text{kVA}_t = \sqrt{(OA + AE)^2 + (CE - DC)^2}$$
$$= \sqrt{1582^2 + 813^2} = 1779 \text{ (vector } OD)$$

and the overall power factor

$$\text{PF}_t = \frac{1582}{1779} = 0.89$$

which is a considerable improvement in power factor. Furthermore, with the added synchronous motor load the line current increases approximately 10%, still within the rating of the transformer. It is interesting to note, however, that if an 750-hp induction motor had been added instead, the line current would have exceeded the transformer-rated current by about 18.5%. Calculation of this has been referred to Problem 12-7.

## 12-11 APPLICATIONS

As we have discussed, synchronous motors offer some definite advantages over other motors. These include constant-speed operation, power factor control, and high operating efficiencies. The high-efficiency operation relates directly to the synchronous motor having a separate source for the required field excitation. We may remember that induction machines must have a small air gap in the magnetic circuit. This will limit the magnetizing current needed to establish the flux. The air gap is made as small as mechanical clearance will allow, generally in the range 0.02 to 0.05 in. Induction machines are excited from a single source, and the excitation current and power component of the current flow in the same lines. The magnetizing current is a large component of the rated current, so that they are found to operate at relatively poor power factors at light loads.

Since the field has its own supply, synchronous motors, usually have larger air gaps than those used with induction motors. In addition, they can operate at unity power factor, resulting in efficiencies for this type of motor in the range 85 to 96%. As such, they are the most efficient of all motors.

Nowadays, the starting and running characteristics of synchronous motors are such that they can compete successfully with induction motors. Because synchronous motors require a dc source and slip rings, the initial cost of the machine is larger than that of a comparably sized induction motor, particularly for small units. In the larger sizes for heavy industrial service, this disadvantage no longer exists; as a matter of fact, there is a horsepower range and a speed range that puts the synchronous motor at an advantage (see Table 12.1 and Fig. 12-25). It is precisely for these reasons that synchronous-motor applications are increasing.

When considering a synchronous motor for a specific service, we need to concern ourselves basically with the following specifications.

**TABLE 12-1** OVERVIEW OF THREE-PHASE SYNCHRONOUS MOTOR APPLICATIONS

| Class (speed) | Hp | $T_{start}$ (% rated $T$) | $I_{start}$ (% rated $I$) | $T_{pull-out}$ (% rated $T$) | $\eta$ (%) | PF | Typical applications |
|---|---|---|---|---|---|---|---|
| Above 500 r/min | 25 to few thousands | 110 up to 170 | 500–700 | Up to 200 | 92–96 | Usually high | Fans, centrifugal pumps, and compressors: PF correction, constant speed |
| Below 500 r/min | Usually above few thousands | Low 30–50 | 200–350 | Up to 180 | 92–96 | High | Directly connected loads, compressors when started unloaded, steel mills, PF correction, constant speed, pulsating loads when flywheel is used |

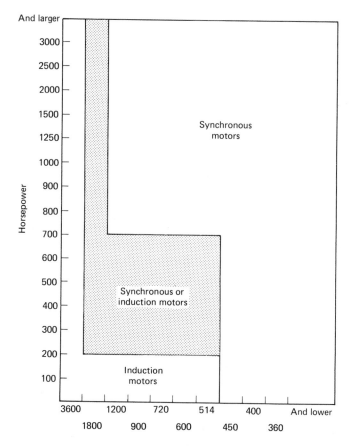

**Figure 12-25** General areas of application of synchronous motors and induction motors.

1. *Starting torque:* the ability of the motor to start the load
2. *Pull-in torque:* the ability to pull into synchronism from induction to synchronous operation
3. *Pull-out torque:* the ability to maintain synchronism under full-load conditions or pulsating loads

As discussed previously, the first two characteristics are associated with the damper winding. It is for this reason that the starting speed–torque characteristic of a synchronous motor is identical to that of an induction motor. The resistance of the damper winding can be made high enough to give good starting characteristics (typically ranging from about 110 to 170% of full-load torque). Pull-in torques are in the range of 120%. The pull-out torque, on the other hand, is a characteristic of the motor and depends on the field strength. This leads us to the connection of the field circuit at startup, since it has an important bearing on the starting of the motor. It may be open circuited or closed through a resistor at startup. Remember, it is the squirrel cage that starts the rotor moving. With the field circuit open at startup, large voltages may be induced in it. We should realize that as long as the rotor is not running at synchronous speed, there is relative motion between the rotating field and the rotor field winding. Thus voltages are induced in the field winding. Such voltages are dangerous and may puncture the insulation of the field.

It is for this reason that the field is normally closed through a resistor during the starting period. The large induced voltage will not exist but the starting torque may be somewhat reduced. Let us see why. The closed field winding acts as a second rotor circuit. Induced currents will flow in the rotor field circuit and by Lenz's law, these induced currents will create a field tending to oppose the main flux through the rotor. Because of the high field circuit inductance, the resultant induced current flow in the rotor winding is limited. Therefore, the reduction in flux through the field poles is limited. On the other hand, there will be some torque developed between the rotating field and the induced currents, tending to aid the starting torque. It is the reduction of flux through the rotor poles that more than counteracts this torque due to induced currents. The overall result is that the starting torque is somewhat reduced.

When synchronous speed is reached, the full excitation should be applied to the field after removal of the shorting resistor.

The types of loads to which synchronous motors are usually well suited fall into three general classes. These are characterized by the specifications that we discussed.

1. In the first category are drives that require relatively low starting and pull-in torques. Applications that meet these requirements are those that permit partial unloading during start-up. Typical examples of machinery using synchronous motors are: compressors, pulp grinders, generator drives, pulp refiner drives, barking drums, chippers, etc.
2. In this class we have applications that require relatively low starting torque, but high pull-in torques. Machines such as pumps and fans generally require

low starting torques. Pumps are usually automatically operated and connected to a system at about the same time as they reach full speed. This requires a motor that develops full-load torque at pull-in.

3. In this group of applications we require high starting and pull-in torques, usually high pull-out torque as well. Applications in this class are: crushers, rubber mills, material handling in rolling mills and steel mills, water, waste and sewage treatment facilities and equipment, etc.

In general, the torque requirements of each type of load should be carefully evaluated prior to setting motor specifications for specific tasks.


## REVIEW QUESTIONS

**12-1.** What is average speed in reference to a synchronous motor?

**12-2.** What determines the number of poles on the rotor of a synchronous motor?

**12-3.** In what manner does a synchronous motor adjust itself to an increasing shaft load?

**12-4.** Name two important characteristics of a synchronous motor not found in induction motors.

**12-5.** What factors decide the selection of a synchronous motor over an induction motor for large-horsepower motors?

**12-6.** How can the speed of a synchronous motor be changed?

**12-7.** What is a damper winding? What is its function, and where is it located?

**12-8.** What is the pull-out torque in reference to a synchronous motor?

**12-9.** Under what operating condition is a synchronous motor referred to as a synchronous condenser?

**12-10.** What do V-curves represent in terms of synchronous motor characteristics?

**12-11.** A synchronous motor is operating at half-load. An increase in its field excitation causes a decrease in armature current. Does the armature current lead or lag the terminal voltage?

**12-12.** Show with the help of diagrams the effect of armature reaction on the main flux of a synchronous motor due to zero PF leading, unity, and zero PF lagging currents.

**12-13.** **(a)** What are the assumptions made in developing a simplified theory of synchronous machines?

**(b)** Draw an equivalent circuit of a synchronous motor showing the resistance of the motor windings, the reactance due to leakage inductance of the windings, and the fictitious reactance introduced to account for armature reaction.

**(c)** What is the synchronous reactance?

**(d)** Draw a vector diagram of a loaded, synchronous motor at some lagging power factor, showing the various voltage drops.

**12-14.** **(a)** Draw the typical curves relating line current and field excitation of a synchronous motor, for no load and full load. Indicate clearly the lagging and leading power factor regions of the curves.

**(b)** A synchronous motor delivering full load is drawing minimum line current from the supply source. If the load is kept constant and the excitation is increased, explain briefly what effect this will have on the power factor.

**12-15.** What is the power factor penalty?

**12-16.** What action can be taken if an industrial consumer's power factor is too low?

**12-17.** A synchronous motor operating at 1/4 load and unity power factor has its load increased to full load. To maintain operation at unity power factor, would the field current have to be adjusted? Explain.

**12-18.** State at least two practical applications where you would use a synchronous motor.


## PROBLEMS

**12-1.** Electrical power is to be supplied to a three-phase 25-Hz system from a three-phase, 60-Hz system through a motor generator set consisting of two directly coupled synchronous machines.
(a) What is the minimum number of poles the motor may have?
(b) What is the minimum number of poles the generator may have?
(c) At what speed in r/min will the set operate?

**12-2.** A 60-Hz synchronous motor is coupled to, and drives, a 50-Hz alternator. How many poles does each machine have, and at what speed does the motor-generator set operate?

**12-3.** (a) What is the synchronous speed of a 12-pole 460-V 60-Hz synchronous motor?
(b) At what speed will it run when the full load is reduced to 50%?
(c) What is the speed if the supply voltage is 380 V?

**12-4.** For the synchronous motor in Example 12-1, assume unity power factor (1.0) and repeat the calculations. (This problem indicates the slightly reduced input power for minimum line current.) Maintain the output power as in the example. What would be the efficiency of the motor in this case?

**12-5.** A three-phase 6600-V Y-connected synchronous motor is tested. The open-circuit and short-circuit data are as follows:
Open circuit:

| $I_F$ (A)       | 40   | 80   | 120  | 160  | 240  |
|-----------------|------|------|------|------|------|
| $V_{L-L}$ (V)   | 3800 | 5800 | 7000 | 7800 | 8700 |

Short circuit: $I_F = 180$ A, $I_A = 460$A
When running as a synchronous motor it takes 80 A at a power factor of 0.6 leading. What field current is necessary? Assume that $R_A$ is negligible.

**12-6.** The motor of Problem 12-5 has a field circuit resistance of 0.5 Ω. The friction and windage losses are 9.65 kW and the magnetic losses are equivalent to 12.0 kW.
(a) What is the efficiency of the motor?
(b) What is the shaft horsepower developed?

**12-7.** In Example 12-7, assume that a 750-hp induction motor was added (otherwise same data) to the system instead of the synchronous motor. Would this overload the transformer? If so, what is the transformer current in this case?

**12-8.** A 2000-V three-phase Y-connected synchronous motor has an effective resistance and synchronous reactance of 0.2 Ω and 2.2 Ω/phase, respectively. The input is 800 kW at normal voltage and the induced line EMF is 2500 V. Calculate the line current and power factor.

**12-9.** A three-phase induction motor having a rating of 100 hp and operating at a power factor 0.8 lagging is in parallel with a three-phase synchronous motor taking 200 kVA at 0.8 power factor leading. The supply voltage is 2400 V. Calculate the line current and power factor of this combined load.

**12-10.** A 1500-hp 6600-V three-phase Y-connected synchronous motor with a synchronous reactance of 6 Ω/phase, delivers full load at unity power factor and normal rated voltage. The armature resistance is negligible. What percentage increase in field excitation current is needed to achieve a PF = 0.866 leading? Assume linear operation and constant torque angle.

**12-11.** A 30-hp 440-V Y-connected synchronous motor is operating at a load that causes the rotor to lag the stator field by 10 electrical degrees. The field excitation is adjusted to produce a generated phase voltage equal to the applied phase voltage.
(a) What is the resultant voltage per phase in the armature?
(b) What angle does it make with the applied phase voltage?

**12-12.** The input to a 13,800-V three-phase, Y-connected synchronous motor is 75 A. The synchronous reactance per phase is 25 Ω. The armature resistance is negligible.
(a) Find the power supplied to the motor and the induced EMF for PF = 0.8 leading.
(b) For PF = 0.8 lagging, what are the induced EMF, the line current, and power supplied to the motor? Assume that the torque angle remains constant at the value determined in part (a).

**12-13.** A 100-hp 500-V three-phase Y-connected synchronous motor has a negligible resistance and a synchronous reactance of 0.8 Ω/phase. For full load and 0.8 power factor leading, calculate:
(a) The total power developed in watts.
(b) The EMF per phase.
(c) The torque angle.
Assume an efficiency of 93% (excluding exciter losses).

**12-14.** A three-phase Y-connected synchronous motor has negligible armature resistance and a synchronous reactance of 12.0 Ω/phase. Input to the motor is 12,000 kW at a line voltage of 13,200 V. The field current is adjusted so that the excitation voltage is 9000 V/phase.
(a) What is the power angle?
(b) At what power factor is the motor operating?
(c) What is its armature current?

**12-15.** A synchronous motor has a Δ-connected armature with a synchronous reactance of 2.4 Ω/phase. This motor takes a line current of 68 A at 220 V, with the field current adjusted to give an excitation voltage of 160 V/phase.
(a) What power does the motor take from the line?
(b) At what power factor is it operating?
Assume the armature resistance to be negligible.

**12-16.** The input to an industrial plant is 1200 kW at a power factor of 0.6 lagging. It is desired to connect a synchronous motor that operates at a leading power factor to 0.8 to the power mains and have it correct the overall power factor to 0.9 lagging.
(a) What should be the power input to the synchronous motor?
(b) If the power factor is not corrected, what is the power factor penalty, assuming the allowed PF is 85% minimum?

**12-17.** A manufacturing plant has a load of 3600 kVA at a lagging power factor of 0.707. A 500-hp synchronous motor having an efficiency of 90% is installed and is used to improve the overall power factor to 0.937 lagging.

(a) Draw the kVA vector diagram to represent these conditions. Label each vector clearly.

(b) Calculate the kilovolt-ampere input rating of the synchronous motor and the power factor at which it must operate.

**12-18.** An industrial factory has a load of 1600 kVA at an average power factor of 0.60 lagging. Calculate:

(a) The kVA input to a synchronous condenser for an overall power factor of unity.

(b) The total kW load.

**12-19.** Repeat Problem 12-18 for an overall power factor of 0.90 lagging.

# 13

# STARTING and CONTROL of AC MOTORS

## 13-1 INTRODUCTION

At this point in our study we examine some ways to start the three-phase ac motors we have discussed and look at some ways to control their speed. Because of their widespread use, we start off with conventional starting methods, particularly those that are in use for induction motors. Synchronous motors use the damper winding as a starting aid, which really is basically a squirrel-cage induction motor. Therefore, the starting characteristics of synchronous motors are similar to those discussed for squirrel-cage induction motors. As we discussed in Chapter 10, the starting current of induction machines is prohibitive. Therefore, all starting methods deal with this problem; however, there are various schemes applied which we will discuss.

Modern machinery consists of a large number of components each assigned a definite function. In general, operated together they serve or carry out a specific production process. The electric motor (for our purpose) has systems of control and transmission, which may consist of shafts, gears, pulleys and belts, and so on, to impart motion to the driven load. These keep the workpiece in motion, and the term *drive* is generally used to designate the equipment that produces the motion. In other words, we can say that an electric drive is a piece of equipment designed to convert electric energy into mechanical energy and to provide electrical control in the process.

Equipment that includes electric drives are electric motors, rectifier units, frequency changers, and converters, to name a few. Such equipment provides versatile control and enables motor operation on a specific torque–speed characteristic.

Modern industry uses a large variety of many machines operating at different and variable speeds. Typical examples are metalcutting machine tools; electric cranes; elevators; equipment used in textile, paper, mining, electronic industries; and many others. All drives used must have some form of speed control to do the job well. Speed control implies intentional change in speed. For instance, elevators, cranes, and conveyor mechanisms must decelerate smoothly and come to a standstill at the required stop point. This chapter provides a brief introduction to some of the common starting methods employed in industry, and modern methods to control motor speeds of electric drives.

## 13-2 STARTING THE SYNCHRONOUS MOTOR

The synchronous motor as such is not self-starting. The rotating field passes the rotor field poles at synchronous speed and because of the rotor inertia and its connected load, it cannot produce acceleration. Therefore, it must be brought up to near or actual synchronous speed before it can operate.

Any method that accomplishes this will do, but there are, of course, practical considerations as well. Sometimes this is done by using a small dc motor mounted on the rotor shaft. In fractional- and small-horsepower motors the reluctance torque is put to use (this is discussed more fully in Chapter 14). But in the large industrial units with which we are concerned in this chapter, the most common method is to use a damper winding for starting purposes. In Section 12-2 we saw what the damper winding looks like and where it is located. Now we will see how it starts the motor.

With a three-phase supply connected to the stator winding, the rotating field will create rotor currents in the damper bars. This action is identical to that of induction motors with squirrel-cage rotors. The result is that torque is developed and the rotor accelerates in the direction of the rotating field.

The synchronous motor with a damper winding will start as a squirrel-cage induction motor and will reach a speed slightly below synchronism. At this small slip the rotor field winding is energized and the motor will pull into synchronism. This necessitates a momentary quick acceleration of the rotor to "catch up" to the stator field. It is for this reason that high-speed synchronous motors are started at greatly reduced loads. The inertia of the rotor plus the load would be too large for the rotor to step into synchronism with the rotating stator field.

When a dc field is applied, opposite poles of stator and rotor fields need not necessarily be opposite each other; they may be slightly ahead of each other. However, sufficient torque is usually developed to accelerate the rotor momentarily to lock in. If it happens that like poles are opposite each other, there will be a repelling force. The rotor will slow down momentarily, but within half a cycle a stator pole of unlike polarity will come by and the attractive force will then pull the rotor up to synchronous speed.

During the instant that like poles are opposite each other, the air-gap flux will be reduced because of a partial cancellation of fields. This causes an inrush of current from the supply lines. To avoid this problem, synchronous motors are often started on reduced voltage and the rotor field applied before full voltage is supplied to the stator. Large line currents are thereby avoided.

As we discussed in Chapter 10, the starting torque of a synchronous motor with a damper winding can be controlled. For instance, increasing the resistance of the damper winding gives increased starting torque. But as we know, this also means that the motor will not approach synchronism as closely since it will run at a larger slip speed. On the other hand, a low-resistance damper winding is more effective for damping oscillations. So there is a trade-off. Under normal operating conditions we should realize that the damper winding carries no current; therefore, it has no influence on the torque. If for some reason the mechanical load fluctuates due to some disturbance, the rotor position with respect to the stator pole axis will fluctuate. This oscillation is normally referred to as *hunting*. This implies relative motion between field and damper bars. According to Lenz's law, currents will be induced in the rotor bars which will have such a direction as to counteract these swings. The net effect is that the currents will have a damping effect on oscillating rotor swings.

### Automatic Starters for Synchronous Motors

Since most synchronous motors are three-phase and are provided with a damper winding, the common practice is to start them as induction motors. Like squirrel-cage motors they may be connected directly across the line or started on reduced voltage. When started on reduced voltage the usual practice is as follows. The starting contactor is closed first, connecting the stator to the reduced voltage, starting the motor. Once up to nearly synchronous speed, the starting contacts are opened and the running contacts are closed, thereby putting the stator to full line voltage. Immediately after, the field circuit contactor closes, connecting it to the field circuit supply.

The actual method employed to provide reduced voltage at startup is similar to that for starting induction motors. As we will see, these include starting resistances or reactances in the stator circuit, or autotransformers may be used. Whatever method is chosen, it is apparent that the specific control problem as related to synchronous motors is the method for transferring control connections to the rotor when reaching synchronous speed. Also, it is usual to short-circuit the field circuit at startup as a precautionary measure. The field winding has many turns and when stationary may have a high voltage induced in it.

As an example of a full-voltage starter for a synchronous motor, Fig. 13-1 shows a controller using a frequency sensitive relay (FSR) which will operate at a selected frequency. The operation of the circuit is as follows. Depressing the START button energizes control relay *CR*. One of its contacts provides a holding circuit for this relay while another energizes the main contactor relay *M*. *M* is a

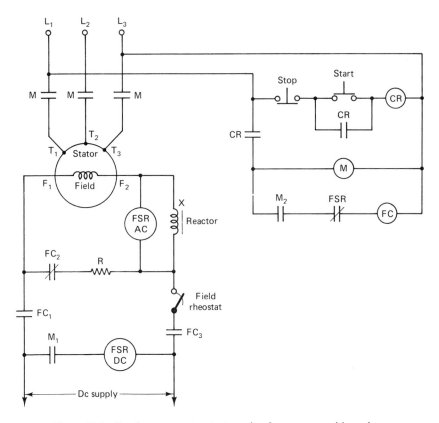

**Figure 13-1** Synchronous motor starter using frequency-sensitive relay.

three-pole line contactor which also has two normally-open interlock contacts, $M_1$ and $M_2$. When contactor $M$ closes, its main contacts ($M$) close ahead of its interlock contacts and energize the stator of the motor. Relay $FSR$ is to close fast enough to open the circuit to relay $FC$ by opening contact $FSR$ before interlock contact $M_2$ closes.

The motor will start as an induction motor with the field winding $F_1-F_2$ short-circuited through the normally-closed contact $FC_2$, discharge resistor R, and frequency-sensitive relay coil $FSR_{ac}$ in parallel with a small reactor $X$. The $FSR$ or polarized frequency relay, as it is also called, has two windings or coils on its relay core, as opposed to one on common relays. One coil $FSR_{ac}$ is operated on ac while the other, $FSR_{dc}$, is supplied with dc voltage through $M_1$. The direction of the flux produced by the dc winding is constant but that of the ac winding will alternate.

At startup of the motor, the frequency of the induced current in the field circuit is high; therefore, the reactance of the reactor $X$ will be high, so that most of the field-induced current flows through winding $FSR_{ac}$. As the motor accelerates, the frequency of the induced current in the field circuit ($F_1-F_2$) decreases. Subsequently, a larger proportion of this current is diverted through reactor $X$, since its reactance decreases in proportion to this frequency.

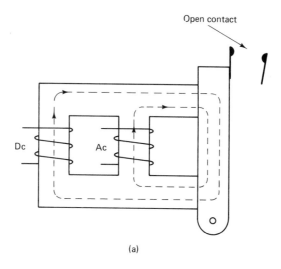

(a)

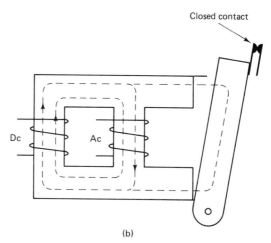

(b)

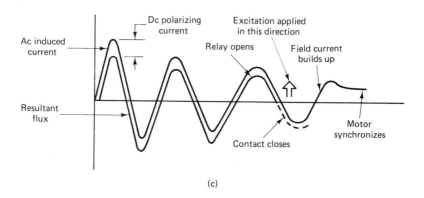

(c)

**Figure 13-2**  Action of frequency-sensitive relay, *FSR*: (a) armature closed, contact open; (b) armature open, contact closed; (c) current and flux waveforms.

Sec. 13-2    Starting the Synchronous Motor                    **363**

When synchronous speed is nearly reached, reactor $X$ virtually shorts out coil $FSR_{ac}$, which causes coil $FSR_{dc}$ to act such that normally-closed contact $FSR$ closes, thereby energizing the field circuit relay $FC$. This in turn connects the dc supply to the field through contacts $FC_1$ and $FC_3$; at the same time, the field rheostat is connected in series with it through the low dc resistance of the reactor coil.

Let us examine the $FSR$ relay operation more closely. At first ac current passes through its ac coil so that both the ac and dc coils are energized. We then have the condition depicted in Fig. 13-2a. As long as this relay remains energized, its contact $FSR$ is open. This prevents relay $FC$ from energizing, thereby preventing the dc supply from energizing the rotor field, since $FC_1$ and $FC_3$ are open.

When the motor starts up, the magnetic force of the $FSR$ ac coil becomes weaker. When the motor reaches its proper speed, the ac field flux is so weak that its contacts return to their normal position (Fig. 13-2b; i.e., contact $FSR$ closes). Field contactor $FC$ then operates and supplies dc to the field through contacts $FC_1$ and $FC_3$; contact $FC_2$ opens, eliminating $R$ in the circuit. Thus $FSR$ takes care of the proper speed. How does it take care of the proper rotor field position with respect to the rotating field? This is due to the relay polarization, which is so designed that its armature drops out when the magnetic fluxes of the two coils have a definite value but in opposition. The ac flux is the variable one; it determines when the armature drops out. The relatively weak dc flux has the effect of offsetting the ac flux; in other words, the resultant flux has a nonzero average value. As the ac flux becomes weaker, the combined strength of the two fluxes keeps the relay closed—but not so when they are opposed. Figure 13-2c shows the instant of closing. The time lag of mechanical operation delays contact closure until the first half of the negative wave. The direct current in the field then builds up properly for synchronization.

## 13-3 DYNAMIC BRAKING OF SYNCHRONOUS MOTORS

Situations may arise where rapid stopping of a synchronous motor is required. In rubber mills, for example, a breakdown may occur due to overloading caused by the rubber jamming the rolls. To protect the operators, the mill must be shut down rapidly. The quickest way to unload is to reverse the rolls, but to do so, the motor must stop. To stop the system, the rotating inertias must dissipate their kinetic energy in the form of heat somewhere in the braking system employed. With mechanical brakes it is in the brake lining or shoe. Other methods involve electrical braking, such as (1) plugging the motor, or (2) removing the power supply and connecting the armature across a resistor bank, the field circuit being kept intact.

In plugging, two stator terminal leads are reversed to reverse the direction of rotation of the synchronously rotating air-gap field. Currents are then induced in the damper winding to produce a torque to stop the rotor. During this interval the field circuit is short-circuited. When standstill is reached, the voltage supply must be disconnected to prevent the motor from accelerating back up to speed but in the reverse direction. The disadvantage of this method involves the severe strains that are imposed on the equipment, and it is for this reason that the second method is commonly used. In this method, the rotational energy of the motor is quickly dissipated in the resistors, since the machine acts as a generator under these conditions. In addition, it is generally found that faster stopping action is achieved.

## 13-4 EXCITATION SYSTEMS FOR SYNCHRONOUS MOTORS

A number of excitation systems for supplying direct current to the field winding of a synchronous motor are in use. The dc supply to the field winding is necessary not only for normal synchronous operation, but also to pull the machine into synchronism at the end of the starting period. Excitation systems are commonly 125 V up to ratings of 50 kW, with higher voltages at the larger ratings. Several methods are in use, such as a direct shaft-connected dc shunt motor.

The output of the exciter (i.e., the field current of a synchronous motor) is varied by adjusting the field rheostat. Another economical excitation system would be a motor-generator set driven by an induction motor, as represented schematically in Fig. 13-3. Such sets are generally employed in low-speed synchronous motor applications where space around the motor is at a premium.

More recent excitation systems employ static exciters using silicon diodes or thyristors. Such units operate at high efficiencies and develop conventional 125- or 250-V dc outputs. Various types of protection are incorporated in these units; for instance, the primary winding of the transformer used to match the static exciter to the line voltage can be fused for short-circuit protection. A typical motor excitation power supply using silicon diodes is in principle illustrated in Fig. 13-4. The operation of the three-phase bridge rectifier is explained in Section 13-7.

As can be seen from Figure 13-4, the dc exciter is eliminated from the motor shaft, thereby circumventing the usual problems associated with dc generators,

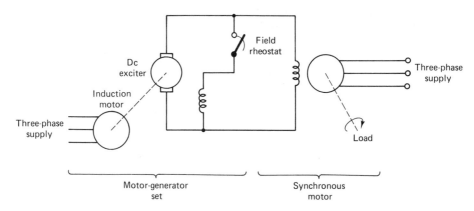

**Figure 13-3** Motor–generator set excitation for synchronous motor.

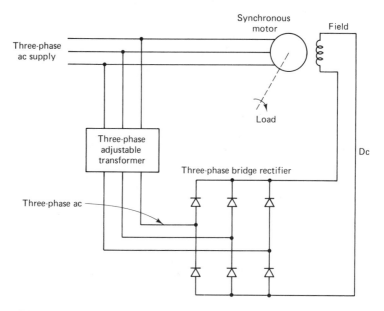

**Figure 13-4**  Elementary static excitation systems for a synchronous motor.

such as sparking at the brushes on the commutator. However, it is still necessary to supply the dc to the synchronous motor field via brushes and slip rings.

To eliminate the slip rings, the brushless excitation system was developed. Such a system is shown in principle in Fig. 13-5. Comparing this diagram to that of Fig. 13-4, we see that the three-phase adjustable transformer has been replaced by an ac alternator which has a three-phase rotating armature and a stationary dc field. Thus the exciter rotates, since it is on the same shaft as the rotor. The dc excitation for the synchronous motor is adjusted by varying the dc excitation to the stationary exciter field, which in turn controls the armature voltage. This voltage is rectified and applied to the field winding of the synchronous motor. Brushless excitation systems have been used extensively in low-power applications as well, where reduced atmospheric pressure intensifies the problem of brush wear, for example in aircraft applications.

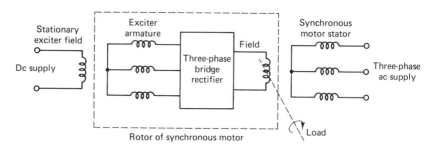

**Figure 13-5**  Component of brushless synchronous motor.

Because of the large starting currents of large motors and the consequent effect on distribution systems (called *voltage flicker*), power utilities have set rules and regulations regarding the size of motors that may be started directly on the line. To limit the objectionable effects of across-the-line starting, various types of reduced-voltage starters have been produced to limit the starting current. But since the starting torque is proportional to the applied voltage, care must be exercised that when starting under reduced voltage, the starting torque is adequate to accelerate the motor and load. Even a modest reduction to 80% of rated voltage will reduce the starting torque to 64% of its nominal value. It should be kept in mind that the problem of providing the proper starting equipment is not always one of merely selecting a starter that satisfies the rating of the motor. The motor itself must be selected to meet the requirements of the particular application.

Although various methods are employed, only some of the generally used starting methods will be discussed in relation to the squirrel-cage motor. Wound-rotor motors are started by inserting resistors in the rotor circuits.

### Full-Voltage Method

Whether or not an induction motor is started across the line (full voltage) depends on the size of motor, the type of application, and the capacity of the distribution system. Most modern induction motors are designed to withstand the mechanical forces and inrush currents associated with full-voltage starting. Although there is then no limitation on the size of the motor that may be started this way, it must be understood that objectionable line-voltage fluctuation (voltage flicker) generally occurs. The circuit diagram for a full voltage automatic starter is shown in Fig. 13-6a.

The motor is started by pressing the momentary-contact start button. When the start button is pressed, the control circuit is energized and causes the main starter relay $M$ to energize. Its auxiliary contacts $M$ close, thereby starting the induction motor, $IM$. When the start button is released and its contact opens, the starter auxiliary contact $M$, which is wired in parallel with the start button, continues to maintain the circuit. Normally, the circuit with this auxiliary contact is called the *holding circuit*. Motor operation ceases by pressing the stop button, in which case contactor M is deenergized. In the event of power interruption or greatly reduced voltage, the contactor coil M will be deenergized, thereby opening the holding circuit. When voltage is restored, the motor cannot restart automatically. The start button must be pressed again. The safety feature provided by this arrangement is called low-voltage protection and is one of the important advantages of magnetic control.

Flexible control is another feature of the magnetic starters. A pushbutton for energizing the main contactor coil can be located any distance from the starter, on the motor or on the machinery being driven. Furthermore, several pushbuttons can be used to provide multiple-location operation of the same starter and motor.

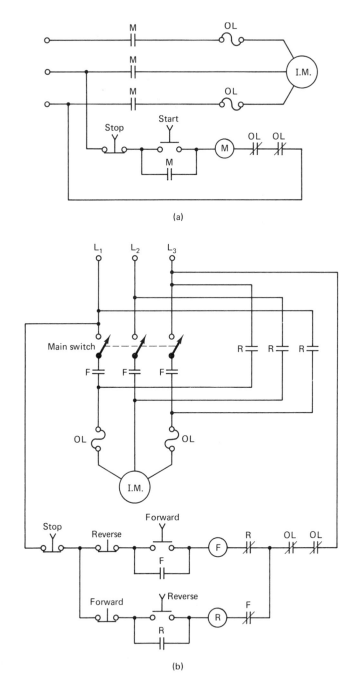

**Figure 13-6** Diagrams of full-voltage automatic starters for three-phase squirrel-cage induction motors: (a) simple starter circuit; (b) circuit includes reverse-run.

Instead of a pushbutton, a relay, timer, or pressure switch, for instance, could be used to provide automatic operation in addition to remote control.

One of the basic characteristics of squirrel-cage motors is that the direction of rotation can be changed by reversing any two of the incoming line-voltage leads. Reversing motor rotation can also be accomplished with a standard magnetic contactor with the proper electrical and mechanical interlocking. This is accomplished in Fig. 13-6b. The interlocking feature prevents both sets of contacts, forward and reverse, from being closed at the same time, even for a short time. There is no possibility of a short circuit then.

When the forward button is pressed in Fig. 13-6b, contactor $F$ is energized through the stop button, upper contacts of the reverse button, and normally closed contact $R$ of the reverse contactor. The three main contacts of the forward contactor close to start the motor. Coil $F$ is kept energized by its holding contact once the start button is released.

As can be noticed, another normally closed contact $F$ opens the reverse-run circuit to ensure that the forward and reverse circuits are electrically interlocked. The starter, apart from being interlocked electrically is generally mechanically interlocked as well. This prevents a short circuit if both forward and reverse pushbuttons were to be accidentically closed together. In addition, overload protection is provided by thermal elements $OL$ placed in the motor leads. Should the motor overheat, these normally closed contacts are opened, thereby deenergizing the contactor, which in turn disconnects the motor from the source.

If the reverse button in Fig. 13-6b is pressed, the forward circuit opens through the normally closed contact of the reverse pushbutton. This deenergizes the forward contactor coil, opening the forward contacts in the motor circuit. The normally closed auxiliary contact $F$ in the reverse circuit is thereby closed. The reversing circuit is thus completed through the momentarily closed lower contacts of the reverse button, the lower contacts of the forward button, and auxiliary contact $F$. Reverse contact or $R$ is now energized and the reverse contacts in the motor circuit are closed. Auxiliary contact $R$ opens in the forward circuit for added safety. The reverse contactor is kept energized by the holding contact $R$ across the reverse pushbutton, until the forward button or stop button is pressed. Note that lines $L_1$ and $L_3$ are interchanged in the reverse-run position as compared to the forward-run mode.

### Y-Δ Starting Method

For this method of starting the ends of each phase winding must be brought out to the terminal box. The reduced voltage for starting is obtained by connecting the winding in star. This means that the phase winding EMF will only be $1/\sqrt{3}$, or 58% of its rated value. When the motor has reached 75 to 80% of rated speed, full-rated voltage is applied to the windings by reconnecting them in delta. The starting torque is reduced to one-third of normal value. If this is sufficient for the particular application, this method should be given due consideration because of its simplicity and therefore, low cost.

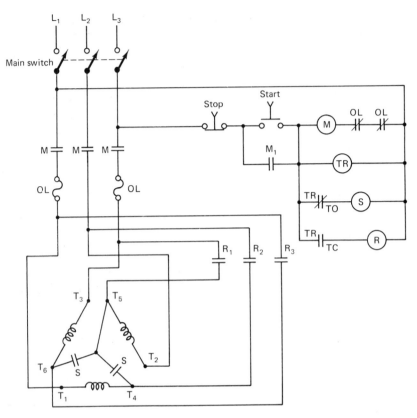

**Figure 13-7** Automatic Y-Δ starter.

As shown in Fig. 13-7, the operation is as follows; when the start button is pressed, the main contactor $M$ energizes, closing its contacts, thereby connecting winding terminals $T_1$, $T_2$, and $T_3$ to the line, and contacts $S$ close to complete the star-point connection. After timing relay $TR$ times out, the $S$ contacts open first, which is followed by the closing of the $R$ contacts. As is readily seen, terminals $T_1$-$T_6$, $T_2$-$T_4$, and $T_3$-$T_5$ are joined, thereby connecting the windings in delta.

### Autotransformer Method

Figure 13-8 shows the connection for an autotransformer to reduce the motor terminal voltage during starting. The taps are generally made at the 50, 65, and 80% points, so that selection of the proper tap can be made for the proper starting torque. The corresponding starting torques will be 25, 42, or 64% of rated torque, respectively. Furthermore, the current, being reduced by the lower starting voltage, is reduced further by the transformer action. Hence the starting current is decreased in proportion to the impressed voltage squared, or to 25, 42, or 64% of full-voltage starting.

The starting sequence is as follows. Pressing the START button causes contactors $S$ and $N$ to be energized closing their respective contacts and start timer

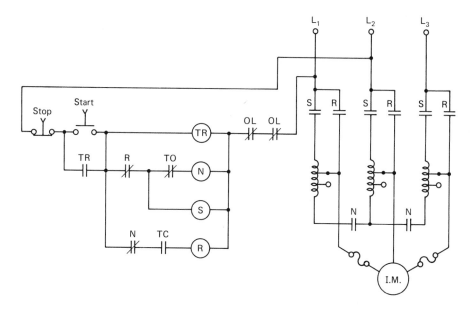

**Figure 13-8** Starter using a three-phase autotransformer.

relay *TR*. This connects the motor to the autotransformer taps. After timing relay *TR* times out, the thermally operated contact *TO* opens, dropping out contactor *N* and closing *TC*, energizing the *R* contactor. Contacts *R* put the induction motor directly in the line and deenergize contactor *S*. The autotransformer is thereby deenergized.

### Primary Resistance Method

This method of starting is a less costly but also a less effective way of reduced starting voltage—by connecting a starting resistor in series with the motor during starting. When the motor has accelerated to about 75% of rated full-load speed, the resistor is shorted-out, connecting the motor across the line.

### Primary Reactor Method

This method of starting uses a reactor instead of a resistor, discussed previously. However, it differs from the primary-resistance starting in its effect on the accelerating torque available. To show this, let us examine Fig. 13-9. As shown, both have been adjusted to have the same starting torque. As the motor comes up to speed, the reactor voltage being out of phase with the motor voltage causes the resultant voltage to be higher for the reactor system than for the resistor system. This is most desirable, especially for accelerating loads for which the torque increases with speed, such as compressors and fans.

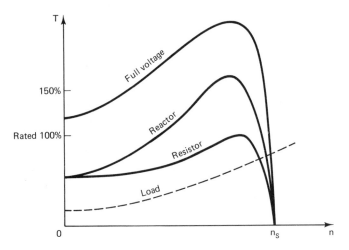

**Figure 13-9** Accelerating torque characteristics for starting methods indicated of starting squirrel-cage induction motors for an assumed load torque.

## Wound-Rotor Method

Secondary resistor starters, as opposed to primary resistor starters used for the automatic acceleration of wound-rotor induction motors, consist of an across-the-line starter and one or more accelerating contacts to shunt out resistance in the

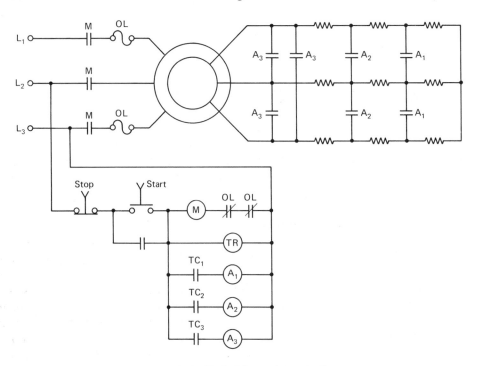

**Figure 13-10** Typical four-point controller.

Starting and Control of AC Motors    Chap. 13

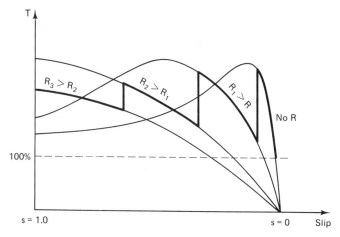

**Figure 13-11**  Starting of wound-rotor induction motor with four-point starter.

secondary rotor circuit. The resistances connected to the slip rings of the motor are meant for starting duty only, and this starter cannot be used for speed regulation. The operation of the accelerating contactors is controlled by a timing relay which provides timed acceleration and eliminates the danger of improper startup procedures. A typical starter diagram with four points of acceleration is shown in Fig. 13-10.

As can be seen, balanced rotor-resistance conditions are maintained by the closing of the successive sets of contacts $A_1$, $A_2$, and $A_3$. Since the final step comprises the full-speed conditions, short-circuiting three pairs of lines helps to reduce contact resistances to zero. Figure 13-11 shows the torque–speed characteristics with the three different external resistances added to the rotor circuit. During startup the resistances are eliminated much in the fashion of a dc motor starter. By proper selection of the starting resistors, any torque requirement can be met within the capabilities of the motor.

## 13-6  INDUCTION MOTOR SPEED CONTROL

The speed of a squirrel-cage induction motor operating from a constant-voltage, constant-frequency supply is essentially constant. It normally runs at a slip of approximately 3 to 5% of synchronous speed. The torque developed varies as the square of the applied voltage and steady-state operation is attained when the motor torque balances the load torque.

Provided that the frequency is fixed, the slip is determined by the load torque and applied voltage, and speed control can be achieved by adjustment of the voltage. Reducing the voltage decreases the rotor speed, but the maximum torque available from the motor is also reduced. Although this provides a wide range of speed control, the ohmic losses become excessive and the efficiency poor, particularly at lower speeds. This is so since at a given slip the motor current is proportional to the voltage, whereas the torque developed varies as the square of the

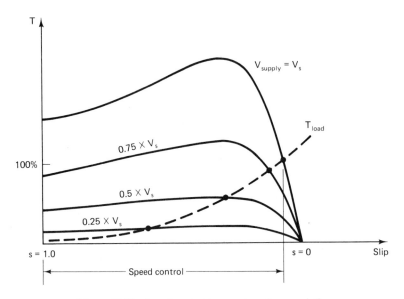

**Figure 13-12** Speed control by supply voltage variation.

voltage. Therefore, at reduced speeds an appreciably larger current is required to develop a large torque. It suggests, therefore, that voltage control methods are more suitable in applications where the torque demand is also lower at increased slips. This, in fact, is true in such applications as fan and pump drives, where the load torque varies approximately as the square of the speed (see Fig. 13-12).

As can be seen, the torque required for starting and high-slip operation is small, resulting in satisfactory operation. However, the operating efficiency is still poor and generally the motor is derated at low speeds to avoid excessive overheating due to reduced ventilation. This method of control is often referred to as *slip control*, and obviously can be done only with induction motors.

Another method for controlling the speed of an induction motor is to vary the supply frequency. Although this is simple enough in principle, the practical implementation is generally not so convenient. With the advent of the thyristor, many sophisticated variable-frequency inverter systems are now available, which will be discussed in principle, shortly. In a variable-frequency system, the supply

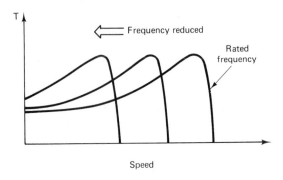

**Figure 13-13** Torque–speed characteristic at different supply frequencies and constant air-gap flux.

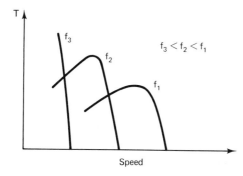

**Figure 13-14** Torque–speed characteristic for different supply frequencies and constant terminal voltage.

frequency is reduced for starting. Figure 13-13 shows the torque–speed characteristics of an induction motor at different stator frequencies. The maximum torque is maintained by keeping the volts/frequency ratio the same so that the air-gap flux is essentially constant.

These characteristics are suitable for driving a constant-torque load at variable speed. If the frequency is varied while the stator applied voltage remains constant, the air-gap flux and stalling torque would decrease with frequency, as illustrated in Fig. 13-14. These characteristics would be more suitable for traction applications, where large torques are needed at startup but smaller torques at higher speeds. An alternative approach is a solid-state system, such as the dc link converter to be described in the following sections.

## 13-7 SOLID-STATE DRIVES

In earlier chapters we have discussed the principles of dc machines, induction motors, and synchronous machines. In Chapter 5 and so far in this chapter, we have concerned ourselves with the conventional starting techniques of these motors. Various circuits have been shown throughout to illustrate how some motors or field excitation systems can be supplied with dc voltage from solid-state rectifiers. No details or specifics were given as to the working of these solid-state circuits. In addition, in those chapters it was mentioned that it is possible to control the motor characteristics, such as speed, torque, and so on.

There are numerous motor applications that require precise control of one or more of the motor output parameters. Adjustable dc voltage supplies or adjustable-speed drives allow the speed of the load to be varied continuously to meet the demands of the process. Depending on the application, there are several drive systems available and the selection depends on such factors as initial cost, reliability, ease of operation and maintenance.

The remainder of this chapter deals with some basic schemes for motor control, with the ultimate goal of discussing induction motor speed control by means of a solid-state inverter drive. Before we examine the more complex thyristor circuits, it may serve us well to discuss the three-phase diode rectifier, because it is simpler.

## Three-Phase Diode Rectifiers

Full-wave, single-phase bridges are satisfactory for motors up to about 5-hp ratings. However, above this size, line currents become excessive and it is necessary to operate the motors on three-phase supplies. The simplest three-phase system has three diodes connected between the source and the load as shown in Fig. 13-15. In effect, we have three single-phase generators connected through diodes to a common load. The generators are 120° out of phase with each other, so that if generator $E_{AN}$ is beginning its cycle at time $t_0$, generator $E_{BN}$ begins its cycle at time $t_4$, which corresponds to one-third of a cycle or 120 electrical degrees. Similarly, $E_{CN}$ begins its cycle at time $t_8$, which corresponds to 240 electrical degrees. Because of the arrangement, whichever generator voltage is most positive will provide the current to the load at any given time. For example, during the time interval $t_1 < t < t_5$, $E_{AN}$ is more positive than either $E_{BN}$ or $E_{CN}$, so that phase $A$ will conduct current to the load during this time interval. At time $t_5$, the situation changes since $E_{BN}$ becomes more positive than $E_{AN}$. This allows current to switch from phase $A$ to phase $B$. Actually, during the time interval $t_5 < t < t_9$, when $E_{BN}$ has the largest value, it provides current to the load and simultaneously back-biases diodes $D_A$ and $D_C$. During the time interval $t_5 < t < t_6$, $E_{AN}$ is positive but smaller than $E_{BN}$, while $E_{CN}$ is negative.

We can draw the equivalent circuit of Fig. 13-16 for the conditions prevailing during time interval $t_5 < t < t_6$. Since $E_{BN}$ is the largest voltage, it causes the current $I_B$ to flow through the resistor $R$. It causes a voltage drop $V_R$ equal to $E_{BN}$. This voltage is greater than $E_{AN}$, so that diode $D_A$ cannot conduct. Because

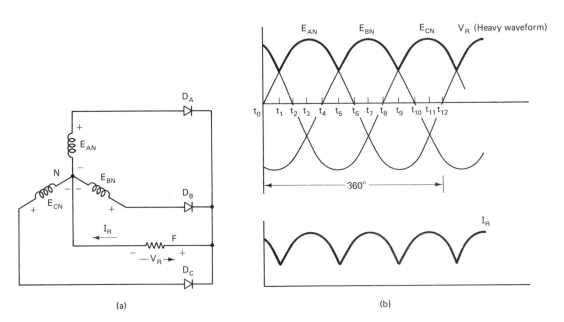

(a)                    (b)

**Figure 13-15**   Three-phase diode rectifier supplying resistive load: (a) circuit diagram; (b) waveforms.

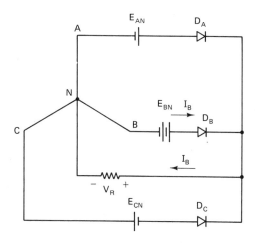

**Figure 13-16**  Equivalent circuit for the time interval $t_5$–$t_6$.

of the reverse voltage of $E_{CN}$, diode $D_C$ cannot conduct, so that current must be supplied only from phase $B$.  The current waveshape is identical to $V_R$, and its value is given by

$$I_R = \frac{V_R}{R}$$

where   $I_R$ = instantaneous value of load current, A
          $V_R$ = instantaneous value of load voltage, V

The peak-to-peak value can easily be determined from Fig. 13-15b.  If we consider the peak value of a sine wave to be 1, then the value at $t_1$, which corresponds to 30°, would be sin 30°, which is 0.5; then the ripple is 50%.  The average or dc value for a three-phase half-wave-rectified sine wave is 0.827 of its maximum value, so that the ratio of ripple to average is 60%.  This is still a high value, but the circuit would require a considerably smaller filter inductor than would the corresponding single-phase rectifier.  If the motor is supplied from a three-phase transformer, only one of its phases is being utilized at any given time.  This is really poor usage of the transformer, since two-thirds of it is not being used; it is said to have a poor *utilization factor*.

The utilization factor can be improved and the ripple reduced by using the three-phase full-wave bridge shown in Fig. 13-17.  In this diagram there are six diodes, only two of which are conducting at any one time.  The voltages $V_{AB}$, $V_{BC}$, and $V_{CA}$ are 120° out of phase with each other, with $V_{AB}$ being the leading voltage.  The voltages are shown in Fig. 13-17b with the rectified portion of each wave shown dashed.  As we can see, the times marked $t_0$, $t_1$, and so on, correspond to 30° of a sine wave.  From the load point of view, the voltage always appears to be positive from $A$ to $B$ due to the action of the rectifiers.  At any instant of time, current will be flowing through two diodes.  For example, during the interval when $V_{AB}$ is positive ($t_2 < t < t_4$), current flows through diodes $D_1$ and $D_5$.  During this interval ($V_{AB} > V_{AC}$), diode $D_6$ is back-biased and cannot conduct.  At time $t_4$, $V_{AC}$ becomes greater than $V_{AB}$ and current will switch from $D_5$ to $D_6$.  The

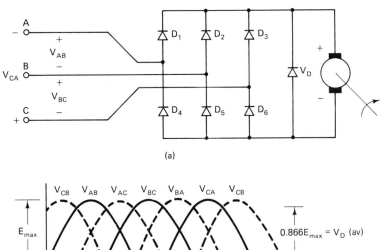

(a)

(b)

**Figure 13-17** Three-phase, full-wave bridge supplying dc motor: (a) circuit diagram; (b) waveforms.

voltage across the load will follow the peaks of the applied voltage. The conducting diodes at any instant are indicated in Fig. 13-18.

From Fig. 13-17b it can be seen that the maximum output voltage occurs at $t_1$, $t_3$, $t_5$, $t_7$, and so on, while the minimums occur at $t_2$, $t_4$, $t_6$, and so on. The voltage waveform for $V_{AB}$ is a sine wave starting at $t_0$ and ending at $t_{12}$, and each time interval then corresponds to 30°. A minimum occurs at $t_2$, which corresponds to 60°, so that if the maximum value of the sine wave is $E_{max}$, the minimum value of the output is $E_{max} \sin 60°$, or $0.866E_{max}$. The peak-to-peak ripple is thus $(1 - 0.866)E_{max}$, or $0.134E_{max}$. The dc can be shown to be $0.955E_{max}$, so that the percent ripple is only $(0.134/0.955)100\%$, or 14%.

### Thyristor Switches

In the preceding section we described how dc motors may be energized directly from ac lines using diodes. A diode allows ac current to flow through it during the portion of the half-cycle when the anode is positive with respect to the cathode. A thyristor or controlled rectifier is a device that will allow current to flow through it for only part of the half-cycle, and by this means can control the average current to the load. A thyristor, originally called a silicon-control rectifier (SCR), is a three-terminal device similar in appearance to a diode, but the third terminal is called the gate, which allows the thyristor to be turned on under separate control.

Figure 13-19a is the symbol for a thyristor, while Fig. 13-19b shows the thyristor connected in a circuit. If the voltage $V_{GK}$ is negative, the thyristor acts like an open circuit under all conditions. If $V_{GK}$ goes slightly positive so that the current into the gate circuit exceeds a small value, of the order of several milliamperes, and $V_{AK}$ is positive, the thyristor will turn on abruptly and stay on as long as $V_{AK}$ is positive and current is flowing from $A$ to $K$. In order for the thyristor to turn off, it is necessary either to allow the current to fall to zero naturally, or to apply a reverse voltage from $K$ to $A$ which will drive the current to zero. Figure 13-20 illustrates the action in a thyristor-controlled circuit. $V_{CK}$ is applied to the gate to produce the gate current $I_G$. When $V_{CK}$ is negative, the thyristor cannot conduct and no current flows, although $V_{AK}$ is positive. At $t_{on}$, $V_{CK}$ goes slightly positive, producing a large enough current ($I_G$) in the gate circuit to cause the thyristor to turn on. At this point, the anode-to-cathode circuit becomes a short circuit and current flows through the resistor. When the thyristor turns on, the gate circuit loses control and has no further effect until the thyristor is turned off. Current will naturally decrease to zero when the line voltage goes to zero at the end of the voltage half-cycle. During the negative portion of the wave, no conduction will occur because $V_{AK}$ is negative. The cycle will repeat itself the next time that $V_{AK}$ and $V_{GK}$ are both positive.

By shifting the point at which $V_{CK}$ first becomes positive, it is possible to change the average value of $V_R$. The relationship between the average output voltage and the firing angle for a sine wave can be seen in Fig. 13-21. The maximum value for the half-wave configuration is $0.314V_{max}$ and it can be varied from this value down to zero by delaying the firing angle. There are many circuits specifically designed to control the firing angle, but it is not the purpose of this book to explore them in detail.

### Thyristor Bridge Circuits

By replacing some or all of the diodes in the circuits of Fig. 13-15 or 13-17 by thyristors, it is possible to construct power sources having a variable dc output. If we replace the diodes of Fig. 13-17a with thyristors, we obtain the circuit shown

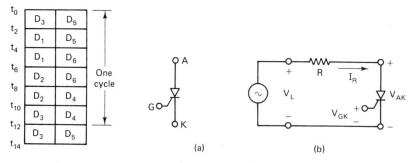

**Figure 13-18** Diode conduction intervals.

**Figure 13-19** Thyristor or silicon-controlled rectifier: (a) symbol ($A$, anode; $G$, gate; $K$, cathode); (b) circuit diagram.

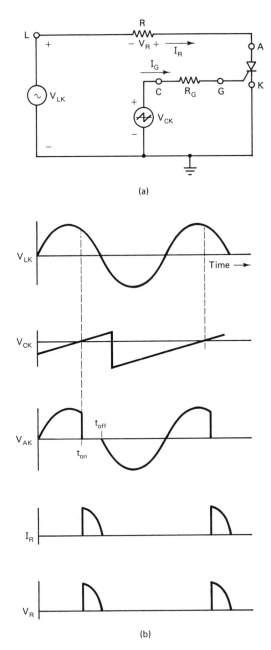

(a)

(b)

**Figure 13-20** Thyristor-controlled resistor load with sawtooth gate signal: (a) circuit diagram; (b) waveforms.

in Fig. 13-22. Each of the thyristor gates is driven from a separate signal source, $E_{GA}$, $E_{GB}$, and $E_{GC}$. Very often this source is a pulser arranged to be synchronized with the line and the three pulsers are 120° electrically out of phase with each other. The point at which the pulse occurs can be delayed, so that the firing angle can be varied and the output controlled. A complete set of waveforms for a delay angle of 60° is shown in Fig. 13-23.

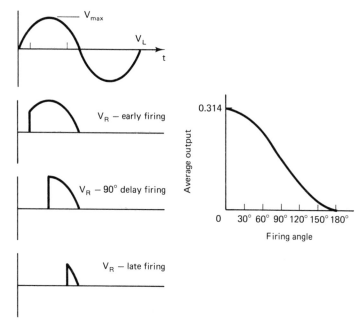

**Figure 13-21** Output voltage versus firing angle for a sine wave.

In ac circuits, it is more convenient to discuss phase angles than actual time. We assume that all the circuits are operating at 60 Hz unless otherwise noted. Since one cycle at 60 Hz is 1/60 s or 16.67 ms, we can convert from time to phase angle using the following:

$$1 \text{ cycle} = 2\pi \text{ radians} = 360° = 16.67 \text{ ms}$$

Therefore,

$$1° = \frac{16.67}{360} \text{ ms}$$

$$1 \text{ radian} = \frac{16.67}{2\pi} \text{ ms}$$

A delay of 60° is then

$$60° \times \frac{16.67}{360} = 2.78 \text{ ms}$$

The action of the circuit of Fig. 13-22 and the resultant waveforms of Fig. 13-23 can be explained as follows. At a phase angle of less than 60°, thyristor $T_A$ cannot conduct since it has not been fired. $E_{GA}$ delivers a pulse at 60° which turns $T_A$ on; it immediately starts to conduct and the voltage $E_{AN}$ is immediately applied to the resistor. Conduction continues until 180° since there is no voltage present to turn $E_{AN}$ off. In the case of the simple diode, $E_{BN}$ would have been present and when $E_{BN}$ became greater than $E_{AN}$ at 150°, $E_{BN}$ would have been applied to

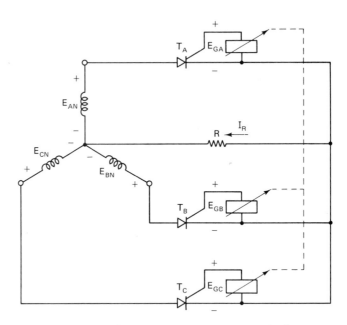

**Figure 13-22**  Thyristor three-phase one-way circuit.

the load.   However, $T_B$ is not turned on until the phase angle is 180°.   The voltage waveform across the load is now a string of partial sine waves with largely varying amplitude but with a reduced dc component.   By adjusting the angle, the dc value can be varied from $0.885E_{max}$ to zero, as shown in Fig. 13-24.   Using a pulse to trigger the circuit will limit the minimum phase angle to 30°.   If we apply a pulse to turn on phase $A$ at, say, 20°, $E_{CN} > E_{AN}$, and $E_{AN}$ will not turn on.   When $E_{AN} > E_{CN}$ the pulse has gone and $E_{AN}$ cannot be applied to the load.   In any event, we cannot effect current transfer from one thyristor to another if we try to turn on one which is at a lower voltage than the other.

This circuit requires three pulses per cycle, one for each thyristor.   The circuit of Fig. 13-17a can be modified to that of Fig. 13-25 by replacing the diodes with thyristors.   This circuit requires six pulses per cycle and will produce the waveforms shown in Fig. 13-26.   There are many variations of these circuits, replacing some thyristors with diodes, but the purpose is the same, namely obtaining a variable torque, speed, or power from the motor.

It should be pointed out that the gate circuits are electrically isolated from each other and that they are synchronized with the line and with each other. Isolation can be accomplished by using isolation transformers or optically coupled devices.   The generation and synchronization of the gate firing signals can be done conveniently, using integrated-circuit packages.   It is no longer necessary to design these circuits from first principles since very sophisticated packages—both analog and digital—are readily available at reasonable prices.

Electronic controllers have one major advantage over older electromechanical devices—they can be programmed to start a motor as well as being used for running. The controller can bring a motor up to speed in a predetermined manner without

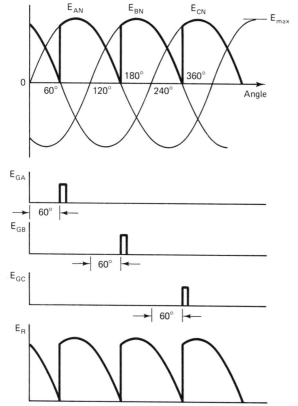

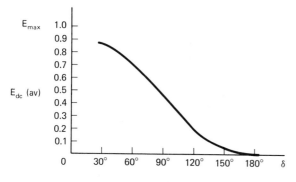

**Figure 13-23** Waveforms for Fig. 13-22 with delay angle of 60°.

drawing excessive line currents, and can also be programmed to shut down automatically if faults are encountered either during startup or while running.

### Frequency Converters

There are basically two types of frequency converter, the dc link converter discussed here, and the cycloconverter. A block diagram of the dc link converter is shown in Fig. 13-27. The three-phase ac supply is first rectified in a three-phase con-

**Figure 13-24** Average dc output voltage versus delay angle or firing angle.

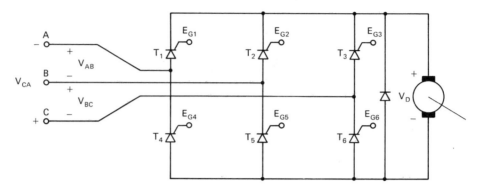

**Figure 13-25** Three-phase thyristor bridge with motor load.

trolled, bridge rectifier circuit. The control circuit controls the rectifier gating, which subsequently sets the dc output voltage $V_o$. This rectified voltage in turn is supplied to the inverter via a filtering network, which converts the resulting dc power to three-phase ac power. The inverter, as will be shown, is switched sequentially, so that alternating waveforms are delivered to the motor. The output waveforms are not sinusoidal, but in practice this does not impair motor operation apart from a minor reduction in speed and efficiency due to the presence of har-

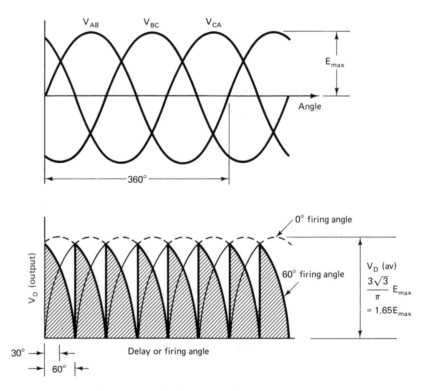

**Figure 13-26** Waveforms for thyristor bridge circuit.

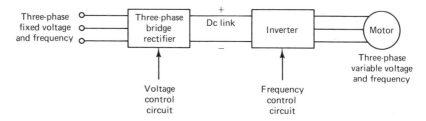

**Figure 13-27** Block diagram for induction motor speed control using a dc link converter.

monics. The output frequency of the inverter is established by the rate at which the thyristors are triggered into conduction. The voltage/frequency ratio is maintained by proper adjustment of the dc level from the rectifier circuit.

As is evident, the dc link converter involves a double power conversion; nevertheless, high efficiencies are generally achieved in practice, on the order of 85 to 95%.

The cycloconverter, on the other hand, converts the network frequency directly to a lower output frequency without intermediate rectification. In principle, the output is composed of segments of the input waveform. Since for efficient operation the output frequency must be at least one-third of the input frequency, it is suitable for low-speed motor operation only if operated from a normal main frequency supply. This type of converter will not be discussed here.

### Voltage Controllers

At this point we can examine the current and voltage waveforms of a single-phase controller as shown in Fig. 13-28. The two thyristors are connected "back to back" and are triggered alternately at identical points on the voltage waveform. For the circuit the delay angle is α degrees and the load is resistive. As shown, the voltage waveform has no dc component, but odd harmonics are present, primarily the third. If, however, the load possesses some inductance as well as resistance, the load current would no longer rise instantaneously at $\omega t = \alpha$, and current flow is delayed beyond the point $\omega t = \pi$. Variation of the firing angle α controls the average power delivered to the load by the ac supply.

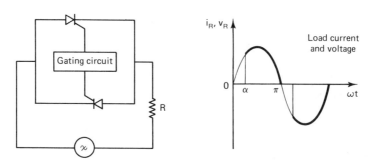

**Figure 13-28** Output waveforms in a resistive load of an ac power controller.

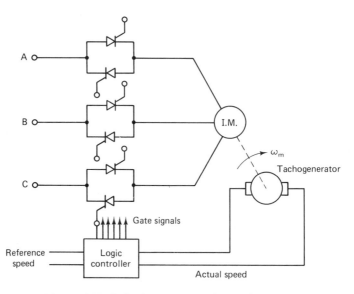

**Figure 13-29** Induction motor speed control system.

A three-phase version of this circuit as used in a closed-loop motor control system is illustrated in Fig. 13-29.

The tachogenerator develops a voltage which depends on the motor speed. This voltage is compared with the reference speed signal and any error (i.e., a difference signal) is translated by the logic controller into a change of firing angle. For instance, if the tachogenerator signal is below that of the reference signal, the conduction periods are increased, resulting in an increase in motor torque, and subsequently in shaft speed.

This type of controller gives some measure of speed adjustment by voltage control, subsequently by increased slips. The drawbacks of voltage control have been discussed, and in practice this form of control is efficient where the torque falls considerably with decreasing speeds.

## 13-8 *VARIABLE-FREQUENCY INDUCTION MOTOR DRIVE*

As indicated by Eq. (10-1), the synchronous speed of an induction machine can be altered by varying the frequency. However, from the transformer equation it can be seen that any decrease in frequency results in a corresponding increase in the magnetic flux (assuming linear behavior). If currents are to remain at their normal values and saturation of the magnetic circuit is to be avoided, the applied voltage must also be varied somewhat proportionally to the frequency.

One of many possible circuits for achieving an adjustable ac voltage whose magnitude is a function of the output frequency is the converter circuit illustrated in Fig. 13-30. Voltage control is obtained by controlling the rectifier gating of the phase-controlled rectifier circuit, which is a three-phase bridge rectifier circuit of the type shown in Fig. 13-25.

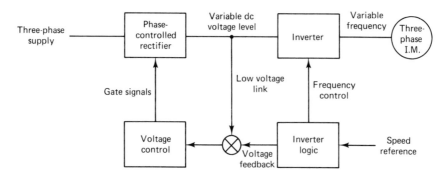

**Figure 13-30**   Voltage-controlled induction motor drive.

The drop in average load voltage with an increase in the firing angle α has been demonstrated.   When the firing delay angle α is increased to 90° it was shown that the average output voltage is zero.

### Inverter Circuit

Figure 13-31 shows the circuit of a three-phase bridge inverter as a means of controlling the speed of a squirrel-cage induction motor by varying the frequency. The dc voltage applied to the inverter is from the controlled rectifier via the filtering network in the dc link.   The capacitor shown is to remind us that the dc input voltage is reasonably constant over each cyclic change in the inverter.   The input voltage V from the controlled rectifier may change over several cycles to match the inverter load requirements.   The gating signals required to produce the three-phase output voltages are shown in Fig. 13-32 together with the resulting line-to-line voltages applied to the motor.   It is readily seen that the fundamental components of these voltages form a balanced three-phase set.   It must be kept in mind that the inverter circuit of Fig. 13-31 is shown in its basic form.   As mentioned earlier, with inductive loads as an induction machine represents, the load current does not rise instantaneously upon firing the thyristor and commutation is delayed; in other words, the current lags the voltage.   This means that current continues to flow for some time even though its thyristor may be turned off.   To circumvent the possibility of this current flowing through the other thyristor in the same leg (e.g., $T_1$–$T_4$), which is now being fired, feedback diodes are normally placed in antiparallel with each thyristor, although not shown in Fig. 13-31.   Many other commutating circuits are employed; however, no other circuits will be discussed here.

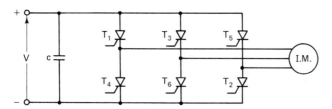

**Figure 13-31**   Basic three-phase bridge inverter circuit.

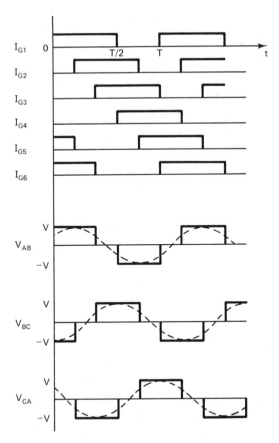

**Figure 13-32** Gating signals and resulting load voltage of inverter circuit in Fig. 13-31.

A feature of the induction motor converter is to hold the inverter voltage/frequency relationship fixed over most of its operating range; as has been mentioned, this is to maintain constant air-gap flux. It may therefore be necessary to increase the voltage somewhat in the lower-frequency range in order to achieve this, since the stator impedance drop becomes a larger component of the total motor voltage. The power entering the inverter is the product of the dc link voltage and current, whereas the motor output power is the product of torque and speed. It can be reasoned that the inverter input current is related solely to the output torque of the motor. This implies that by controlling the dc link current, the torque is controlled directly. This type of control is generally referred to as constant torque as opposed to constant horsepower control. With reference to Fig. 13-30, the reference voltage feedback signal is compared to the dc link voltage and any resulting error signal fed to voltage control of the rectifier gating. This control, in turn, assures that the required volts/frequency ratio is obtained.

By changing any of the two leads to the three-phase induction motor, the phase sequence is reversed, thereby reversing the direction of rotation. One of the principal advantages of the circuit discussed, apart from its speed-control capability, is that the direction of rotation can be reversed by reversing the firing sequence of the thyristors in the inverter circuit, thereby eliminating the need to switch power leads.

## REVIEW QUESTIONS

**13-1.** What are drives?

**13-2.** In what manner does a synchronous motor develop a starting torque?

**13-3.** How does the resistance of a damper winding affect the starting torque of a synchronous motor?

**13-4.** Why is it desirable to short-circuit the dc field when a synchronous motor is started?

**13-5.** Briefly describe the phenomenon of "hunting" in a synchronous motor.

**13-6.** With reference to Fig. 13-1, describe the operation of the circuit.

**13-7.** Explain briefly how quick stopping of a synchronous motor may be accomplished.

**13-8.** Describe two methods to obtain the dc excitation field supply for a synchronous motor.

**13-9.** Why are the larger induction motors not started with full voltage applied at startup? How are they usually started?

**13-10.** State at least three starting methods for induction motors; describe one of them.

**13-11.** What is the effect on the efficiency of a wound-rotor induction motor when external rotor resistances are added?

**13-12.** Describe the effect on the normal torque–speed characteristic of an induction motor produced by:
(a) Halving the applied voltage with normal frequency.
(b) Halving both the applied voltage and frequency.
Sketch the associated torque–speed characteristics in their approximate relative positions with respect to the normal one. Neglect the effect of stator resistance and leakage reactance.

**13-13.** When starting an induction motor by the star-delta method, what are the values of inrush current and starting torque in terms of starting at full voltage?

**13-14.** A 5-hp motor is started by means of a full-voltage automatic starter as depicted in Fig. 13-6a. If this circuit is now considered for a 15-hp motor, what components should be changed?

**13-15.** A 250-hp squirrel-cage induction motor is to be started by means of an automatic starter. Select a "best" circuit and describe it. If this motor is sized to the load and full-load starting is desired, what factors must be considered for this application?

**13-16.** Modify the circuit in Fig. 13-10 to obtain a three-point controller.

**13-17.** In the dynamic braking mode, the torque of the induction motor varies greatly. It is low at the instant of braking but high when the motor approaches standstill. Why is this so?

**13-18.** Discuss briefly the three general speed-control methods for the induction motor.

**13-19.** Describe how the speed of an induction motor can be controlled by varying the voltage. Is the operating range extensive?

**13-20.** Describe a way to control the speed of an induction motor by varying the frequency. What quantity should remain relatively constant over most of the operating range? Explain.

# 14

# SPECIAL-PURPOSE
# MOTORS and DEVICES

## 14-1 INTRODUCTION

The purpose of this chapter is to introduce the reader to a wide variety of small (less than ¼ hp to 1/2000 hp) machines that are used for special applications as well as general purpose. The construction of these machines is peculiar and unique; for example, permanent-magnet fields, brushless solid-state switching, printed-circuit-board armatures, linear moving magnetic fields, and so on. Many of these machines are used in combination with electronic amplifiers to provide precise speed control, positional control, and speed acceleration sensing.

Conventional position-control and speed-regulation schemes involve servomotors in closed-loop feedback systems. Their applications range from control within computers to manipulation of space satellites. Servomotors and stepper motors are sometimes described as instrument motors. Recent innovations have applied pulse-width modulators (PWM) to small specialized machines for sophisticated electric drives. Modern aircraft are equipped with brushless permanent-magnet generators and motors. The outstanding features of permanent-magnet motors are the linear speed–torque and current–torque characteristic curves. As a result of this linearity, precise speed–torque control can be achieved by applying suitable armature voltage.

In previous chapters we have discussed general-purpose motors. A *general-purpose motor* may be defined as one that is not restricted to a specific application, but is suitable for general use under normal service conditions as specified by manufacturing standards.

In contrast, a *definite-purpose motor* is designed for use in specific applications or for use under particular service conditions. Definite-purpose motors may have

390

standard ratings and provide standard operating characteristics and construction features. However, a *special-purpose motor* will not have standard operating characteristics or be of standard construction.

In the following sections we describe the more common small machines for definite-purpose and specialized use as well as for general-purpose use. Essential principles of generator and motor action have been described in earlier chapters. In this chapter we present unique operating features, different construction, and useful applications for the most common variety of small machines.

## 14-2 PERMANENT-MAGNET DC MOTORS

Permanent-magnet dc motors have their pole fields supplied by permanent magnets. The semicircular permanent magnets are assembled on the stator frame to provide one or more field pole pairs. Most permanent magnets are composed of ceramic, Alnico, or rare earth elements (see Fig. 14-1a).

Without a field-wound structure, permanent-magnet dc motors become more compact in size and tend to have higher operating efficiency. Elimination of field-coil power loss results in cooler operation with totally enclosed construction. Alnico-type permanent magnets are least affected by higher temperatures.

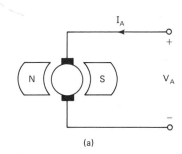

(a)

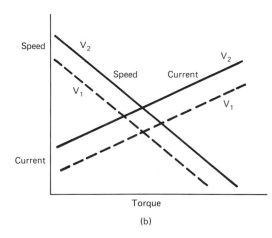

(b)

**Figure 14-1** Permanent-magnet dc motor: (a) circuit diagram; (b) speed and current versus torque.

If permanent-magnet motors are not overloaded, demagnetization of the air-gap flux will not occur and the speed–torque and current–torque characteristic curves remain essentially linear. All speed–torque control is achieved by adjustment of the armature voltage. Below rated load, the operating characteristics are linear and directly proportional to the armature voltage. The operating characteristics for two armature voltages are illustrated in Fig. 14-1b.

In summary, permanent-magnet motors provide comparatively simple and reliable dc drives where high efficiency, high stall torque, and linear speed–torque operating characteristics are desirable. Other advantages are the reduced physical size and availability of totally enclosed frames.

## 14-3 DC SERVOMOTORS

DC servomotors are used in control systems primarily in position- and velocity-control applications. The types that we examine are small machines below a 2-hp rating. In addition, we focus our attention on a typical application for a small permanent-magnet dc motor. Applications vary from 1-A motors developing 0.0023 N · m used in photography cameras to 2-hp motors used in antenna drives and moving equipment in automatic warehouses.

In velocity-control devices, for fixed voltage, torque is relatively constant with speed, similar to most large dc motors. The major difference for a servomotor is that it will usually be better built and will have very little torque ripple, as it will have a large number of commutator bars. It will normally be operated from a variable-voltage source, in contrast to most machines, which operate at a fixed voltage.

Position-control devices are used primarily in high-torque, low-speed applications. The torque–speed characteristic is much steeper than that of a conventional dc motor since the device produces very high torque at standstill but relatively little torque at high speed. Motors used primarily for position control are also called *torque motors*.

A set of operating characteristics for a 2-hp direct-drive dc motor is shown in Fig. 14-2. Motors specifically designed for direct-drive applications can be used over a very wide range of speed and torque without requiring the use of gears. These motors can be brought to a dead stop within three revolutions from 1000 r/min. Since the dc servomotor will be run over a wide speed and voltage range, there is no fixed voltage curve, and it then becomes necessary to determine the power dissipation in the motor (Fig. 14-2). Note that there are three different regions in which the motor can operate:

1. Continuous-duty zone
2. Safe-acceleration zone
3. Intermittent zone

The specific conditions under which the motor can operate in zones 2 and 3 are specified in manufacturers' data sheets.

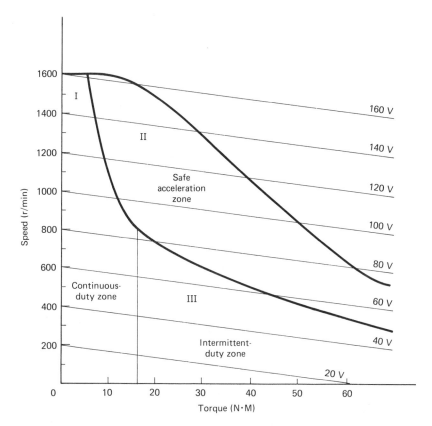

**Figure 14-2** Torque–speed characteristic of a 2-hp direct-drive dc servomotor.

Figure 14-3 shows the torque–speed characteristic of a dc servomotor used in a position-control device. This would be an application where an output has to move at very low speed and is essentially stationary. One example is a crystal-pulling machine used in the making of pure silicon for semiconductors. Motor speed must be very precisely controlled, speed variations must be very carefully controlled, and the speeds are very low.

Note the curves in Fig. 14-3. The particular data were obtained from a permanent-magnet motor. The torque–speed curves are extremely linear since the magnetics—in this situation ceramic magnets—are virtually unaffected by armature reaction. Machines that use Alnico magnets are not as inherently resistant to demagnetization effects as are ceramic magnets. Special fittings in Alnico magnet machines make them nearly as linear as ceramic magnets.

Servomotors are usually higher-resistance machines than conventional motors, since they may require high currents at relatively high voltages in order to produce high torque. For this reason, and also due to the fact that only a small amount of cooling may be obtained from air motion through the machine, the frame must be comparatively large. This allows the heat to be dissipated over a larger surface area. The motor is also usually completely enclosed, so that in many cases it is necessary to use a fan if operation is continuous in this region.

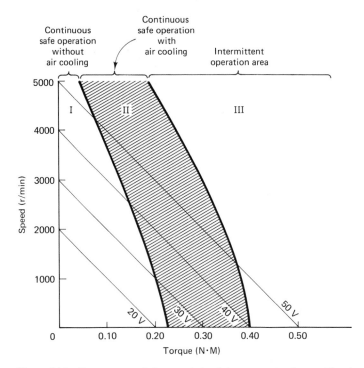

**Figure 14-3**  Torque–speed characteristic of dc servomotor for positional control.

The motor can momentarily be overloaded and run at high dissipation, but certain temperature ratings must not be exceeded.   Again, it is necessary to consult actual manufacturers' literature for exact ratings.

### EXAMPLE 14-1

As an example of the use of Fig. 14-3, we consider the following.   The torque of a given process has the following torque–speed characteristic:

$$T = K_1\omega$$

where    $\omega$ = speed, rad/s
       $T$ = torque, N · m
      $K_1$ = 96.1 × $10^{-6}$ N · m/rad/s
         = $10^{-4}$ N · m/(r/min)

Find:

(a) The motor voltage when the load is rotating at 2000 r/min.
(b) The starting torque available at the voltage determined in part (a).
(c) The maximum speed that the motor can attain, driving the given load if no external cooling is used.
(d) The power delivered to the load.
(e) The maximum torque that the motor can deliver to the load, assuming that it can be air cooled.

**SOLUTION**

Figure 14-3 has been redrawn as Fig. 14-4 to show the load curve superimposed on the motor characteristic curve. From this we can see that:

(a) At 2000 r/min, the load torque is $200 \times 10^{-3}$ N · m or 0.2 N · m. For a speed of 2000 r/min and a torque of 0.2 N · m, the voltage is approximately 40 V.

(b) The starting torque would then be $400 \times 10^{-3}$ N · m or 0.4 N · m.

(c) If we wish to avoid overheating, the maximum speed obtainable occurs at the point where the load line meets the zone I–zone II boundary. This occurs at about 1900 r/min, 36 V.

(d) From part (a), the load torque is 0.2 N · m. The speed is 2000 r/min.

$$P_1 = \omega T = \frac{2\pi \times 2000}{60} \times 0.2 = 41.9 \text{ W}$$

(e) Maximum torque with air cooling occurs at the zone II–zone III boundary and is about $300 \times 10^{-3}$ N · m or 0.3 N · m. The speed is 3000 r/min. At this operating point, the power output is

$$P_2 = \frac{3000 \times 2\pi}{60} \times 0.3 = 94 \text{ W}$$

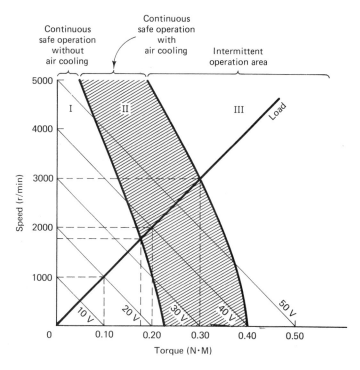

**Figure 14-4** Solution for Example 14-1: torque–speed characteristic of dc servomotor.

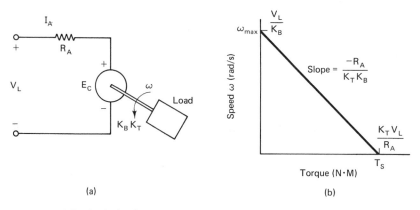

**Figure 14-5** Analysis of permanent-magnet dc motor: (a) equivalent circuit; (b) speed versus torque characteristic.

## 14-4 ANALYSIS OF THE PERMANENT-MAGNET DC MOTOR

A permanent-magnet dc servomotor can be represented by the equivalent circuit shown in Fig. 14-5a. In the diagram the various quantities are defined as follows:

$$V_L = \text{line voltage, V}$$

$$I_A = \text{armature current, A}$$

$$\omega = \text{rotational velocity, rad/s}$$

$$K_B = \text{counter EMF constant, V/rad/s}$$

$$K_T = \text{torque constant, N·m/A}$$

$$E_C = \text{counter EMF}$$

The counter EMF and armature current are expressed by

$$E_C = K_B\omega \tag{14-1}$$

$$I_A = \frac{V_L - E_C}{R_A} \tag{14-2}$$

The motor torque is given by

$$T = K_T I_A$$

$$T = \frac{K_T(V_L - E_C)}{R_A} = \frac{K_T(V_L - K_B\omega)}{R_A} \tag{14-3}$$

The curve of speed versus torque will be a straight line as indicated by Eq. (14-3). Transposing yields

$$TR_A = K_T V_L - K_T K_B \omega$$

and

$$\omega = \frac{K_T V_L - T R_A}{K_T K_B} = \frac{V_L}{K_B} - \frac{T R_A}{K_T K_B} \qquad (14\text{-}4)$$

This linear relationship of speed versus torque is illustrated by Fig. 14-5b.
At standstill $\omega = 0$; then from Eq. (14-4),

$$T = T_S = \frac{K_T V_L}{R_A}$$

The maximum speed occurs when $T = 0$:

$$\omega_{max} = \frac{V_L}{K_B}$$

The electrical power that will be converted to mechanical power is

$$P_1 = E_C I_A = K_B \omega I_A$$

The mechanical power to the load is

$$P_2 = \omega T = \omega K_T I_A$$

Ideally, $P_1 = P_2$. Then

$$K_B \omega I_A = \omega K_T I_A$$

and

$$K_B = K_T \qquad K_B = K_C$$

Note that this equivalence is true only for the SI system of units.

## 14-5 MOVING-COIL ARMATURE MOTORS

Moving-coil armature-type motors are constructed of permanent-magnet field poles on the stator with either a shell or hollow-cup armature winding, or a printed-circuit disk armature winding. Commutator segments are connected to the armature windings, which are assembled on the rotor. Within the armature structure there is no iron or steel. Thus the armature is of extremely light weight, resulting in very low inertia. Also, the armature windings possess low inductance and correspondingly low electrical time constants. As a consequence, moving-coil armature-type motors are used in applications requiring rapid response due to low inertia and high acceleration. The high acceleration is produced by the high starting torque, which in turn is the result of a high value of armature current. Because of the armature design, this high value of armature current must be limited to short durations, to prevent overheating.

Shell-type armature windings composed of insulated copper or aluminum coils are assembled and bonded using resin to form a hollow cup. The hollow-cup winding is connected to the commutator assembly and mounted on the rotor shaft as illustrated in Fig. 14-6a.

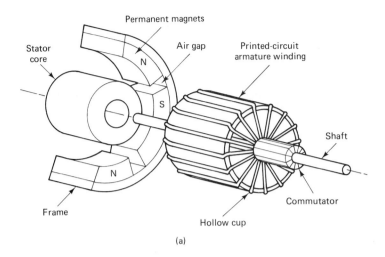

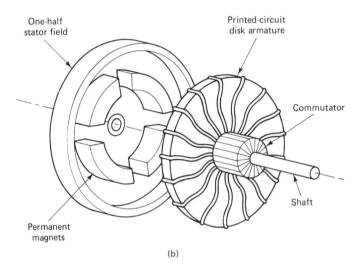

**Figure 14-6** Moving-coil armature motors: (a) hollow-cup armature motor; (b) printed-circuit disk armature motor.

Printed-circuit armature windings composed of laminated or etched tracings of copper form the conductors on a fiberglass disk which are connected to the commutator assembly. The printed-circuit winding and commutator segments are mounted on the shaft and assembled as shown in Fig. 14-6b.

## 14-6 BRUSHLESS DC MOTORS

Where there is a need for a dc motor in which the hazards produced by sparking or arcing at the brushes and commutator segments must be eliminated, the brushless dc motor is finding increased application. Essentially, the brushless dc motor

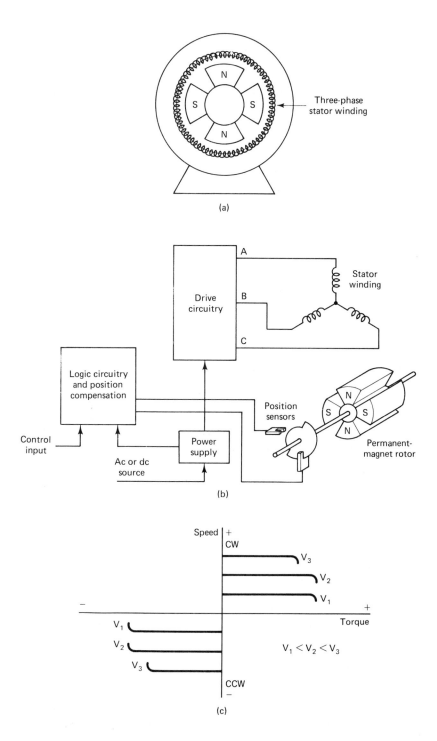

**Figure 14-7** Brushless dc motor: (a) construction; (b) control and drive circuitry; (c) torque–speed characteristic.

consists of a permanent-magnet field rotor and stator armature windings which are commutated by means of electronic switching. Effective rotation of the armature field is achieved by a change of current directions in the armature windings by means of switching power transistors. To provide synchronism of the permanent-magnet fields of the rotor with the rotating armature fields of the stator, shaft position sensors are used to initiate the proper switching times. The electronic circuitry for brushless dc motors may be quite complicated; however, the general principles are illustrated in Fig. 14-7. As shown, a three-phase brushless dc motor control system consists of three groups of armature stator coils which are excited by electronic driver circuitry controlled by the position sensors. The number of field poles is determined by the permanent-magnet field structure. Electronic control of both magnitude and switching rate of armature current in the stator windings will determine the inherent speed–torque charactertistics of the particular brushless dc motor control system. With decreased costs and increased reliability of semiconductors, brushless dc motors with their linear operating characteristics and a wide speed range are becoming competitive with conventional dc drives.

## 14-7 DC TACHOMETERS

Dc tachometers are precision small dc generators designed to develop a voltage output directly proportional to angular velocity. Many tachometers are applied to feedback control systems in which the tachometer supplies an accurate voltage-sensing signal to the closed-loop system of speed regulation. Generally, tachometers are permanent-magnet generators with suitable numbers of commutator segments and armature coils. The armature design is such that output voltage ripple over the desired speed range is negligible. Tachometers are designed for continuous operation.

## 14-8 STEPPER MOTORS

A stepper motor is in principle an electromechanical actuator that responds to a command in electrical input by displacing its mechanical output member by a fixed angular displacement. It can be controlled by simple electronic and drive circuitry and with proper interface equipment can be controlled by a digital computer. Stepper motors can therefore be employed where digital input pulses are converted into analog shaft-output motions. Each shaft revolution can be related to a number of identical, discrete steps, each triggered by a single pulse. The sense of rotation depends on the switching sequence and the rotor speed varies according to the rate of switchings.

Stepper motors are classified as permanent-magnet, reluctance-, and inductor-type machines. They may be wound as machines of any number of phases, the most popular being two and four phases. Because of their widespread use, only the permanent-magnet and variable-reluctance-type stepper motor will be discussed.

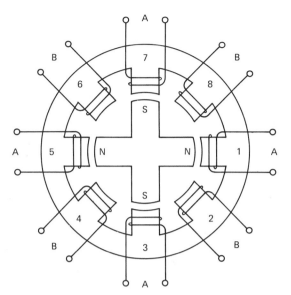

**Figure 14-8** Permanent-magnet stepper motor.

### *Permanent-Magnet Stepper Motors*

A typical two-phase stepper motor with a four-pole permanent-magnet rotor is illustrated in Fig. 14-8. When the stator is energized, the rotor is held in one of the eight positions, where the air gap is at a minimum and the torque then depends on the strength of the magnetic field. Since the opposite poles of the rotor are of the same magnetic polarity, the most satisfactory motor performance is obtained if the stator poles are similarly energized but of opposite poles. For example, in Fig. 14-8, poles 1 and 5 should have south polarity while poles 3 and 7 should be north, resulting in the polarization given in Fig. 14-9a. If in addition to energizing

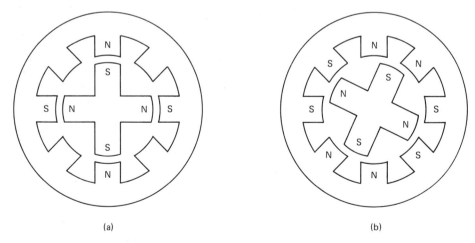

**Figure 14-9** Operating principle of stepper motor: (a) phase *A* energized; (b) phases *A* and *B* energized.

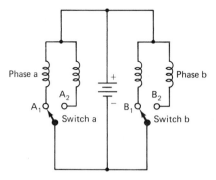

**Figure 14-10** Bifilar stator windings to achieve two-phase excitation.

coils $A$, phase coils $B$ are energized with the polarities shown in Fig. 14-9b, the result will be that the rotor advances $22\frac{1}{2}°$ in a clockwise rotation. Deenergizing the phase $a$ winding will result in a further $22\frac{1}{2}°$ movement. Reversing the phase $a$ winding current will produce an additional $22\frac{1}{2}°$ movement; and so on.

Two-phase excitation of the stator windings to accomplish these polarization patterns discussed with only one winding on each pole is difficult to implement with solid-state switches since winding-current reversals are continually necessary. To circumvent this difficulty, it is usual to employ double coils in each phase, bifiliar wound.

The connection of the windings is such that for one polarization on the $A$ poles the windings are connected to $A_1$ in Fig. 14-10, and the other windings of the $A$ poles, for reverse polarization, are connected to $A_2$. The phase $B$ windings are similarly connected. Of course, only one winding per phase should be ener- gized at any time. The switching indicated in the circuit would be electronic.

As explained, rotation of the stepper-motor rotor is accomplished by switching the polarities of the stator field to the next step position, causing the rotor to follow. Naturally, the rotor is always somewhat behind the stator field switching, since it tries to catch up and align itself, north to south, with the stator field. This situation may be illustrated by Fig. 14-11. As shown, with the rotor and stator poles aligned, there is zero torque. As soon as an angular displacement occurs between the two field poles, there exists a restoring torque tending to realign the field, thereby providing motor action. This developed torque will be maximum when the fields are displaced by $90°$ from one another, and is referred to as the *holding* or *pull-in torque*. The motor cannot drive this load; therefore, a more meaningful value of

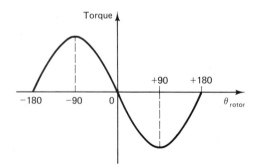

**Figure 14-11** Stepper motor torque versus rotor displacement angle.

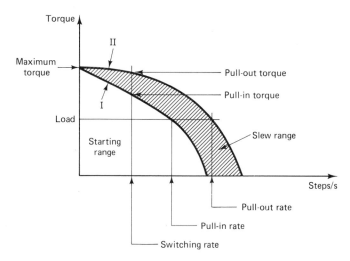

**Figure 14-12** Stepper motor characteristics. I, Pull-in torque versus switching rate; II, pull-out torque versus switching rate.

the motor capacity is the *pull-out* or *running torque*. Figure 14-12 shows the variation of pull-in and pull-out torque as a function of the switching rate. The pull-in torque depends on the total inertia and is a measure at which the motor can start without losing steps. The pull-out torque is a plot of the allowable load that can be supplied at its maximum stepping rate after it has reached its speed.

The area between the curves is the unstable range of the motor, called the *slew range*, in which the motor may tend to fall out of step and stop. The difference between pull-in and pull-out rates is due to the inertia of the rotor. Without this inertia, the curves would coincide.

### Variable-Reluctance Stepper Motors

A typical four-phase variable-reluctance-type stepper motor is illustrated in Fig. 14-13. As each phase is energized in turn, the soft-iron rotor will align itself with the energized poles. As shown, phase *A* is energized and the rotor has assumed a position to minimize the reluctance in the circuit. If phase *A* is now deenergized and phase *B* simultaneously energized, the rotor will again assume a state of minimum reluctance, and in doing so torque is developed. This angular rotation of the shaft per step, the *stepping angle*, is inversely related to the number of teeth and phases, by

$$\text{step angle } \alpha = \frac{360°}{\text{teeth} \times \text{phases}} \tag{14-5}$$

Energizing phase *D* instead of *B* would have resulted in opposite angular motion, but of course, the step angle remains the same.

Since the rotor continuously seeks a position to minimize the circuit's reluctance, this motor is designated as a variable-reluctance stepper motor. Although the operating principle of variable-reluctance and permanent-magnet mo-

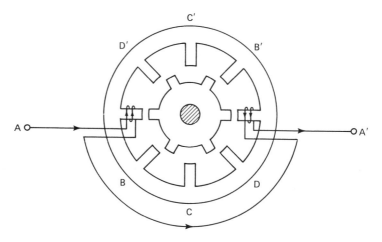

**Figure 14-13** Four-phase variable-reluctance stepper motor. Only phase *A* winding is shown for clarity.

tors is similar, there is one notable difference: the variable-reluctance motor does not have any residual torque to hold the rotor at one position when turned off.

The operating characteristics of the variable-reluctance motor are similar to those displaced in Fig. 14-12 for the permanent-magnet type; however, the torque produced is generally lower. Once driven into the slew rate by increasing the switching rate, the motors cannot be stopped or reversed instantaneously without losing synchronism. As can be seen from Eq. (14-5), the step angle of a motor is determined by the number of poles and teeth. Typical step angles are 15°, 5°, and 2°, although fractional-degree step angles are possible.

Typical applications of stepper motors include paper drives on printers, X-Y positioning systems for machine tools, digitally controlled valves, and many other instances where precise control of rotary motion is desired.

## 14-9 *LINEAR INDUCTION MOTORS*

The basic principles of producing linear motion in linear induction motors (LIMs) are similar to those that produce angular motion in squirrel-cage induction motors. This is true to such an extent that the LIM is often developed from the induction motor. Therefore, let us consider the motor of Fig. 14-14a, which consists of two parts. The stator has the conventional three-phase winding and provides a sinusoidally distributed magnetic field in the air gap which rotates at synchronous speed, related to the number of poles and line frequency, as discussed earlier, namely

$$n_s = \frac{120f}{p} \tag{14-6}$$

When the rotor rotates at a slower than synchronous speed, electromotive forces are induced in the squirrel-cage rotor and induced currents are created in the rotor bars. The forces arising between these currents and the rotating magnetic field drag the rotor in the direction of the field. This principle holds when the rotor is replaced by a continuous sheet of conducting material, as illustrated in Fig. 14-14b. If the motor is now split and unrolled, we obtain the motor shown in Fig. 14-14c. Note that we have now a gliding flux instead of a rotating one. Thus linear motion is the result, hence the name. Also, because of this, the stator is now referred to as the primary member and the rotor as the secondary member, although in principle the moving member could be either one.

Furthermore, if the magnetic core of the rotor is removed from the conducting sheet (which serves as a conducting path for the magnetic flux), the linear induction motor developed takes on the form shown in Fig. 14-14d. Assuming that the secondary is fixed, the primary will be moved by the forces acting on it and subsequently leave the air gap; hence the secondary is shown extended, since otherwise

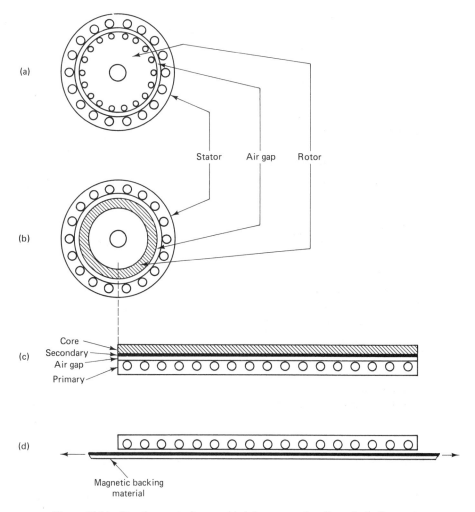

**Figure 14-14** Development of a one-sided, long secondary linear induction motor from a squirrel-cage induction motor.

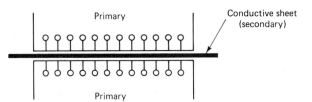

Primary

Conductive sheet
(secondary)

Primary

**Figure 14-15** Double-sided LIM.

the magnetic field would no longer be continuous. Although here we have shown the secondary as the extended member, in principle either one could be extended, depending on the particular application, but not both. Which member is selected to be extended depends on the particular application. For instance, in transportation systems the vehicle carries the primary while the track constitutes the secondary. In this case power is supplied to the vehicle by means of collectors running on rails or wires.

An alternative form of an LIM is one that has two similar primary members straddling the conducting sheet acting as secondary. Such an arrangement is illustrated in Fig. 14-15 and is referred to as *double-sided* whereas the configuration of Fig. 14-14d is *singled-sided*. Large LIMs generally have three-phase windings and find application for material handling, transportation, and liquid-metal pumps. Smaller LIMs are generally used as curtain pullers, sliding doors, conveyer applications, and so on.

### Operating Characteristics

As Eq. (14-6) shows, the gliding field moves through two pole pitches during one cycle of the ac supply. Since the synchronous speed or synchronous velocity of an LIM is dependent on the pole pitch $\tau$, it is a very important parameter in machine operation. The synchronous velocity $v_s$ is determined as

$$v_s = 2\tau f \quad \text{m/s} \tag{14-7}$$

where $f$ is the supply frequency. To provide continuous force, thus motion, there must exist relative velocity between the primary and secondary. Expressing the velocity of the moving member by $v$, the slip of an LIM is

$$s = \frac{v_s - v}{v_s} \tag{14-8}$$

In addition, the coil shape is very important in relation to secondary induced-current patterns and leakage reactances. Figure 14-16 shows the induced currents for a section of the secondary. The secondary currents travel in closed, quasi-rectangular paths. Only the currents traversing the width of the secondary will produce force, since they are perpendicular to the moving field. Due to the finite width of the machine, the distribution of magnetic forces is not uniform laterally across the secondary. The field tapers off to zero at either side of the machine. This phenomenon is called the *lateral edge effect* and is encountered in all LIMs. There is a *longitudinal edge effect* as well, due to the finite length, which stems from the fact that the primary units must be wound on and off the primary yoke

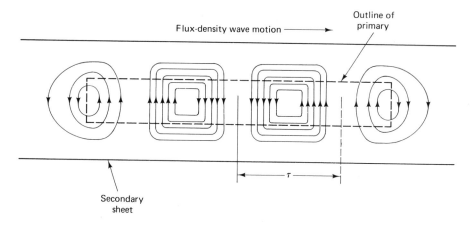

Flux-density wave motion ⟶

Outline of primary

τ

Secondary sheet

**Figure 14-16** Idealized sketch of secondary induced current paths.

at either end. Therefore, the magnetic flux density wave must build up and tail off to zero at either end of the machine. These end effects cause the motor to have a higher effective resistance with consequent lower efficiency. Also, the maximum thrust that the motor can produce is reduced.

Under load, the motor moves at a speed that coincides with the intersection of the force and load characteristic. For many applications these speeds are too high and some form of speed regulation is needed. This can be accomplished by frequency or pole changes. The first method is inconvenient or too expensive; the second requires elaborate switching to change the interconnections of the coils. Pole changing, although less expensive, adds to the size of the machine and only discrete speed values are obtainable. The best type of control is to use a velocity-feedback signal to modulate the power to the motor. In this way, the force–speed characteristic of the motor can be modified so that it intersects with the time-varying load line at the same velocity point (see Fig. 14-17). Strictly speaking, it is the force that is controlled rather than the speed. At all times the motor force is matched to the load; therefore, constant velocity is maintained.

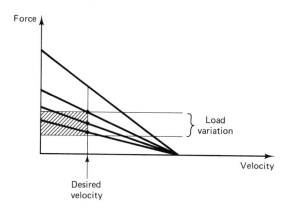

Force

Load variation

Velocity

Desired velocity

**Figure 14-17** Velocity control of LIM.

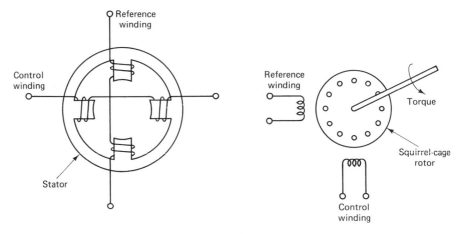

**Figure 14-18** Two-phase servomotor.

## 14-10 AC SERVOMOTORS

The ac servomotor used in position-control systems is basically a two-phase induction motor. The stator has two field windings wound on pole pieces and at right angles to each other. The rotor is an ordinary squirrel-cage type. The arrangement of the stator construction is shown in Fig. 14-18.

One of the windings, called the *reference winding*, is connected to the ac supply; the other, the *control winding*, is connected to some other voltage supply, being of the same frequency as the reference winding. If a voltage phase difference exists between the two voltages applied to the stator windings, a revolving field will be created and the rotor will rotate in the fashion discussed. The sense of direction will depend on the relative phase difference between the two voltages. Thus the design enables the motor speed or stalled torque to be controlled by regulating one phase, keeping the other fully energized.

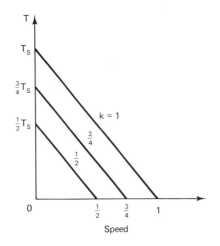

**Figure 14-19** Idealized torque–speed characteristics of a servomotor ($k = 1$ is maximum control voltage).

The motor characteristics differ considerably from those of ordinary motors, since the torque–speed curve is approximately linear. By reducing the voltage to the control winding, a series of speed curves can be obtained, as shown in Fig. 14-19. To assure high torques and acceleration, the rotor length is relatively large compared to its diameter. Also, the rotor resistance is high to prevent the motor from running single-phase after the control voltage has been reduced to zero.

## 14-11 AC TACHOMETER GENERATORS

To measure the angular velocity of a shaft, small two-phase machines may be used, such as an alternator with a permanent magnet. Its output can then be measured with a voltmeter that is calibrated in terms of speed. Unfortunately, in such a device the measured output voltage not only varies with speed, but its frequency is speed dependent as well. In many applications, such as the servo systems using synchros, the input and output variables are at a fixed frequency and only the voltages change. If the output from a tachometer is to be added directly as an ac signal, its output frequency must also be fixed. Thus it is clear that a conventional alternator cannot be used in such an application. To overcome this frequency dependency upon speed, a machine known as a *drag-cup generator* is often employed, illustrated schematically in Fig. 14-20. Basically, the tachometer generator consists of an input winding energized from an ac source. This winding is so arranged as not to induce a voltage in a separate output winding. A copper or

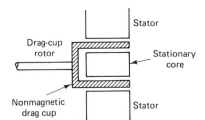

**Figure 14-20** Drag-cup tachometer; rotor cross section.

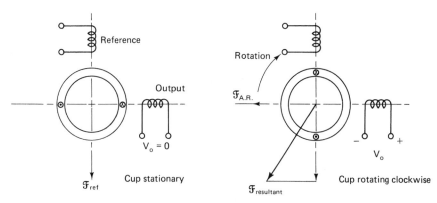

**Figure 14-21** Field relations in drag-cup tachometer.

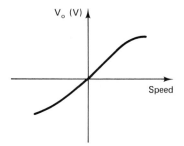

**Figure 14-22** Drag-cup tachometer output voltage $V_o$ versus speed characteristic.

aluminum sleeve, the drag-cup rotor, runs in the air gap and is supported by a shaft. By transformer action eddy currents are induced in the rotor, which in turn set up an armature reaction flux at right angles to the field set up by the reference winding (see Fig. 14-21).

The resultant field, being shifted in space in the direction of rotation, will induce a voltage in the output winding. If the field due to the reference winding is constant, the induced armature currents, and hence its created armature flux, will depend directly on the rotational speed of the motor. Figure 14-22 shows the voltage–speed characteristic for the tachometer.

## 14-12 ELECTROMAGNETIC CLUTCHES

In the operation of controlled devices it is often desirable to be able quickly to engage or disengage a motor from the devices it is driving. This is the function of a clutch. Although there are numerous types of clutches, only two of the most commonly used clutches are described.

### Friction Disk Clutch

The basic principle of this device is illustrated in Fig. 14-23. The input shaft and the clutch disk fastened to it are rotated by the motor. The second disk is fastened to the output shaft, which in turn is connected to the actuated device. As shown

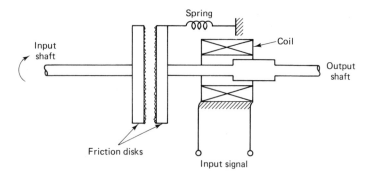

**Figure 14-23** Operating principle of a friction clutch which engages upon energizing coil.

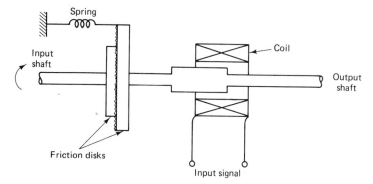

**Figure 14-24** Friction disk clutch which disengages upon energizing coil.

in the figure, the two disks are separated and the output shaft does not rotate. As the two disks are pressed together, the output shaft is made to rotate because of the friction between the surfaces of the disks. When separated, the output shaft is uncoupled from the input shaft and rotation ceases.

There are various ways possible to bring the two clutch disks together; the usual one is electromagnetically, by employing an electromagnet. As shown in Fig. 14-23, the disks are kept separated by a spring. Upon actuating the coil, the magnetic field set up by the coil current will actuate one of clutch disks against the spring action, causing it to engage with the other. Interrupting the coil current causes the spring force to disengage the clutch disks.

The reverse arrangement, where the disks are normally engaged, is shown in Fig. 14-24. Here, upon energizing the coil the magnetic field will create a force to pull the disks apart, thereby uncoupling the shaft against the spring action. When the input signal has been removed, the spring force will keep the disks normally engaged.

It should be observed in Figs. 14-23 and 14-24 that the input and output shafts refer to the clutch arrangement. In actual practice, the motor and controlled device shafts are attached to the input and output shafts of the clutch, respectively, via a shaft coupler.

### Eddy Current Clutch

The electromagnetic clutch discussed has two stable positions; that is, it is an on-off device; input coupled or not coupled. When coupled, the output shaft rotates at the same speed as the input shaft.

Many industrial operations, however, require that the speed of the actuated device be controlled, a function that the friction clutch cannot provide. There are several types of clutches available that enable us to perform this function, one of which is the eddy current clutch. The basic principle on which this device is based is as follows. Figure 14-25 shows a simple electromagnet arrangement consisting of two U-shaped iron bars inside a coil and mounted on a shaft. Upon energizing the coil with a dc current through a slip ring arrangement, magnetic poles are created, setting up a flux as shown. This arrangement is now inserted in a soft-

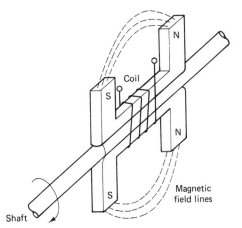

**Figure 14-25** Eddy current clutch. Magnetic sleeve omitted for clarity.

iron sleeve or drum, preserving a small air gap, and the magnet arrangement rotated. The moving poles, and hence the rotating magnetic flux linking the drum, set up eddy currents in the drum. These eddy currents create a secondary field that will interact with the rotating magnet field. The resulting force is in such a direction as to cause the drum to follow the magnet field (see Fig. 14-26). It can now be imagined that the magnet arrangement is coupled to the input shaft and the drum to the output shaft, with no mechanical connection between the two shafts.

The torque created by the force between the magnet field and eddy current field can be controlled by varying the dc excitation current of the coil, thereby controlling the rotational speed of the drum. Because eddy currents are created only when there is relative motion between the poles and the drum, it is inherent

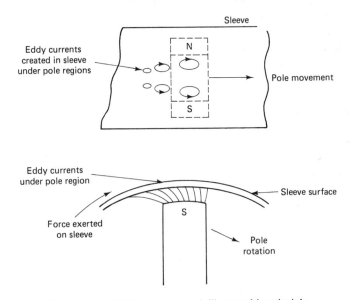

**Figure 14-26** Eddy current clutch illustrated in principle.

Special-Purpose Motors and Devices    Chap. 14

in the device operation that the rotational speed of the output shaft cannot be synchronized with that of the input shaft. In other words, eddy current clutches require slip to function, which normally does not exceed a value of 5% at rated torque. It is interesting to note that the torque–speed characteristic of an eddy current clutch is similar to that of an induction motor. It develops sufficient peak torque to prevent stalling in case of momentary overload, and its characteristic can be influenced by selection of drum material. This is analogous to adding resistance in the rotor circuit of wound-rotor induction motors.

## REVIEW QUESTIONS

**14-1.** Describe the types of generators and motors that may be found in modern aircraft. Why are they used in this application?

**14-2.** What are the differences among general-purpose motors, definite-purpose motors, and special-purpose motors?

**14-3.** Explain the characteristics of the permanent magnets required for dc motors.

**14-4.** Why are the speed–torque and current–torque characteristic curves of permanent-magnet motors extremely linear? Draw the circuit diagram and the characteristic curves.

**14-5.** Explain how speed–torque control is adjusted in permanent-magnet dc motors. Draw the operating characteristics.

**14-6.** What are dc servomotors? Describe the operating characteristics that make them suitable for position- and velocity-control applications. Describe several unique construction features of this motor.

**14-7.** Describe the construction of a moving-coil armature motor. Clearly state the advantages of this type of motor and give several applications for it.

**14-8.** Describe the construction of a printed-circuit disk armature motor. Clearly state the advantages of this motor and give several applications for it.

**14-9.** State the major reasons for using brushless dc motors. Are there any disadvantages in their use? Give several applications for their use.

**14-10.** Describe the operating characteristics of dc stepper motors. Give several good applications for their use.

**14-11.** What material property is essential in the rotor construction of a hysteresis motor? Why does the torque–speed characteristic deviate from the ideal curve in practical motors?

**14-12.** Give a reason for the low efficiencies encountered in:
(a) Hysteresis motors.
(b) Reluctance motors.

**14-13.** What is meant by the *slew range* in reference to a stepper motor? Why is operation in this range unstable?

**14-14.** How can the speed of an LIM be controlled?

**14-15.** An ac tachometer generator has an output voltage of which the frequency is constant. How is this accomplished?

**14-16.** Describe the principle of operation in eddy current clutches. Why does the output shaft rotate at slip speed compared to the input shaft? How is this slip controlled?

**14-17.** Describe how braking action can be provided by an eddy current clutch.

## PROBLEMS

**14-1.** A stepper motor with six pole pairs has two control windings. Determine the step angle for this motor.

**14-2.** A linear induction motor wound with seven poles has a pole pitch of 180 mm. (*Note:* The number of poles need not be even; why?) Determine the synchronous velocity of this machine when operating from a 25-Hz supply.

**14-3.** Describe a two-phase induction motor for use in ac servo systems. State any features that make this machine different from the conventional induction motor.

**14-4.** If the stepper motor of Problem 14-1 has bifiliar stator windings, what is the stepping angle?

**14-5.** Determine the switching sequence of the windings in Problem 14-1 to achieve counterclockwise rotation.

**14-6.** If the rotor in Problem 14-1 drives an output shaft through a reduction gearbox having a ratio of 30:1, what is the output shaft's stepping angle?

**14-7.** What should be the switching rate in Problem 14-1 to obtain an output shaft speed of 1 r/s?

**14-8.** By the use of formula (14-5), determine the stepping angle of the permanent-magnet stepper motor of Fig. 14-13. What should the switching rate be to obtain an output shaft speed of 1 r/s?

**14-9.** A small dc permanent-magnet motor draws 1 A at 12 V and runs at 600 r/min. The armature resistance is $1 \Omega$. What value of resistance should be added to the armature circuit to drop the speed to 450 r/min. Assume that the armature current remains constant.

**14-10.** The torque–speed characteristic of a dc servo motor used as a position-control device is as given in Fig. 14-27. The particular data shown were obtained from a permanent-magnet motor. The characteristics are linear since the magnets in this case are ceramics and are virtually unaffected by armature reaction. It is used to drive a load having a torque–speed characteristic as $T = K_1 n$, where $n$ is the speed in r/min, $T$ is the torque in N · m, and $K_1 = 64 \times 10^{-6}$ N · m/rad/s $= 0.67 \times 10^{-4}$ N · m/(r/min). Determine:
  **(a)** The motor voltage when the load is rotating at 2000 r/min.
  **(b)** $T_{start}$ at the voltage determined in part (a).
  **(c)** $n_{max}$ if no external cooling is used.
  **(d)** The power to the load in part (a).
  **(e)** $T_{max}$ the motor can deliver if air cooled.

**14-11.** The torque–speed characteristic curve of a dc servomotor is given in Fig. 14-27. It is to be used to drive a load having a torque–speed relationship given by

$$T = 4 \times 10^{-6} \omega^2$$

where $T$ is the torque in N · m and $\omega$ is the speed in rad/s.

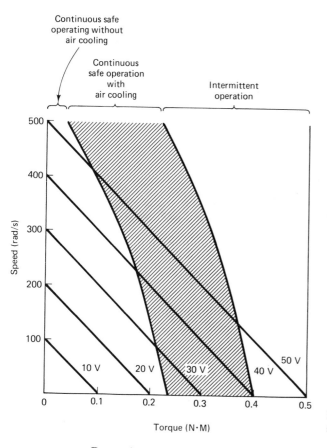

Speed (rad/s)

Continuous safe
operating without
air cooling

Continuous
safe operation
with
air cooling

Intermittent
operation

10 V    20 V    30 V    40 V    50 V

Torque (N·M)

**Figure 14-27**   Torque versus speed dc servomotor.

Determine:

**(a)** The maximum voltage that may be applied to the motor if it is not fan-cooled.

**(b)** The speed at which the motor will operate if the applied voltage is 25 V.

**(c)** The power output of the motor if it is driving the load at 250 rad/s.

**(d)** The range of the voltage applied if the motor is to be fan cooled and operate within safe limits.

**14-12.** In problem 14-9 the armature current was assumed constant with the added resistance in the armature circuit;

**(a)** If this assumption is not made, what would be the corresponding motor speed?

**(b)** If the output power is constant, what resistance value is required to reduce the speed to 450 r/min.?

# APPENDIX A
# COMMON SYMBOLS

| Symbol | Name | Unit |
|--------|------|------|
| $a$ | parallel paths in armature acceleration | |
| $a$ | turns ratio | |
| $A$ | area | |
| $A$ | unit of current | amperes |
| At | unit of magnetomotive force | ampere-turns |
| $B$ | magnetic flux density | teslas |
| $B$ | susceptance | siemens |
| $C$ | capacitance | farads |
| C | coulombs | |
| °C | Celsius | |
| $d$ | distance | meters |
| $D$ | diameter | meters |
| $e, E$ | source voltage | volts |
| $f$ | frequency | hertz |
| $f, F$ | force | newtons |
| F | unit of capacitance | farads |
| $F_m$ | magnetomotive force | ampere-turns |
| $G$ | conductance | siemens |
| h | hours | |
| $H$ | magnetizing force or magnetic field intensity | ampere-turns/meter |
| H | henry | |
| $i, I$ | current | amperes |
| J | joules | |
| $k, K$ | constant of proportionality | |
| $l$ | length | meters |

| Symbol | Name | Unit |
|--------|------|------|
| $L$ | inductance | henrys |
| m | meters | |
| $M$ | mass | kilograms |
| $M$ | mutual inductance | henrys |
| $n$ | rotational speed | revolutions/minute |
| $N$ | number of turns | |
| N | newtons | |
| $p$ | number of poles | |
| $P$ | average power | watts |
| $Q$ | reactive power | reactive volt-amperes |
| $r$ | radius | meters |
| $R$ | resistance | ohms |
| $R_m$ | reluctance | ampere-turns/weber |
| $s$ | slip | |
| s | seconds | |
| $S$ | apparent power | volt-amperes |
| S | siemens | |
| $t$ | time | seconds |
| $T$ | period | seconds |
| $T$ | torque | newton-meters |
| T | teslas | |
| $v, V$ | voltage or potential drop | volts |
| $V$ | volume | cubic meters |
| V | volts | |
| $W$ | energy | joules |
| W | watts | |
| Wb | webers | |
| $X$ | reactance | ohms |
| $Y$ | admittance | siemens |
| $Z$ | impedance | ohms |
| $\delta$ (delta) | delay or firing angle | degrees |
| $\Delta$ (delta) | incremental change | |
| $\theta$ (theta) | impedance angle | degrees |
| $\mu$ (mu) | absolute permeability | henrys/meter |
| $\pi$ (pi) | ratio of circumference to diameter (constant = 3.141592654 . . .) | |
| $\rho$ (rho) | resistivity | ohm-meters |
| $\sigma$ (sigma) | conductivity | siemens/meter |
| $\tau$ (tau) | time constant | seconds |
| $\phi$ (phi) | magnetic flux | webers |
| $\phi$ (phi) | power factor angle | degrees |
| $\omega$ (omega) | angular velocity | radians/second |
| $\Omega$ (omega) | resistance | ohms |

# APPENDIX B
# SI UNITS and
# CONVERSION FACTORS

### BASIC SYSTEME INTERNATIONAL UNITS

| | |
|---|---|
| Length, l | meters (m) |
| Mass, $m$ | kilograms (kg) |
| Time, $t$ | seconds (s) |
| Current, $i$ | amperes (A) |
| Temperature, $T$ | kelvin (K) |
| Luminous intensity, $I$ | candela (cd) |

### CONVERSION FACTORS

| | |
|---|---|
| Length: | 1 inch (in.) = 2.5400 centimeters (cm) |
| | 1 foot (ft) = 0.3048 meter (m) |
| | 1 mile (mi) = 1.6093 kilometers (km) |
| Mass: | 1 ounce (oz) = 28.3495 grams (g) |
| | 1 pound (lb) = 0.4536 kilogram (kg) |
| Temperature: | kelvin (K) = degrees Celsius (°C) + 273.15 |
| Volume: | 1 liter (L) = 1000 cubic centimeters (cm³) |
| | = 61.0237 cubic inches (in.³) |
| | = 0.2642 U.S. gallon (gal) |
| Force: | 1 pound (lbf) = 4.4482 newtons (N) |
| | 1 newton (N) = 100,000 dynes (dyn) |

| | |
|---|---|
| Torque: | 1 newton-meter = 0.7376 pound-foot (lbf · ft) |
| Energy: | 1 foot-pound force (ft · lbf) = 1.3558 joules (J) |
| | 1 joule (J) = 1 watt-second (W · s) = $10^7$ ergs |
| | 1 British thermal unit (Btu) = 1054 |
| | joules = 0.252 kilocalorie (kcal) |
| | 1 kilowatt-hour (kWh) = 3412 Btu |
| Power: | 1 watt (W) = 1 joule/second |
| | 1 horsepower (hp) = 746 watts = 550 foot-pounds |
| | force per second |
| Magnetic flux: | 1 weber (Wb) = $10^8$ lines or maxwells |
| Magnetic flux density: | 1 tesla (T) = 1 weber/meter$^2$ |
| | = 64,516 lines/square inch |
| | = $10^4$ gauss |
| Magnetizing force: | 1 ampere-turn/inch = 39.3701 ampere-turns/meter |
| | 1 oersted = 79.5775 ampere-turns/meter |

# APPENDIX C
# COMPUTER SOLUTIONS
# of ELECTRICAL
# PROBLEMS

The digital computer makes it possible to solve many circuit problems with relative ease. One of the attractive features of a computer is its ability to solve a circuit problem repetitively, with changes in input data. Also, nonlinear problems may readily be solved using iterative methods. As an example, consider the problem of Example 1-9. In the original problem, driving MMF is given as 1400 At, and it is required to find the resulting air-gap flux. In the solution, an iterative method was used, since the sheet steel has a nonlinear characteristic and the reluctance is a function of the flux. The method of solution is to assume a flux density and calculate the required MMF to produce this flux. In essence, we guess at a solution by first estimating the air-gap flux and then calculate the required MMF to produce this flux. If the answer is incorrect, we take another guess and we can make the solution converge to the correct answer.

If the circuit is at all complex, the process becomes tedious for all except the most trivial problems. By writing a computer program, the initial effort may be somewhat greater, but the repeated calculations are all done quite readily by the machine.

Computer solutions to many of the problems in this text have been provided. They are meant to show the power of this approach and are not to be viewed as the best or only solution for a given problem.

BASIC was selected as the programming language, as it is a very simple language, easy to program, and readily available to most students of engineering and technology.

## Computer Solution 1   Example 1-9

Computer Solution 1 was written to solve Example 1-9.   It is a comparatively straightforward problem and the use of a computer solution is probably "techno-logical overkill," but it will serve to illustrate the method.

In this problem (see Fig. 1-21), it is required to calculate the flux in the air gap produced by a given MMF.   The program steps are the following.   In lines 20 to 33, the data for the $B-H$ curve are entered as part of the program and lines 5, 6, and 40 are the physical dimensions of the circuit.

The calculation begins by selecting the first two points of the $B-H$ curve: $B = 0.2$ T and $H = 50$ At/m.   The circuit reluctance and the total flux are then calculated in lines 40 to 90 and it is then possible to determine $NI$, the number of ampere turns required to produce the air-gap flux corresponding to the first point on the $B-H$ curve.   The values of $NI$, $B$, and $H$ are then displayed on the screen in line 91.

This process is repeated for all seven sets of data entered in lines 20 and 33. It is then possible to estimate the actual air-gap flux by interpolating between values displayed on the screen.

Obviously, this is a very simple solution to a rather simple problem.   The next step, and a more sophisticated approach, is to eliminate the necessity of interpolating between data points.

```
ready.

  5    a = .0004                              Mechanical dimensions
  6    l = .001
 10    dim b(7),h(7)
 20    data .2,.4,.6,.8,1.,1.2,1.4
 30    fori = 1to7
 31    read b(i)                              B-H curve data
 32    next i
 33    data 50,70,90,120,180,310,880
 34    for i = 1to7
 35    read h(i)
 36    nexti
 37    fori = 1to7
 40    ls = 1.4
 42    rs = ls*h(i)/((b(i)*a))
 70    rg = l/((4*3.14e - 7)*a)               Reluctance calculations
 72    r = rs + rg
 90    ni = b(i)*a*r                          NI = φ × R
 91    printni;b(i);h(i)                      Output
105    next i
```

## Computer Solution 2   Example 1-9

Computer Solution 2 can best be understood by referring to Fig. CS2.   The program starts at the initial values of $B$ and $H$, namely $B_1$ and $H_1$.   It calculates the value of $NI$ to produce the resulting flux, as before.   However, this time it compares

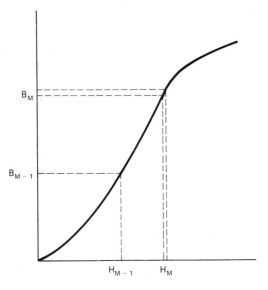

**Figure CS2**  Interpolation routine.

this value of *NI* with the given value of 1400 At.  If *NI* is less than 1400, it goes to the next point on the curve and repeats the calculation.  Eventually, it will reach a point where the calculated value of *NI* will exceed 1400.  Call this point *M* and the corresponding values of input data are $B_m$ and $H_m$.  The program will then interpolate values of *B* and *H* between $B_m$ and $H_m$ and the previous data points $B_{m-1}$ and $H_{m-1}$.  It does this by taking steps of

$$\frac{B_m - B_{m-1}}{100} \quad \text{and} \quad \frac{H_m - H_{m-1}}{100}$$

backward from $B_m H_m$.  That is, it takes 1% steps downward from the point where it first overshot and calculates (*NI* = 1400) at each point.  If this quantity is positive, it proceeds to the next lower point.  At the first point where the calculation goes negative, calculation ceases and the machine prints out the values of *NI*, *B*, and *H*.  The answer is then correct to within 1% of the required value.

```
5     a=.0004
6     l=.001
10    dim b(7),h(7)
20    data .2,.4,.6,.8,1.,1.2,1.4
30    fori=1to7
31    read b(i)
32    next i
33    data 50,150,200,300,480,800,1700
34    for i=1to7                          Identical to previous solution
35    read h(i)
36    nexti
37    fori=1to7
40    ls=1.4
42    rs=ls*h(i)/((b(i)*a))
```

```
70    rg = l/((4*3.14e − 7)*a)
72    r = rs + rg
90    ni = b(i)*a*r
91    if(ni − 1400)>0 then 106                    Comparison

105   next i
106   b = b(i)
107   h = h(i)
108   n = 1
140   b1 = b − .01*b*n
145   h1 = h − n*h*.01                            Interpolation routine
150   rs = (ls*h1)/(b1*a)
155   r1 = rg + rs
160   nit = b1*a*r1
170   n = n + 1
180   if(nit − 1400)>0 then140
190   printnit,b1,h1
```

## Computer Solution 3   Example 1-9

Computer Solution 3 is a more general solution to the problem and requires input from the keyboard.   It requires that seven sets of values of $B$ and $H$ be entered from the keyboard in answer to the questions

VALUES OF B ARE?

VALUES OF H ARE?

It will calculate all of the quantities required, and then print out the information as follows:

NUMBER OF A.T. = XXXX AMP TURNS

INDUCTION IS =    YYYY A.T./M

FLUX DENSITY IS = ZZZZ TESLA

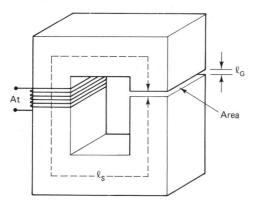

**Figure CS3**   Magnetic circuit for Computer Solution 3, Example 1-9.

```
1       print"3"
2       input"area in sq.m =";a
3       input"length of air gap in m =";l          Input dimensions
4       input"length of steel ="; ls
5       input"no of a.t.  in";at
8       dim b(10),h(10)
9       print"values of b are"
15      input b(1),b(2),b(3),b(4),b(5),b(6),b(7)
33      print"values of h are"                      Input B–H values
35      input h(1),h(2),h(3),h(4),h(5),h(6),h(7)
36      fori = 1to7
42      rs = ls*h(i)/((b(i)*a))
70      rg = l/((4**1e−7*a))                         Calculate reluctance
72      r = rs + rg
90      ni = b(i)*a*r
91      if(ni − 1400)>0 then 106                     Check for limit
105     next i
106     b = b(i)
107     h = h(i)
108     n = 1
140     b1 = b − .01*b*n
145     h1 = h − n*h*.01
150     rs = (ls*h1)/(b1*a)
155     r1 = rg + rs
160     nit = b1*a*r1
170     n = n + 1
180     if(nit − 1400)>0 then140
181     fori = 1to3000:nexti
185     print"3"
190     print"no.of a.t. =";tab(20);nit;tab(32);"ampturns"
199     print"induction is =";tab(20);h1;tab(32);"a.t./m"
200     print"flux density is";tab(20);b1;tab(32);"tesla"
```

### Computer Solution 4

This is a more general example which further demonstrates the power of the computer in solving problems (Fig. CS4). In this problem, 10 sets of data are read in for $B$ and $H$. After these values are read in, the screen is cleared and all the physical dimensions are requested in meters. Also, the last line of input data, "FLUX," is the desired value of the air-gap flux.

Computer Solution 4 calculates the MMF on the center leg for a given flux density in the air gap. It uses the same procedure as outlined in Computer Solution 3 for linear interpolation. When calculations are completed it displays the values of $H$ in sections 1, 2, and 3, as well as the lengths of each section as read in. The last line displayed is the number of ampere-turns required for the solution. Upon completion of the data display, the program will return to the beginning and request new input data.

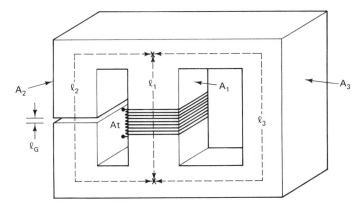

**Figure CS4**   Magnetic circuit for Computer Solution 4

```
1      print"3"
2      dimb(10),h(10)
3      print"values of b are"
4      input b(1),b(2),b(3),b(4),b(5),b(6),b(7),b(8),b(9),b(10)
5      print"values of h are"
6      input h(1),h(2),h(3),h(4),h(5),h(6),h(7),h(8),h(9),h(10)
10     print"3"
20     input"a1";a1
30     input"a2";              a2
40     input"a3";              a3
50     input"l1";              l1
60     input"l2";              l2
70     input"l3";              l3
80     input"lg";lg
85     input"flux";                  phi
140    bg = phi/a2
145    hg = bg/(4*3.14e − 07)
150    for i = 1to10
160    del2 = bg − b(i)
165    j = i
166    b2 = b(i)
170    if del2< =0 then185
180    next i
185    h2 = ((h(j) − h(j − 1))*(bg − b(j − 1))/(b(j) − b(j − 1))) + h(j − 1)
190    h3a = ((h2*l2) + (hg*lg))/l3
200    for k = 1 to 10
210    del3 = h3a − h(k)
220    l = k
225    h(3) = h(k)
230    if del3< =0then250
240    next k
250    b3 = ((b(k) − b(k − 1))*(h3 − h(k − 1))/(h(k) − h(k − 1))) + b(k − 1)
255    phi1 = b3*a3 + b2*a2
```

```
260    b1a=phi1/a1
270    for i=1to10
280    del4=b1a-b(i)
285    m=i
286    b1=b1a
290    if del4<=0then300
295    next i
300    h1=((h(m)-h(m-1))*(b1a-b(m-1))/(b(m)-b(m-1)))+h(m-1)
310    at=h1*l1+h3*l3
320    print"h1";h1;"l1";l1
330    print"h2";h2;"l2";l2
340    print"h3";h3;"l3";l3
345    print"at";at
350    go to 20
```

# APPENDIX D
# TWO-WATTMETER
# METHOD

Power systems in North America are three-phase systems and operate at 60 Hz, as discussed in Chapter 6. Synchronous generators are configured in a wye (Y) connection, usually with the neutral grounded (four-wire system: three line conductors and a neutral). Otherwise, the generator windings are delta (Δ) connected (three-wire system: three line conductors).

The total power in three-phase circuits can easily be measured using the two-wattmeter method shown in Fig. D-1. Wattmeter $A$ is connected with its current coil in line $A$ and with the potential coil across lines $A$ and $B$. Wattmeter $C$ is connected with its current coil in line $C$ and with the potential coil across lines $C$ and $B$. (*Note:* The two wattmeters are sensing two line currents and two voltages with respect to the third line.)

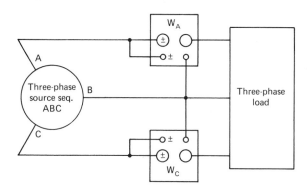

**Figure D-1** Two-wattmeter method for measuring power and power factor.

The instantaneous powers are indicated by:

$$\text{Wattmeter } A: W_A = v_{AB}i_A = (v_{AN} - v_{BN})i_A$$

$$\text{Wattmeter } C: W_C = v_{CB}i_C = (v_{CN} - v_{BN})i_C$$

The total instantaneous power is

$$W_A + W_C = v_{AN}i_A + v_{CN}i_C - v_{BN}(i_A + i_C)$$

For a three-wire load:

$$i_A + i_B + i_C = 0$$

$$i_A + i_C = -i_B$$

Therefore,

$$W_A + W_C = v_{AN}i_A + v_{CN}i_C + v_{BN}i_B$$

$$= p_A + p_C + p_B$$

which is the total instantaneous power in a three-phase load assumed to be a wye-connected ungrounded connection. With rms or effective values of voltages and currents, the two-wattmeter method measures the total average three-phase power of any balanced or unbalanced load.

The power factor of any balanced three-phase load may be determined from the readings of the two wattmeters. From the phasor diagrams of voltages and currents, the average powers indicated by the two wattmeters are

$$W_A = V_{AB}I_A \cos (30° + \phi)$$

$$= V_{L-L}I_L \cos (30° + \phi)$$

$$W_C = V_{CB}I_C \cos (30° - \phi)$$

$$= V_{L-L}I_L \cos (30° - \phi)$$

Note that

$$V_{L-L}I_L = \frac{S_{3\phi}}{\sqrt{3}}$$

where $V_{L-L}$ = line-to-line voltage
$I_L$ = line current
$\phi$ = power factor angle
$S_{3\phi}$ = three-phase apparent power

$$\text{power factor (PF)} = \cos \phi$$

$$= \frac{P_{3\phi}}{S_{3\phi}}$$

Next we consider the ratio of the two wattmeter readings $W_A/W_C$. From the power triangle,

$$\tan\phi = \frac{Q_{3\phi}}{P_{3\phi}} = \frac{\sqrt{3}(W_C - W_A)}{W_A + W_C}$$

$$= \frac{\sqrt{3}\,(1 - W_A/W_C)}{1 + W_A/W_C}$$

Therefore,

$$\text{power factor} = \cos\left[\arctan\frac{\sqrt{3}(1 - W_A/W_C)}{1 + W_A/W_C}\right]$$

This relationship of power factor and the ratio of the two wattmeter readings is plotted as a power factor chart, illustrated in Fig. D-2.

For phase sequence $ABC$ and the circuit of Fig. D-1, it can be shown by means of phasor diagrams that wattmeter $A$ will indicate a zero reading at 0.5 power factor lagging (inductive). As the power factor of the three-phase load decreases below 0.5, wattmeter $A$ would give a negative indication, and therefore its current leads must be reversed to provide an indication that must be considered a negative value. Conversely, for the circuit of Fig. D-1, if wattmeter $C$ indicates a zero reading, the power factor is 0.5 leading (capacitive).

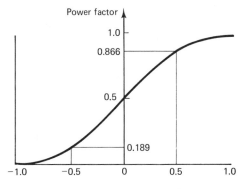

Figure D-2   Power factor chart.

# APPENDIX E
# ELECTRICAL MACHINE
# and TRANSFORMER
# DATA

*Direct-Current Motors*

| Motor rating (hp) | Voltage (V) | Typical full-load efficiency (%) |
|---|---|---|
| $\frac{1}{4}$ | 120 | 54 |
| $\frac{1}{3}$ | 120 | 58 |
| $\frac{1}{2}$ | 120 | 60 |
| $\frac{3}{4}$ | 120 | 63 |
| 1 | 120 | 66 |
| $1\frac{1}{2}$ | 120 | 71 |
| 2 | 120 | 73 |
| 3 | 120 | 75 |
| 5 | 120 | 78 |
| $7\frac{1}{2}$ | 120/240 | 80 |
| 10 | 120/240 | 82 |
| 15 | 120/240 | 85 |
| 20 | 120/240 | 86 |
| 25 | 120/240 | 87 |
| 30 | 240/500 | 88 |
| 40 | 240/500 | 88 |
| 50 | 240/500 | 89 |
| 60 | 240/500 | 89 |
| 75 | 240/500 | 90 |
| 100 | 500 | 90 |
| 125 | 500 | 91 |
| 150 | 500 | 92 |
| 200 | 500 | 92 |

### Three-Phase Synchronous Motors
(Operating at unity power factor)

| Motor rating (hp) | Voltage (V) | Typical full-load efficiency (%) |
|---|---|---|
| 25 | 230/440/575 | 86 |
| 30 | 230/440/575 | 86 |
| 40 | 230/440/575 | 87 |
| 50 | 230/440/575 | 87 |
| 60 | 440/575/ 2300 | 88 |
| 75 | 440/575/ 2300 | 88 |
| 100 | 440/575/ 2300 | 89 |
| 125 | 440/575/ 2300 | 89 |
| 150 | 440/575/ 2300 | 90 |
| 200 | 440/575/ 2300 | 90 |

### Three-Phase Induction Motors
(Operating at all voltages)

| Motor rating (hp) | Typical full-load efficiency (%) | Typical full-load pf (%) |
|---|---|---|
| 1 | 72 | 78 |
| $1\frac{1}{2}$ | 75 | 79 |
| 2 | 76 | 79 |
| 3 | 81 | 82 |
| 5 | 82 | 86 |
| $7\frac{1}{2}$ | 84 | 81 |
| 10 | 84 | 85 |
| 15 | 88 | 83 |
| 20 | 89 | 84 |
| 25 | 90 | 86 |
| 30 | 90 | 87 |
| 40 | 90 | 83 |
| 50 | 92 | 84 |
| 60 | 90 | 81 |
| 75 | 92 | 81 |
| 100 | 92 | 84 |
| 125 | 92 | 90 |
| 150 | 92 | 91 |
| 200 | 93 | 89 |

Single-phase transformers
secondary voltage 120/240 volts

| kVA Rating | 2.4 kV | | 7.2 kV | | 13.8 kV | | 34.4 kV | | Typical full-Load efficiency |
|---|---|---|---|---|---|---|---|---|---|
| | R ohms | X ohms | R ohms | X ohms | R ohms | X ohms | R ohms | X ohms | Percent |
| 3 | 30.7 | 42.2 | 362 | 432 | 635 | 1777 | 10,260 | 17,750 | 97.0 |
| 10 | 8.1 | 11.5 | 82.9 | 114 | 248 | 457 | 2,840 | 5,440 | 98.2 |
| 25 | 3.7 | 5.1 | 29.0 | 41.5 | 129 | 175 | 1,041 | 2,220 | 98.4 |
| 50 | 1.8 | 2.6 | 13.5 | 22.8 | 53.3 | 95.2 | 402 | 1,160 | 98.7 |
| 100 | 0.81 | 1.8 | 5.7 | 13.0 | 34.3 | 57.1 | 178 | 592 | 98.8 |
| 333 | 0.21 | 0.76 | 1.6 | 6.2 | 13.2 | 25.3 | 39.1 | 181 | 99.0 |
| 500 | 0.13 | 0.54 | 1.0 | 4.9 | 3.8 | 18.7 | 26.0 | 133 | 99.0 |

The heading for the impedance columns is: Typical leakage impedance valves referred to high voltage winding

Three-Phase Transformers
Typical Resistance and Reactance Ohmic
Values
Referred to High Voltage Winding Based
on 10,000 kVA (Self-Cooled) Rating

| H.V. winding | L.V. winding | Ohmic values | |
|---|---|---|---|
| kV | kV | R | X |
| 15 | 15 | 0.10 | 1.50 |
| 34.5 | 15 | 0.55 | 8.40 |
| 46 | 25 | 1.00 | 16.3 |
| 69 | 34.5 | 2.28 | 40.5 |
| 115 | 34.5 | 6.48 | 132 |
| 230 | 46 | 27.5 | 714 |

Typical full-load efficiency is 98.5 to 99.8 percent.

# ANSWERS TO SELECTED PROBLEMS

**Chapter 1**
**1.1** 4.58A
**1.3** 1.38A
**1.5** 17.0A
**1.7** 1801At
**1.9** 1.82A
**1.11** 1942At 2119
**1.14** **(a)** 1.84A
    **(b)** 80N
    **(c)** 0.144A

**Chapter 2**
**2.1** 2.88N left
**2.2** 0.024N
**2.3** **(a)** 1.8ft.lb
    **(b)** 73 effective conductors

**Chapter 3**
**3.1** 233.3V 1543r/min
**3.6** **(a)** 58 Ω
    **(b)** 340V
    **(c)** 75 Ω

**3.9** **(a)** 268.8V
    **(b)** 84.5%
**3.12** **(b)** 220v
    **(c)** 214 Ω
**3.15** **(a)** **(i)** 0.1 Ω, 103.3 Ω
       **(ii)** 10.0%
**3.16** **(a)** 154.8V
    **(b)** 151.0V
**3.17** **(b)** 224At
**3.18** 83.9%
**3.19** 248.7V
**3.20** **(a)** 712.1V
    **(b)** 676.5V
**3.21** **(a)** **(i)** 200 Ω
       **(ii)** 300 Ω
    **(b)** 25 Ω

**Chapter 4**
**4.1** **(a)** 8.8 kW
    **(b)** 8.8 hp
**4.3** **(a)** 16kW
    **(b)** 67 N.m
    **(c)** 78%

**4.5 (a)** 127.0 rad/s
  **(b)** 126.7 rad/s
**4.6 (a)** 165.8A
  **(b)** 76 kW
  **(c)** 1.4 kW
  **(d)** 593 N.m
**4.9 (a)** 3.1 kW
  **(b)** 5.6%
**4.11 (a)** 160W
  **(b)** 78%
**4.13 (a) (i)** 53.8 rad/s 1002 N.m
    **(ii)** 132 rad/s 178 N.m
  **(b)** 4.83 Ω
**4.15 (a)** 225.8V
  **(b)** 7.0 kW 80%
  **(c)** 6.3%
  **(d)** 33.5 N.m
  **(e)** 71 turns
**4.17** 2.15 Ω
**4.19** 1242r/min, 54.7 ft.lb
**4.21 (a)** 34.5A
  **(b)** 87%
  **(c)** 276 rad/s
  **(d)** 5.4%

## Chapter 5
**5.1** 114.8 A

## Chapter 6
**6.1** 900 rpm
**6.3 (a)** 4839 W
  **(b)** 14,520 W
**6.5** 262.4 A
**6.7** 2875 V
**6.9 (a)** 15.2°
  **(b)** 26.6° I lags V.

## Chapter 7
**7.1** 976 V/phase
**7.3** $V_{L-L}$ = 614 V
**7.5** 0.957
**7.7 (a)** 38.8 V/coil, E/phase =
    1552 V
  **(b)** $V_{L-L}$ = 2688 V

**7.9** 37.4 percent
**7.11 (a)** 21.0 ohms
  **(b)** 81.5 percent
  **(c)** 24.2°
  **(d)** 749.83 kW
  **(e)** $V_{RATED}$ = 2656V, kVA
    same as rated
**7.13 (a)** 60 poles
  **(b)** 108 MW
  **(c)** 4518 A
  **(d)** 111.34 MW
  **(e)** 8860 kN-m
**7.15** 73.9 percent
**7.17 (a)** 13.86 ohms, 13.79 ohms
  **(b)** 28.9 percent, −13.4 percent

## Chapter 8
**8.1 (a)** 5.63 mWb
  **(b)** 1200 V
**8.3 (a)** 1.458 ohms
  **(b)** 0.091 ohms
**8.5** 4512 V
**8.7 (a)** 11.62 ohms, 33.1 ohms,
    31.0 ohms
  **(b)** 4.5 percent
**8.9** 98.04 percent, 98.48 percent
**8.11** 89.94 percent
**8.13** 2.23 percent
**8.15** 93.7 percent

## Chapter 9
**9.1** 4 turns
**9.3** 57,8 kVA
**9.5** 17.8 A
**9.7** $kVA_1$ = 12, $kVA_2$ = 8
**9.9 (a)** 33
  **(b)** 18.3 kVA, 15.88 kW
  **(c)** 48.1 A
  **(d)** $I_s$ = 48.1 A, $I_p$ = 1.46 A
**9.11 (a)** 3333kVA, $V_p$ = 13.2kV,
    $V_s$ = 66kV, $I_p$ = 253A,
    $I_s$ = 50.5A

(c) 3333kVA, $V_p = 7.62$kV,
$V_s = 38.1$kV, $I_p =$
437.5A, $I_s = 87.5$A

## Chapter 10
**10.1** 564 rpm
**10.3** (a) 6-p
  (b) 5 percent
  (c) 3 Hz
  (d) 60 rpm 1140 rpm, same speed
**10.5** 7 V
**10.9** (a) 148.4A, 0.896
  (b) 123.4 Hp, 1007 N-M
  (c) 89.5 percent
**10.11** 497 N-M, 581A
**10.13** (a) $X_e = 1.34$ ohms, $R_R =$
    0.30 ohms, $X_M = 19.85$
    ohms
  (b) 18.2A, 0.87
  (c) 92.7 percent, 1720 rpm
**10.15** (a) 52 N-M
  (b) 379V
  (c) 87.3A
  (d) 110V

## Chapter 11
**11.1** only c is a fractional Hp motor
**11.3** 57.8 percent
**11.5** 88.4 F
**11.7** 144 F, 9 F, 720 V minimum

**11.9** (a) 925 r/min
  (b) 4.48 N-M

## Chapter 12
**12.1** (a) 24
  (b) 10
  (c) 300 r/min
**12.3** $n_s = 600$ r/min
**12.5** 198A
**12.7** overloaded by 356kVA, $I_L =$
    2957A
**12.9** 58.5A, 0.965 leading
**12.11** (a) 44.3V, 84.9°
**12.13** (a) 80.215 kW
  (b) 352.2V
  (c) 12.1°
**12.15** (a) 16.687kW
  (b) 0.644 lagging
**12.17** (a) $kVA_{SM} = 1652$, $pf_{SM} =$
    0.25 leading
**12.19** $kVA_{SM} = 636$

## Chapter 14
**14.1** 15 degrees
**14.2** 9 m/s
**14.6** 0.5 degree
**14.9** 2.75
**14.10** (a) 34V
  (b) 0.35 N-M
  (c) 2500 r/min
  (d) 27.2 W
  (e) 113 W

# REFERENCES

1. S. A. Boctor, *Electric Circuit Analysis*, Prentice-Hall, Englewood Cliffs, N.J. (to be published 1987).

2. R. G. Kloeffler, R. M. Kerchner, and J. L. Brenneman, *Direct-Current Machinery*, revised edition, Macmillan, New York, 1950.

3. A. E. Fitzgerald, C. Kingsley, Jr., and S. D. Umans, *Electric Machinery*, 4th ed., McGraw-Hill, New York, 1983.

4. G. R. Slemon and A. Straughen, *Electric Machines*, Addison-Wesley, Reading, Mass., 1980.

5. R. A. Millermaster, *Harwood's Control of Electric Motors*, 4th ed., Wiley-Interscience, New York, 1970.

6. R. A. Pearman, *Power Electronics: Solid State Motor Control*, Reston, Reston, Va.,1980.

7. D. Platnick, Brushless D.C. Motor, doctoral dissertation, University of Rochester, Rochester, N.Y., 1965.

8. O. I. Elgerd, *Basic Electric Power Engineering*, Addison-Wesley, Reading, Mass., 1977.

9. I. L. Kosow, *Electric Machinery and Transformers*, Prentice-Hall, Englewood Cliffs, N.J., 1972.

10. V. Del Toro, *Electromechanical Devices for Energy Conversion and Control Systems*, Prentice-Hall, Englewood Cliffs, N.J., 1968.

11. F. P. Sullivan, Assistant Editor, Electric Generators, a *Power* special report, March 1966.

12. C. Bodine, *Small Motor, Gear Motor, and Control Handbook*, 4th ed., Bodine Electric Co., Chicago, 1978.

# INDEX

# A

# I

# K

# L

# M